Production Note
Cornell University Library produced this volume to replace the irreparably deteriorated original. It was scanned using Xerox software and equipment at 600 dots per inch resolution and compressed prior to storage using CCITT Group 4 compression. The digital data were used to create Cornell's replacement volume on paper that meets the ANSI Standard Z39.48-1984. The production of this volume was supported in part by the Commission on Preservation and Access and the Xerox Corporation. 1992.

ELEMENTS

OF

GEOMETRY AND TRIGONOMETRY

FROM THE WORKS OF

A. M. LEGENDRE

ADAPTED TO THE COURSE OF MATHEMATICAL INSTRUCTION
IN THE UNITED STATES

BY CHARLES DAVIES, LL.D.
AUTHOR OF A FULL COURSE OF MATHEMATICS

EDITED BY

J. HOWARD VAN AMRINGE, A.M., Ph.D.
PROFESSOR OF MATHEMATICS IN COLUMBIA COLLEGE

NEW YORK ·:· CINCINNATI ·:· CHICAGO
AMERICAN BOOK COMPANY

DAVIES'S MATHEMATICAL SERIES

FOR ELEMENTARY SCHOOLS
Davies's Primary Arithmetic
Davies's Intellectual Arithmetic
Davies's First Book in Arithmetic
Davies's Standard Arithmetic
Davies's Practical Arithmetic
Davies's Complete Arithmetic

FOR SECONDARY SCHOOLS
Davies's University Arithmetic
Davies's New Elementary Algebra
Davies's Bourdon's Algebra
Davies's Elementary Geometry and
    Trigonometry
Davies's Legendre's Geometry and
    Trigonometry

FOR COLLEGES AND ADVANCED STUDENTS
Davies's University Algebra
Davies's Analytical Geometry
Davies's Analytical Geometry and
    Calculus
Davies's Descriptive Geometry
Davies's Elements of Surveying

D. LEGENDRE GEOM. AND TRIG.

W. P. 4

PREFACE.

O F the various treatises on Elementary Geometry which have appeared during the present century, that of **M. Legendre** stands pre-eminent. Its peculiar merits have won for it not only a European reputation, but have also caused it to be selected as the basis of many of the best works on the subject that have been published in this country.

In the original treatise of **Legendre,** the propositions are not enunciated in general terms, but by means of the diagrams employed in their demonstration. This departure from the method of **Euclid** is much to be regretted. The propositions of Geometry are general truths, and ought to be stated in general terms, without reference to particular diagrams. In the following work, each proposition is first enunciated in general terms, and afterward with reference to a particular figure, that figure being taken to represent any one of the class to which it belongs. By this arrangement, the difficulty experienced by beginners in comprehending abstract truths is lessened, without in any manner impairing the generality of the truths evolved.

The term *solid,* used not only by **Legendre,** but by many other authors, to denote a limited portion of space, seems calculated to introduce the foreign idea of matter into a science which deals only with the abstract properties and relations of figured space. The term *volume* has been introduced in its place, under the belief that it corresponds more exactly to the idea intended. Many other departures have been made from the original text, the value and utility of which have been made manifest in the practical tests to which the work has been subjected.

In the present edition, numerous changes have been made, both in the Geometry and in the Trigonometry. The definitions have been carefully revised — the demonstrations have been harmonized, and, in many instances, abbreviated — the principal object being to simplify the subject as much as possible, without departing from the general plan. These changes are due to Professor Peck, of the Department of Pure Mathematics and Astronomy in Columbia College. For his aid, in giving to the work its present permanent form, I tender him my grateful acknowledgments.

The edition of **Legendre**, referred to in the last paragraph, will not be altered in form or substance; and yet, Geometry must be made a more practical science. To attain this object, without deranging a system so long used, and so generally approved, an Appendix has been prepared and added to **Legendre**, embracing many Problems of Geometrical construction, and many applications of Algebra to Geometry.

It would be unjust to those giving instruction, to add to their daily labors, the additional one, of finding appropriate solutions to so many difficult problems: hence, a Key has been made for the use of Teachers, in which the best methods of construction and solution are fully given.

CHARLES DAVIES.

FISHKILL-ON-HUDSON, *June*, 1875.

NOTE. — The edition of **Legendre** referred to in the foregoing preface was prepared by the late Professor Davies the year before his lamented death. The present edition is the result of a careful re-examination of the work, into which have been incorporated such emendations, in the way of greater clearness of expression or of proof, as could be made without altering it in form or substance.

Practical exercises have been placed at the end of the several books, and comprise additional theorems, problems, and numerical exercises upon the principles of the Book or Books preceding. They will, it is hoped, be found of service in accustoming students, early in and throughout their course, to make for themselves practical application of geometric principles, and constitute, in addition, a large body of review and test questions for the convenience of teachers.

The Trigonometry has been carefully revised throughout, to simplify the discussions and to make the treatment conform in every particular to the latest and best methods.

It is believed that in clearness and precision of definition, in general simplicity and rigor of demonstration, in orderly and logical development of the subject, and in compactness of form, **Davies' Legendre** is superior to any work of its grade for the general training of the logical powers of pupils, and for their instruction in the great body of elementary geometric truth.

J. H. VAN AMRINGE,
Editor of Davies' Course of Mathematics.

COLUMBIA COLLEGE, N. Y., *June*, 1885.

CONTENTS.

GEOMETRY.

BOOK VIII.

BOOK IX.

PLANE TRIGONOMETRY.

INTRODUCTION.

PLANE TRIGONOMETRY.

ANALYTICAL TRIGONOMETRY.

SPHERICAL TRIGONOMETRY.

MENSURATION.

CONTENTS.

LOGARITHMIC TABLES.

ELEMENTS

OF

GEOMETRY.

INTRODUCTION.

DEFINITIONS OF TERMS.

1. QUANTITY is any thing which can be increased, diminished, and measured.

To measure a thing, is to find out how many times it contains some other thing, of the same kind, taken as a standard. The assumed standard is called the *unit of measure.*

2. In GEOMETRY, there are four species of quantity, viz.: LINES, SURFACES, VOLUMES, and ANGLES. These are called GEOMETRICAL MAGNITUDES.

Since the unit of measure of a quantity is of the same kind as the quantity measured, there are four kinds of units of measure, viz.: *Units of Length, Units of Surface, Units of Volume,* and *Units of Angular Measure.*

3. GEOMETRY is that branch of Mathematics which treats of the properties, relations, and measurement of the Geometrical Magnitudes.

4. In Geometry, the quantities considered are generally represented by means of the straight line and curve. The operations to be performed upon the quantities, and the relations between them, are indicated by signs, as in Analysis.

The following are the principal signs employed:

The *Sign of Addition,* $+$, called *plus:*
Thus, $A + B$, indicates that B is to be added to A.

The *Sign of Subtraction,* $-$, called *minus:*
Thus, $A - B$, indicates that B is to be subtracted from A.

The *Sign of Multiplication,* $\times$:
Thus, $A \times B$, indicates that A is to be multiplied by B.

The *Sign of Division,* $\div$:
Thus, $A \div B$, or, $\dfrac{A}{B}$, indicates that A is to be divided by B.

The *Exponential Sign:*
Thus, A^3, indicates that A is to be taken three times as a factor, or raised to the third power.

The *Radical Sign,* $\sqrt{}$:
Thus, $\sqrt{A}$, $\sqrt[3]{B}$, indicate that the square root of A, and the cube root of B, are to be taken.

When a compound quantity is to be operated upon as a single quantity, its parts are connected by a vinculum or by a parenthesis:
Thus, $\overline{A + B} \times C$, indicates that the sum of A and B is to be multiplied by C; and $(A + B) \div C$, indicates that the sum of A and B is to be divided by C.

A number written before a quantity, shows how many times it is to be taken.

Thus, $3(A + B)$, indicates that the sum of A and B is to be taken three times.

The *Sign of Equality,* $=$:
Thus, $A = B + C$, indicates that A is equal to the sum of B and C.

The expression, $A = B + C$, is called an equation. The part on the left of the sign of equality is called the *first member;* that on the right, the *second member.*

The *Sign of Inequality,* $<$:

Thus, $\sqrt{A} < \sqrt[3]{B}$, indicates that the square root of A is less than the cube root of B. The opening of the sign is towards the greater quantity.

The sign, $\therefore$ is used as an abbreviation of the word *hence,* or *consequently.*

The symbols, 1°, 2°, etc., mean 1st, 2d, etc.

5. The general truths of Geometry are deduced by a course of logical reasoning, the premises being definitions and principles previously established. The course of reasoning employed in establishing any truth or principle is called *a demonstration.*

6. A THEOREM is a truth requiring demonstration.

7. An AXIOM is a self-evident truth.

8. A PROBLEM is a question requiring solution.

9. A POSTULATE is a self-evident Problem.

Theorems, Axioms, Problems, and Postulates, are all called *Propositions.*

10. A LEMMA is an auxiliary proposition.

11. A COROLLARY is an obvious consequence of one or more propositions.

12. A SCHOLIUM is a remark made upon one or more propositions, with reference to their connection, their use, their extent, or their limitation.

13. An HYPOTHESIS is a supposition made, either in the statement of a proposition, or in the course of a demonstration.

14. Magnitudes are equal to each other, when each contains the same unit an equal number of times.

15. Magnitudes are equal *in all respects*, when they may be so placed as to coincide throughout their whole extent; they are equal *in all their parts* when each part of one is equal to the corresponding part of the other, when taken either in the same or in the reverse order.

ELEMENTS OF GEOMETRY.

BOOK I.

ELEMENTARY PRINCIPLES.

DEFINITIONS.

1. GEOMETRY is that branch of Mathematics which treats of the properties, relations, and measurements of the Geometrical Magnitudes.

2. A POINT is that which has position, but not magnitude.

3. A LINE is that which has length, but neither breadth nor thickness.

Lines are divided into two classes, *straight* and *curved*.

4. A STRAIGHT LINE is one which does not change its direction at any point.

5. A CURVED LINE is one which changes its direction at every point.

When the sense is obvious, to avoid repetition, the word *line*, alone, is commonly used for *straight line;* and the word *curve*, alone, for *curved line.*

6. A line made up of straight lines, not lying in the same direction, is called a *broken line.*

7. A SURFACE is that which has length and breadth without thickness.

Surfaces are divided into two classes, *plane* and *curved surfaces*.

8. A PLANE is a surface, such, that if any two of its points be joined by a straight line, that line will lie wholly in the surface.

9. A CURVED SURFACE is a surface which is neither a plane nor composed of planes.

10. A PLANE ANGLE is the amount of divergence of two straight lines lying in the same plane.

Thus, the amount of divergence of the lines AB and AC, is an angle. The lines AB and AC are called *sides*, and their common point A, is called the *vertex*. An angle is designated by naming its sides, or sometimes by simply naming its vertex; thus, the above is called the angle BAC, or simply, the angle A.

11. When one straight line meets another, the two angles which they form are called *adjacent angles*. Thus, the angles ABD and DBC are adjacent.

12. A RIGHT ANGLE is formed by one straight line meeting another so as to make the adjacent angles *equal*. The first line is then said to be *perpendicular* to the second.

13. An OBLIQUE ANGLE is formed by one straight line meeting another so as to make the adjacent angles *unequal*.

Oblique angles are subdivided into two classes, *acute angles*, and *obtuse angles*.

14. An ACUTE ANGLE is less than a right angle.

15. An OBTUSE ANGLE is greater than a right angle.

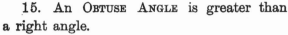

16. Two straight lines are *parallel*, when they lie in the same plane and can not meet, how far soever, either way, both may be produced. They then have the *same direction*.

17. A PLANE FIGURE is a portion of a plane bounded by lines, either straight or curved.

18. A POLYGON is a plane figure bounded by straight lines.

The bounding lines are called *sides* of the polygon. The broken line, made up of all the sides of the polygon, is called the *perimeter* of the polygon. The angles formed by the sides are called *angles* of the polygon.

19. Polygons are classified according to the number of their sides or angles.

A Polygon of three sides is called a *triangle;* one of four sides, a *quadrilateral;* one of five sides, a *pentagon;* one of six sides, a *hexagon;* one of seven sides, a *heptagon;* one of eight sides, an *octagon;* one of ten sides, a *decagon;* one of twelve sides, a *dodecagon,* &c.

20. An EQUILATERAL POLYGON is one whose sides are all equal.

An EQUIANGULAR POLYGON is one whose angles are all equal.

A REGULAR POLYGON is one which is both equilateral and equiangular.

21. Two polygons are *mutually equilateral,* when their sides, taken in the same order, are equal, each to each: that is, following their perimeters in the same direction, the first

side of the one is equal to the first side of the other, the second side of the one to the second side of the other, and so on.

22. Two polygons are *mutually equiangular*, when their angles, taken in the same order, are equal, each to each.

23. A DIAGONAL of a polygon is a straight line joining the vertices of two angles, not consecutive.

24. A BASE of a polygon is any one of its sides on which the polygon is supposed to stand.

25. Triangles may be classified with reference to either their sides, or their angles.

When classified with reference to their sides, there are two classes: *scalene* and *isosceles.*

1st. A SCALENE TRIANGLE is one which has no two of its sides equal.

2d. An ISOSCELES TRIANGLE is one which has two of its sides equal.

When all of the sides are equal, the triangle is EQUILATERAL.

When classified with reference to their angles, there are two classes: *right-angled* and *oblique-angled.*

1st. A RIGHT-ANGLED TRIANGLE is one that has one right angle.

The side opposite the right angle is called the *hypothenuse.*

2d. An OBLIQUE-ANGLED TRIANGLE is one whose angles are all oblique.

If one angle of an oblique-angled triangle is obtuse, the triangle is said to be OBTUSE-ANGLED. If all of the angles are acute, the triangle is said to be ACUTE-ANGLED.

26. Quadrilaterals are classified with reference to the relative directions of their sides. There are then two classes; the *first class* embraces those which have no two sides parallel; the *second class* embraces those which have at least two sides parallel.

Quadrilaterals of the first class, are called *trapeziums.*

Quadrilaterals of the second class, are divided into two species: *trapezoids* and *parallelograms.*

27. A TRAPEZOID is a quadrilateral which has only two of its sides parallel.

28. A PARALLELOGRAM is a quadrilateral which has its opposite sides parallel, two and two.

There are two varieties of parallelograms: *rectangles* and *rhomboids.*

1st. A RECTANGLE is a parallelogram whose angles are all right angles.

A SQUARE is an equilateral rectangle.

2d. A RHOMBOID is a parallelogram whose angles are all oblique.

A RHOMBUS is an equilateral rhomboid.

2

29. SPACE is indefinite extension.

30. A VOLUME is a limited portion of space, combining the three dimensions of length, breadth, and thickness.

AXIOMS.

1. Things which are equal to the same thing, are equal to each other.

2. If equals are added to equals, the sums are equal.

3. If equals are subtracted from equals, the remainders are equal.

4. If equals are added to unequals, the sums are unequal.

5. If equals are subtracted from unequals, the remainders are unequal.

6. If equals are multiplied by equals, the products are equal.

7. If equals are divided by equals, the quotients are equal.

8. The whole is greater than any of its parts.

9. The whole is equal to the sum of all its parts.

10. All right angles are equal.

11. Only one straight line can be drawn joining two given points.

12. The shortest distance from one point to another is measured on the straight line which joins them.

13. Through the same point, only one straight line can be drawn parallel to a given straight line.

POSTULATES.

1.   A straight line can be drawn joining any two points.

2.   A straight line may be prolonged to any length.

3.   If two straight lines are unequal, the length of the less may be laid off on the greater.

4.   A straight line may be bisected; that is, divided into two equal parts.

5.   An angle may be bisected.

6.   A perpendicular may be drawn to a given straight line, either from a point without, or from a point on the line.

7.   A straight line may be drawn, making with a given straight line an angle equal to a given angle.

8.   A straight line may be drawn through a given point, parallel to a given line.

N O T E .

In making references, the following abbreviations are employed, viz.: A. for Axiom; B. for Book; C. for Corollary; D. for Definition; I. for Introduction; P. for Proposition; Prob. for Problem; Post. for Postulate; and S. for Scholium. In referring to the same Book, the number of the Book *is not* given; in referring to any other Book, the number of the Book *is* given.

PROPOSITION I. THEOREM.

If a straight line meets another straight line, the sum of the adjacent angles is equal to two right angles.

Let DC meet AB at C: then is the sum of the angles DCA and DCB equal to two right angles.

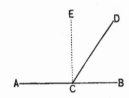

At C, let CE be drawn perpendicular to AB (Post. 6); then, by definition (D. 12), the angles ECA and ECB are both right angles, and consequently, their sum is equal to *two right angles.*

The angle DCA is equal to the sum of the angles ECA and ECD (A. 9); hence,

$$DCA + DCB = ECA + ECD + DCB;$$

But,   ECD + DCB is equal to ECB (A. 9); hence,

$$DCA + DCB = ECA + ECB.$$

The sum of the angles ECA and ECB, is equal to two right angles; consequently, its equal, that is, the sum of the angles DCA and DCB, must also be equal to two right angles; *which was to be proved.*

Cor. 1. If one of the angles DCA, DCB, is a right angle, the other must also be a right angle.

Cor. 2. The sum of the angles BAC, CAD, DAE, EAF, formed about a given point on the same side of a straight line BF, is equal to two right angles. For, their sum is equal to the sum of the

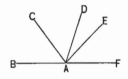

angles EAB and EAF; which, from the proposition just demonstrated, is equal to two right angles.

DEFINITIONS.

If two straight lines intersect each other, they form four angles about the point of intersection, which have received different names, with respect to each other.

1°. ADJACENT ANGLES are those which lie on the same side of one line, and on opposite sides of the other; thus, ACE and ECB, or ACE and ACD, are adjacent angles.

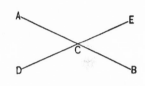

2°. OPPOSITE, or VERTICAL ANGLES, are those which lie on opposite sides of both lines; thus, ACE and DCB, or ACD and ECB, are opposite angles. From the proposition just demonstrated, the sum of any two adjacent angles is equal to two right angles.

PROPOSITION II. THEOREM.

If two straight lines intersect each other, the opposite or vertical angles are equal.

Let AB and DE intersect at C: then are the opposite or vertical angles equal.

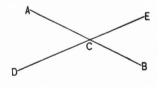

The sum of the adjacent angles ACE and ACD, is equal to two right angles (P. I.): the sum of the adjacent angles ACE and ECB, is also equal to two right angles. But things which are equal to the same thing, are equal to each other (A. 1); hence,

$$ACE + ACD = ACE + ECB;$$

Taking from both the common angle ACE (A. 3), there remains,

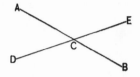

$$ACD = ECB.$$

In like manner, we find,

$$ACD + ACE = ACD + DCB;$$

and, taking away the common angle ACD, we have,

$$ACE = DCB.$$

Hence, *the proposition is proved.*

Cor. 1. If one of the angles about C is a right angle, all of the others are right angles also. For, (P. I., C. 1), each of its adjacent angles is a right angle; and from the proposition just demonstrated, its opposite angle is also a right angle.

Cor. 2. If one line DE, is perpendicular to another AB, then is the second line AB perpendicular to the first DE. For, the angles DCA and DCB are right angles, by definition (D. 12); and from what has just been proved, the angles ACE and BCE are also right angles. Hence, the two lines are mutually perpendicular to each other.

Cor. 3. The sum of all the angles ACB, BCD, DCE, ECF, FCA, that can be formed about a point, is equal to four right angles.

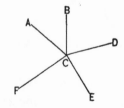

For, if two lines are drawn through the point, mutually perpendicular to each other, the sum of the angles which they form is equal to four right angles, and it is also equal to the sum of the given angles (A. 9). Hence, the sum of the given angles is equal to four right angles.

PROPOSITION III. THEOREM.

If two straight lines have two points in common, they coincide throughout their whole extent, and form one and the same line.

Let A and B be two points common to two lines: then the lines coincide throughout.

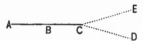

Between A and B they must coincide (A. 11). Suppose, now, that they begin to separate at some point C, beyond AB, the one becoming ACE, and the other ACD. If the lines do separate at C, one or the other must change direction at this point; but this is contradictory to the definition of a straight line (D. 4): hence, the supposition that they separate at any point is absurd. They must, therefore, coincide throughout; *which was to be proved.*

Cor. Two straight lines can intersect in only one point.

NOTE.—The method of demonstration employed above, is called the *reductio ad absurdum*. It consists in assuming an hypothesis which is the contradictory of the proposition to be proved, and then continuing the reasoning until the assumed hypothesis is shown to be false. Its contradictory is thus proved to be true. This method of demonstration is often used in Geometry.

PROPOSITION IV.   THEOREM.

If a straight line meets two other straight lines at a common point, making the sum of the contiguous angles equal to two right angles, the two lines met form one and the same straight line.

Let DC meet AC and BC at C, making the sum of the angles DCA and DCB equal to two right angles: then is CB the prolongation of AC.

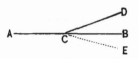

For, if not, suppose CE to be the prolongation of AC; then is the sum of the angles DCA and DCE equal to two right angles (P. I.): consequently, we have (A. 1),

$$DCA + DCB = DCA + DCE;$$

Taking from both the common angle DCA, there remains

$$DCB = DCE,$$

which is impossible, since a part can not be equal to the whole (A. 8). Hence, CB must be the prolongation of AC; *which was to be proved.*

PROPOSITION V.   THEOREM.

If two triangles have two sides and the included angle of the one equal to two sides and the included angle of the other, each to each, the triangles are equal in all respects.

In the triangles ABC and DEF, let AB be equal to DE,

AC to DF, and the angle A to the angle D: then are the triangles equal in all respects.

For, let ABC be ap-
plied to DEF, in such a
manner that the angle
A shall coincide with the
angle D, the side AB
taking the direction DE,
and the side AC the

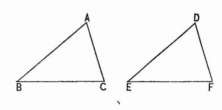

direction DF. Then, because AB is equal to DE, the vertex B will coincide with the vertex E; and because AC
is equal to DF, the vertex C will coincide with the vertex
F; consequently, the side BC will coincide with the side
EF (A. 11). The two triangles, therefore, coincide throughout, and are consequently equal in all respects (I., D. 15);
which was to be proved.

PROPOSITION VI. THEOREM.

*If two triangles have two angles and the included side of the
one equal to two angles and the included side of the other,
each to each, the triangles are equal in all respects.*

In the triangles ABC and DEF, let the angle B be
equal to the angle E, the
angle C to the angle F,
and the side BC to the
side EF: then are the
triangles equal in all respects.

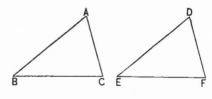

For, let ABC be applied to DEF in such a manner
that the angle B shall coincide with the angle E, the side
BC taking the direction EF, and the side BA the direc-

tion ED. Then, because BC is equal to EF, the vertex C will coincide with the vertex F; and because the angle C is equal to the angle F, the side CA will take the direction FD. Now, the vertex A being at the same time on the lines ED and FD, it must be at their intersection D (P. III., C.): hence, the triangles coincide throughout, and are therefore equal in all respects (I., D. 15); *which was to be proved.*

PROPOSITION VII. THEOREM.

The sum of any two sides of a triangle is greater than the third side.

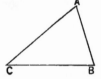

Let ABC be a triangle: then will the sum of any two sides, as AB, BC, be greater than the third side AC.

For, the distance from A to C, measured on any broken line AB, BC, is greater than the distance measured on the straight line AC (A. 12): hence, the sum of AB and BC is greater than AC; *which was to be proved.*

Cor. If from both members of the inequality,

$$AC < AB + BC,$$

we take away either of the sides AB, BC, as BC, for example, there remains (A. 5),

$$AC - BC < AB;$$

that is, *the difference between any two sides of a triangle is less than the third side.*

Scholium. In order that any three given lines may represent the sides of a triangle, the sum of any two must be greater than the third, and the difference of any two must be less than the third.

PROPOSITION VIII. THEOREM.

if from any point within a triangle two straight lines are drawn to the extremities of any side, their sum is less than that of the two remaining sides of the triangle.

Let O be any point within the triangle BAC, and let the lines OB, OC, be drawn to the extremities of any side, as BC: then the sum of BO and OC is less than the sum of the sides BA and AC.

Prolong one of the lines, as BO, till it meets the side AC in D; then, from Prop. VII., we have,

$$OC < OD + DC;$$

adding BO to both members of this inequality, recollecting that the sum of BO and OD is equal to BD, we have (A. 4),

$$BO + OC < BD + DC.$$

From the triangle BAD, we have (P. VII.),

$$BD < BA + AD;$$

adding DC to both members of this inequality, recollecting that the sum of AD and DC is equal to AC, we have,

$$BD + DC < BA + AC.$$

But it was shown that BO + OC is less than BD + DC; still more, then, is BO + OC less than BA + AC; *which was to be proved.*

PROPOSITION IX. THEOREM.

If two triangles have two sides of the one equal to two sides of the other, each to each, and the included angles unequal, the third sides are unequal; and the greater side belongs to the triangle which has the greater included angle.

In the triangles BAC and DEF, let AB be equal to DE, AC to DF, and the angle A greater than the angle D: then is BC greater than EF.

Let the line AG be drawn, making the angle CAG equal to the angle D (Post. 7); make AG equal to DE, and draw GC. Then the triangles AGC and DEF have two sides and the included angle of the one equal to two sides and the included angle of the other, each to each; consequently, GC is equal to EF (P. V.).

Now, the point G may be without the triangle ABC, it may be on the side BC, or it may be within the triangle ABC. Each case will be considered separately.

1°. When G is without the triangle ABC.

In the triangles GIC and AIB, we have, (P. VII.),

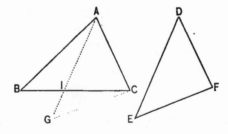

$$GI + IC > GC, \quad \text{and} \quad BI + IA > AB;$$

whence, by addition, recollecting that the sum of BI and IC is equal to BC, and the sum of GI and IA, to GA, we have,

$$AG + BC > AB + GC.$$

Or, since AG = AB, and GC = EF, we have,

$$AB + BC > AB + EF.$$

Taking away the common part AB, there remains (A. 5),

$$BC > EF.$$

2°. When G is on BC.

In this case, it is obvious that GC is less than BC; or since GC = EF, we have,

$$BC > EF.$$

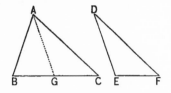

3°. When G is within the triangle ABC.

From Proposition VIII., we have,

$$BA + BC > GA + GC;$$

or, since GA = BA, and GC = EF, we have,

$$BA + BC > BA + EF.$$

Taking away the common part AB, there remains,

$$BC > EF.$$

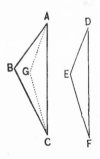

Hence, in each case, BC is greater than EF; *which was to be proved.*

Conversely: If in two triangles ABC and DEF, the side AB is equal to the side DE, the side AC to DF, and BC greater than EF, then is the angle BAC greater than the angle EDF.

For, if not, BAC must either be equal to, or less than, EDF. In the former case, BC would be equal to EF (P. V.), and in the latter case, BC would be less than EF; either of which would contradict the hypothesis: hence, BAC must be greater than EDF.

PROPOSITION X. THEOREM.

If two triangles have the three sides of the one equal to the three sides of the other, each to each, the triangles are equal in all respects.

In the triangles ABC and DEF, let AB be equal to DE, AC to DF, and BC to EF: then are the triangles equal in all respects.

For, since the sides AB, AC, are equal to DE, DF, each to each, if the angle A were greater than D, it would follow, by the last Proposition, that the side BC would be greater than EF; and if the angle A were less than

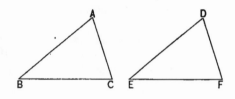

D, the side BC would be less than EF. But BC is equal to EF, by hypothesis; therefore, the angle A can neither be greater nor less than D: hence, it must be equal to it. The two triangles have, therefore, two sides and the included angle of the one equal to two sides and the included angle of the other, each to each; and, consequently, they are equal in all respects (P. V.); *which was to be proved.*

Scholium. In triangles, equal in all respects, the equal sides lie opposite the equal angles; and conversely.

PROPOSITION XI. THEOREM.

In an isosceles triangle the angles opposite the equal sides are equal.

Let BAC be an isosceles triangle, having the side AB equal to the side AC: then the angle C is equal to the angle B.

Join the vertex A and the middle point D of the base
BC. Then, AB is equal to AC, by hypothesis, AD common, and BD equal to DC, by con-
struction: hence, the triangles BAD,
and DAC, have the three sides of the
one equal to those of the other, each to
each; therefore, by the last Proposition,
the angle B is equal to the angle C;
which was to be proved.

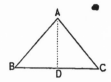

Cor. 1. An equilateral triangle is equiangular.

Cor. 2. The angle BAD is equal to DAC, and BDA to
CDA: hence, the last two are right angles. Consequently,
*a straight line drawn from the vertex of an isosceles tri-
angle to the middle of the base, bisects the angle at the
vertex, and is perpendicular to the base.*

PROPOSITION XII. THEOREM.

*If two angles of a triangle are equal, the sides opposite to
them are also equal, and consequently, the triangle is
isosceles.*

In the triangle ABC, let the angle
ABC be equal to the angle ACB: then
is AC equal to AB, and consequently,
the triangle is isosceles.

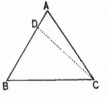

For, if AB and AC are not equal,
suppose one of them, as AB, to be the
greater. On this, take BD equal to AC (Post. 3), and
draw DC. Then, in the triangles ABC, DBC, we have
the side BD equal to AC, by construction, the side BC
common, and the included angle ACB equal to the included
angle DBC, by hypothesis: hence, the two triangles are equal

in all respects (P. V.).   But this is impossible, because
a part can not be equal to the whole (A. 8): hence, the
hypothesis that AB and AC are unequal, is false.   They
must, therefore, be equal; *which was to be proved.*

Cor.   An equiangular triangle is equilateral.

PROPOSITION XIII.   THEOREM.

*In any triangle, the greater side is opposite the greater
angle; and, conversely, the greater angle is opposite the
greater side.*

In the triangle ABC, let the angle
ACB be greater than the angle ABC:
then the side AB is greater than the
side AC.

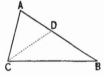

For, draw CD, making the angle BCD
equal to the angle B (Post. 7): then, in
the triangle DCB, we have the angles DCB and DBC equal:
hence, the opposite sides DB and DC are equal (P. XII.).
In the triangle ACD, we have (P. VII.),

$$AD + DC > AC;$$

or, since   DC = DB,   and   AD + DB = AB,   we have,

$$AB > AC;$$

which was to be proved.

Conversely: Let AB be greater than AC: then the angle
ACB is greater than the angle ABC.

For, if ACB were less than ABC, the side AB would
be less than the side AC, from what has just been proved;
if ACB were equal to ABC, the side AB would be equal
to AC, by Prop. XII.; but both conclusions contradict

the hypothesis: hence, ACB can neither be less than, nor equal to, ABC; it must, therefore, be greater; *which was to be proved.*

PROPOSITION XIV.   THEOREM.

From a given point only one perpendicular can be drawn to a given straight line.

Let A be a given point, and AB a perpendicular to DE: then can no other perpendicular to DE be drawn from A.

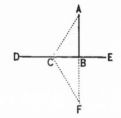

For, suppose a second perpendicular AC to be drawn. Prolong AB till BF is equal to AB, and draw CF. Then, the triangles ABC and FBC have AB equal to BF, by construction, CB common, and the included angles ABC and FBC equal, because both are right angles: hence, the angles ACB and FCB are equal (P. V.). But ACB is, by a hypothesis, a right angle: hence, FCB must also be a right angle, and consequently, the line ACF must be a straight line (P. IV.). But this is impossible (A. 11). The hypothesis that two perpendiculars can be drawn is, therefore, absurd; consequently, only one such perpendicular can be drawn; *which was to be proved.*

If the given point is on the given line, the proposition is equally true. For, if from A two perpendiculars AB and AC could be drawn to DE, we should have BAE and CAE each equal to a right angle; and consequently, equal to each other; which is absurd (A. 8).

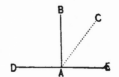

2

PROPOSITION XV.   THEOREM.

If from a point without a straight line a perpendicular is let fall on the line, and oblique lines are drawn to different points of it:

1°. *The perpendicular is shorter than any oblique line.*

2°. *Any two oblique lines that meet the given line at points equally distant from the foot of the perpendicular, are equal.*

3°. *Of two oblique lines that meet the given line at points unequally distant from the foot of the perpendicular, the one which meets it at the greater distance is the longer.*

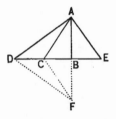

Let A be a given point, DE a given straight line, AB a perpendicular to DE, and AD, AC, AE oblique lines, BC being equal to BE, and BD greater than BC. Then AB is less than any of the oblique lines, AC is equal to AE, and AD greater than AC.

Prolong AB until BF is equal to AB, and draw FC, FD.

1°. In the triangles ABC, FBC, we have the side AB equal to BF, by construction, the side BC common, and the included angles ABC and FBC equal, because both are right angles: hence, FC is equal to AC (P. V.). But, AF is shorter than ACF (A. 12): hence, AB, the half of AF, is shorter than AC, the half of ACF; *which was to be proved.*

2°. In the triangles ABC and ABE, we have the side BC equal to BE, by hypothesis, the side AB common, and the included angles ABC and ABE equal, because both are right angles: hence, AC is equal to AE; *which was to be proved.*

3°. It may be shown, as in the first case, that AD is equal to DF. Then, because the point C lies within the triangle ADF, the sum of the lines AD and DF is greater than the sum of the lines AC and CF (P. VIII.): hence, AD, the half of ADF, is greater than AC, the half of ACF; *which was to be proved.*

Cor. 1. The perpendicular is the shortest distance from a point to a line.

Cor. 2. From a given point to a given straight line, only two equal straight lines can be drawn; for, if there could be more, there would be at least two equal oblique lines on the same side of the perpendicular; which is impossible.

PROPOSITION XVI. THEOREM.

If a perpendicular is drawn to a given straight line at its middle point:

1°. *Any point of the perpendicular is equally distant from the extremities of the line*

2°. *Any point, without the perpendicular, is unequally distant from the extremities.*

Let AB be a given straight line, C its middle point, and EF the perpendicular. Then any point of EF is equally distant from A and B; and any point without EF, is unequally distant from A and B.

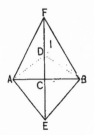

1°. From any point of EF, as D, draw the lines DA and DB. Then DA and DB are equal (P. XV.): hence, D is equally distant from A and B; *which was to be proved.*

2°. From any point without EF, as I, draw IA and IB. One of these lines, as IA, will cut EF in some point D; draw DB. Then, from what has just been shown, DA and DB are equal; but IB is less than the sum of ID and DB (P. VII.); and because the sum of ID and DB is equal to the sum of ID and DA, or IA, we have IB less than IA: hence, I is unequally distant from A and B; *which was to be proved.*

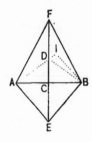

Cor. If a straight line, EF, has two of its points, E and F, each equally distant from A and B, it is perpendicular to the line AB at its middle point.

PROPOSITION XVII. THEOREM.

If two right-angled triangles have the hypothenuse and a side of the one equal to the hypothenuse and a side of the other, each to each, the triangles are equal in all respects.

Let the right-angled triangles ABC and DEF have the hypothenuse AC equal to DF, and the side AB equal to DE: then the triangles are equal in all respects.

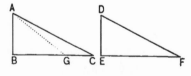

If the side BC is equal to EF, the triangles are equal, in accordance with Proposition X. Let us suppose then, that BC and EF are unequal, and that BC is the longer. On BC lay off BG equal to EF, and draw AG. The triangles ABG and DEF have AB equal to DE, by hypothesis, BG equal to EF, by construction, and the angles B and E equal, because both are right angles; consequently, AG is equal to DF (P. V.). But, AC is equal to DF, by hypothesis: hence, AG and AC are equal, which is impossible (P. XV.). The hypothesis that BC and EF are unequal, is, therefore, absurd: hence, the triangles have all their sides equal, each to each, and are, consequently, equal in all respects; *which was to be proved.*

PROPOSITION XVIII. THEOREM.

If two straight lines are perpendicular to a third straight line, they are parallel.

Let the two lines AC, BD, be perpendicular to AB: then they are parallel.

For, if they could meet in a point O, there would be two perpendiculars OA, OB, drawn from the same point to the same straight line; which is  impossible (P. XIV.): hence, the lines are parallel; *which was to be proved.*

DEFINITIONS.

If a straight line EF intersect two other straight lines AB and CD, it is called a *secant*, with respect to them. The eight angles formed about the points of intersection have different names, with respect to each other.

1°. INTERIOR ANGLES ON THE SAME SIDE, are those that lie on the same side of the secant and *within* the other two lines. Thus, BGH and GHD are interior angles on the same side.

2°. EXTERIOR ANGLES ON THE SAME SIDE are those that lie on the same side of the secant and *without* the other two lines.   Thus, EGB and DHF are exterior angles on the same side.

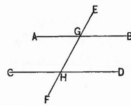

3°. ALTERNATE ANGLES are those that lie on opposite sides of the secant and *within* the other two lines, but not adjacent.   Thus, AGH and GHD are alternate angles.

4°. ALTERNATE EXTERIOR ANGLES are those that lie on opposite sides of the secant and *without* the other two lines.   Thus, AGE and FHD are alternate exterior angles.

5°. OPPOSITE EXTERIOR AND INTERIOR ANGLES are those that lie on the same side of the secant, the one *within* and the other *without* the other two lines, but not adjacent.   Thus, EGB and GHD are opposite exterior and interior angles.

PROPOSITION XIX.   THEOREM.

If two straight lines meet a third straight line, making the sum of the interior angles on the same side equal to two right angles, the two lines are parallel.

Let the lines KC and HD meet the line BA, making the sum of the angles BAC and ABD equal to two right angles; then KC and HD are parallel.

Through G, the middle point of AB, draw GF perpendicular to KC, and prolong it to E.

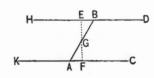

The sum of the angles GBE and GBD is equal to two right

angles (P. I.); the sum of the angles FAG and GBD is equal to two right angles, by hypothesis: hence (A. 1),

$$GBE + GBD = FAG + GBD.$$

Taking away the common part GBD, we have the angle GBE equal to the angle FAG. Again, the angles BGE and AGF are equal, because they are vertical angles (P. II.): hence, the triangles GEB and GFA have two of their angles and the included · side equal, each to each; they are, therefore, equal in all respects (P. VI.): hence, the angle GEB is equal to the angle GFA. But, GFA is a right angle, by construction; GEB must, therefore, be a right angle: hence, the lines KC and HD are perpendicular to EF, and are, therefore, parallel (P. XVIII.); *which was to be proved.*

Cor. 1. If two straight lines are cut by a third straight line, making the alternate angles equal to each other, th·ⁱ two straight lines are parallel.

Let the angle HGA be equal to GHD. Adding to both the angle HGB, we have,

$$HGA + HGB = GHD + HGB.$$

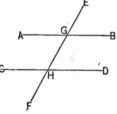

But the first sum is equal to two right angles (P. I.): hence, the second sum is also equal to two right angles; therefore, from what has just been shown, AB and CD are parallel.

Cor. 2. If two straight lines are cut by a third, making the opposite exterior and interior angles equal, the two straight lines are parallel. Let the angles EGB and GHD be equal: Now EGB and AGH are equal, because they are vertical angles (P. II.); and consequently, AGH and GHD are equal: hence, from *Cor.* 1, AB and CD are parallel.

PROPOSITION XX.   THEOREM.

*If a straight line intersects two parallel straight lines, the
sum of the interior angles on the same side is equal to
two right angles.*

Let the parallels AB, CD, be cut by the secant line
FE : then the sum of HGB and GHD is equal to two right
angles.

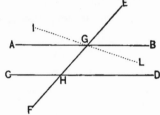

For, if the sum of HGB
and GHD is not equal to two
right angles, let IGL be drawn,
making the sum of HGL and
GHD equal to two right angles ;
then IL and CD are parallel
(P. XIX.) ; and consequently, we have two lines, GB,
GL, drawn through the same point G and parallel to
CD, which is impossible (A. 13) : hence, the sum of HGB
and GHD is equal to two right angles ; *which was to be
proved.*

In like manner, it may be proved that the sum of HGA
and GHC is equal to two right angles.

Cor. 1. If HGB is a right angle, GHD is a right angle
also : hence, *if a line is perpendicular to one of two parallels, it is perpendicular to the other also.*

Cor. 2. *If a straight line intersects two parallels, the alternate angles are equal.*

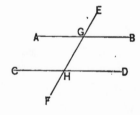

For, if AB and CD are parallel,
the sum of BGH and GHD is equal
to two right angles ; the sum of
BGH and HGA is also equal to two
right angles (P. I.) : hence, these
sums are equal.   Taking away the

common part BGH, there remains the angle GHD equal to
HGA. In like manner, it may be shown that BGH and
GHC are equal.

Cor. 3. *If a straight line intersects two parallels, the
opposite exterior and interior angles are equal.* The angles
DHG and HGA are equal, from what has just been shown.
The angles HGA and BGE are equal, because they are vertical: hence, DHG and BGE are equal. In like manner, it
may be shown that CHG and AGE are equal.

Scholium. Of the eight angles formed by a line cutting
two parallel lines obliquely, the four acute angles are equal,
and so, also, are the four obtuse angles.

PROPOSITION XXI. THEOREM.

*If two straight lines intersect a third straight line, making
the sum of the interior angles on the same side less
than two right angles, the two lines will meet if sufficiently produced.*

Let the two lines CD, IL, meet the line EF, making
the sum of the interior angles HGL, GHD, less than two
right angles: then will IL and CD meet if sufficiently
produced.

For, if they do not meet, they
must be parallel (D. 16). But, if
they were parallel, the sum of the
interior angles HGL, GHD, would
be equal to two right angles
(P. XX.), which contradicts the
hypothesis: hence, IL, CD, will meet if sufficiently produced; *which was to be proved.*

Cor. It is evident that IL and CD will meet on that side of EF, on which the sum of the two angles is less than two right angles.

PROPOSITION XXII. THEOREM.

If two straight lines are parallel to a third line, they are parallel to each other.

Let AB and CD be respectively parallel to EF: then are they parallel to each other.

For, draw PR perpendicular to EF; then is it perpendicular to AB, and also to CD (P. XX., C. 1): hence, AB and CD are perpendic-

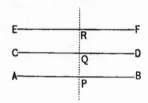

ular to the same straight line, and consequently, they are parallel to each other (P. XVIII.); *which was to be proved.*

PROPOSITION XXIII. THEOREM.

Two parallels are every-where equally distant.

Let AB and CD be parallel: then are they every-where equally distant.

From any two points of AB, as F and E, draw FH and EG perpendicular to CD; they are also perpendicular to AB (P. XX., C. 1), and measure the distance between

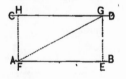

AB and CD, at the points F and E. Draw also FG. The lines FH and EG are parallel (P. XVIII.): hence, the alternate angles HFG and FGE are equal (P. XX., C. 2). The lines AB and CD are parallel, by hypothesis: hence, the alternate angles EFG and FGH are equal. The triangles FGE and FGH have, therefore, the angle HGF equal to GFE, GFH equal to FGE, and the side FG common; they are, therefore, equal in all respects (P. VI.): hence, FH is equal to EG; and consequently, AB and CD are every-where equally distant; *which was to be proved.*

PROPOSITION XXIV. THEOREM.

If two angles have their sides parallel, and lying either in the same or in opposite directions, they are equal.

1°. Let the angles ABC and DEF have their sides parallel, and lying in the same direction: then are they equal.

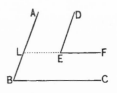

Prolong FE to L. Then, because DE and AL are parallel, the exterior angle DEF is equal to its opposite interior angle ALE (P. XX., C. 3); and, because BC and LF are parallel, the exterior angle ALE is equal to its opposite interior angle ABC: hence, DEF is equal to ABC; *which was to be proved.*

2°. Let the angles ABC and GHK have their sides parallel, and lying in opposite directions: then are they equal.

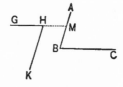

Prolong GH to M. Then, because KH and BM are parallel, the exterior angle GHK is equal to its opposite interior angle HMB; and because HM and BC are parallel, the angle HMB is equal to its alternate angle MBC (P. XX., C. 2): hence, GHK is equal to ABC; *which was to be proved.*

Cor. The opposite angles of a parallelogram are equal.

PROPOSITION XXV. THEOREM.

In any triangle, the sum of the three angles is equal to two right angles.

Let CBA be any triangle: then the sum of the angles C, A, and B, is equal to two right angles.

For, prolong CA to D, and draw AE parallel to BC.

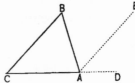

Then, since AE and CB are parallel, and CD cuts them, the exterior angle DAE is equal to its opposite interior angle C (P. XX., C. 3). In like manner, since AE and CB are parallel, and AB cuts them, the alternate angles ABC and BAE are equal: hence, the sum of the three angles of the triangle BAC is equal to the sum of the angles CAB, BAE, EAD; but this sum is equal to two right angles (P. I., C. 2); consequently, the sum of the three angles of the triangle, is equal to two right angles (A. 1); *which was to be proved.*

Cor. 1. Two angles of a triangle being given, the third may be found by subtracting their sum from two right angles.

Cor. 2. If two angles of one triangle are respectively equal to two angles of another, the two triangles are mutually equiangular.

Cor. 3. In any triangle, there can be but one right angle; for if there were two, the third angle would be zero. Nor can a triangle have more than one obtuse angle.

Cor. 4. In any right-angled triangle, the sum of the acute angles is equal to a right angle.

Cor. 5. Since every equilateral triangle is also equi-angular (P. XI., C. 1), each of its angles is equal to the third part of two right angles; so that, if the right angle is expressed by 1, each angle of an equilateral triangle is expressed by $\frac{2}{3}$.

Cor. 6. In any triangle ABC, the exterior angle BAD is equal to the sum of the interior opposite angles B and C. For, AE being parallel to BC, the part BAE is equal to the angle B, and the other part DAE, is equal to the angle C.

PROPOSITION XXVI. THEOREM.

The sum of the interior angles of a polygon is equal to two right angles taken as many times, less two, as the polygon has sides.

Let ABCDE be any polygon; then the sum of its interior angles A, B, C, D, and E, is equal to two right angles taken as many times, less two, as the polygon has sides.

From the vertex of any angle A, draw diagonals AC, AD. The polygon will be divided into as many triangles, less two, as it has sides, having the point A for a common vertex, and for bases, the sides of the polygon, except the two which form the angle A. It is evident, also, that the sum of the angles of these triangles does not differ from the sum of the angles of the polygon: hence, the sum of the angles of the polygon is equal to two right angles, taken as many times as there are triangles; that is, as many times, less two, as the polygon has sides; *which was to be proved.*

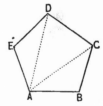

Cor. 1. The sum of the interior angles of a quadrilateral is equal to two right angles taken twice; that is, to four right angles. If the angles of a quadrilateral are equal, each is a right angle.

Cor. 2. The sum of the interior angles of a pentagon is equal to two right angles taken three times; that is, to six right angles: hence, when a pentagon is equiangular, each angle is equal to the fifth part of six right angles, or to $\frac{6}{5}$ of one right angle.

Cor. 3. The sum of the interior angles of a hexagon is equal to eight right angles: hence, in the equiangular hexagon, each angle is the sixth part of eight right angles, or $\frac{4}{3}$ of one right angle.

Cor. 4. In any equiangular polygon, any interior angle is equal to twice as many right angles as the figure has sides, less four right angles, divided by the number of angles.

PROPOSITION XXVII. THEOREM.

The sum of the exterior angles of a polygon is equal to four right angles.

Let the sides of the polygon ABCDE be prolonged, in the same order, forming the exterior angles *a, b, c, d, e*; then the sum of these exterior angles is equal to four right angles.

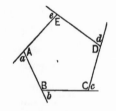

For, each interior angle, together with the corresponding exterior angle, is equal to two right angles (P. I.); hence, the sum of all the interior and exterior angles is equal to two right angles taken

as many times as the polygon has sides. But the sum of
the interior angles is equal to two right angles taken as
many times, less two, as the polygon has sides: hence, the
sum of the exterior angles is equal to two right angles
taken twice; that is, equal to four right angles; *which
was to be proved.*

PROPOSITION XXVIII. THEOREM.

*In any parallelogram, the opposite sides are equal, each to
each.*

Let ABCD be a parallelogram: then
AB is equal to DC, and AD to BC.

For, draw the diagonal BD. Then,
because AB and DC are parallel, the
angle DBA is equal to its alternate
angle BDC (P. XX., C. 2); and, because AD and BC are
parallel, the angle BDA is equal to its alternate angle
DBC. The triangles ABD and CDB, have, therefore, the
angle DBA equal to CDB, the angle BDA equal to DBC, and
the included side DB common; consequently, they are
equal in all respects: hence, AB is equal to DC, and AD to
BC; *which was to be proved.*

Cor. 1. A diagonal of a parallelogram divides it into
two triangles equal in all respects.

Cor. 2. Two parallels included between two other parallels, are equal.

Cor. 3. If two parallelograms have two sides and the
included angle of the one, equal to two sides and the
included angle of the other, each to each, they are equal.

PROPOSITION XXIX. THEOREM.

If the opposite sides of a quadrilateral are equal, each to each, the figure is a parallelogram.

In the quadrilateral ABCD, let AB be equal to DC, and AD to BC: then is it a parallelogram.

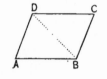

Draw the diagonal DB. Then, the triangles ADB and CBD, have the sides of the one equal to the sides of the other, each to each; and therefore, the triangles are equal in all respects: hence, the angle ABD is equal to the angle CDB (P. X., S.); and consequently, AB is parallel to DC (P. XIX., C. 1). The angle DBC is also equal to the angle BDA, and consequently, BC is parallel to AD: hence, the opposite sides are parallel, two and two; that is, the figure is a parallelogram (D. 28); *which was to be proved.*

PROPOSITION XXX. THEOREM.

If two sides of a quadrilateral are equal and parallel, the figure is a parallelogram.

In the quadrilateral ABCD, let AB be equal and parallel to DC: then the figure is a parallelogram.

Draw the diagonal DB. Then, because AB and DC are parallel, the angle ABD is equal to its alternate angle CDB. Now, the triangles ABD and CDB have the side DC equal to AB, by hypothesis, the side DB common, and the included angle ABD equal to BDC, from what has just been shown:

hence, the triangles are equal in all respects (P. V.);
and consequently, the alternate angles ADB and DBC are
equal. The sides BC and AD are, therefore, parallel, and
the figure is a parallelogram; *which was to be proved.*

Cor. If two points are taken at equal distances from a
given straight line, and on the same side of it, the straight
line joining them is parallel to the given line.

PROPOSITION XXXI. THEOREM.

*The diagonals of a parallelogram divide each other into
equal parts, or mutually bisect each other.*

Let ABCD be a parallelogram, and AC,
BD, its diagonals: then AE is equal to EC,
and BE to ED.

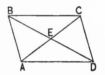

For, the triangles BEC and AED, have
the angles EBC and ADE equal (P. XX.,
C. 2), the angles ECB and DAE equal, and the included
sides BC and AD equal: hence, the triangles are equal in
all respects (P. VI.); consequently, AE is equal to EC, and
BE to ED; *which was to be proved.*

Scholium. In a rhombus, the sides AB, BC, being
equal, the triangles AEB, EBC, have the sides of the one
equal to the corresponding sides of the other; they are,
therefore, equal in all respects: hence, the angles AEB,
BEC, are equal, and therefore, the two diagonals bisect each
other at right angles.

EXERCISES.

1. Show that the lines which bisect (*halve*) two vertical angles, form one and the same straight line.

2. Given two lines, BE and AD; join B with D and A with E, and show that BD + AE is greater than BE + AD. (P. VII.)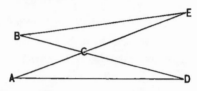

3. One of the two interior angles on the same side, formed by a straight line meeting two parallels, is one-half of a right angle; what is the other angle equal to?

4. The sum of two angles of a triangle is $\frac{4}{5}$ of a right angle; what is the other angle equal to?

5. One of the acute angles of a right-angled triangle is $\frac{2}{3}$ of a right angle; what is the other?

6. Show that the line which bisects the exterior vertical angle of an isosceles triangle is parallel to the base of the triangle. (P. XXV., C. 6; P. XIX., C. 1.)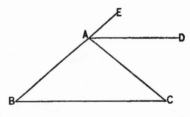

7. The sum of the interior angles of a polygon is 12 right angles; what is the polygon?

8. What is the sum of the interior angles of a heptagon equal to?

9. The sum of five angles of a given equiangular polygon is 8 right angles; what is the polygon?

10. What part of a right angle is an angle of an equiangular decagon?

11. How many sides has a polygon in which the sum of the interior angles is equal to the sum of the exterior angles?

12. Construct a square, having given one of its diagonals.

NOTE 1.—The *complement* of an angle is the difference between that angle and a right angle; thus, EOB is the complement of AOE.

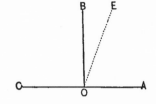

NOTE 2.—The *supplement* of an angle is the difference between that angle and two right angles; thus, EOC is the supplement of AOE.

13. An angle is $\frac{4}{5}$ of a right angle; what is its complement? and what its supplement?

14. Show that any two adjacent angles of a parallelogram are supplements of each other.

15. Show that if two parallelograms have one angle in each equal, their remaining angles are equal each to each.

16. Show that if two sides of a quadrilateral are parallel and two opposite angles equal, the figure is a parallelogram.

17. Show that if the opposite angles of a quadrilateral are equal, each to each, the figure is a parallelogram.

18. Show that the lines which bisect the angles of any quadrilateral form, by their intersection, another quadrilateral, the opposite angles of which are supplements of each other. [Twice the angle B is equal to the sum of the angles CDE and DEF.]

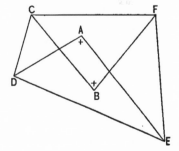

BOOK II.

DEFINITIONS.

1. THE RATIO of one quantity to another of the same kind. is the quotient obtained by dividing the second by the first. The first quantity is called the ANTECEDENT, and the second, the CONSEQUENT.

2. A PROPORTION is an expression of equality between two equal ratios. Thus,

$$\frac{B}{A} = \frac{D}{C},$$

expresses the fact that the ratio of A to B is equal to the ratio of C to D. In Geometry, the proportion is written thus,

$$A : B :: C : D,$$

and read, A is to B, as C is to D.

3. A CONTINUED PROPORTION is one in which several ratios are successively equal to each other; as,

$$A : B :: C : D :: E : F :: G : H, \&c.$$

4. There are four terms in every proportion. The first and second form the *first couplet*, and the third and fourth,

the *second couplet.* The first and fourth terms are called *extremes;* the second and third, *means,* and the fourth term, a *fourth proportional* to the three others. When the second term is equal to the third, it is said to be a *mean proportional* between the extremes. In this case, there are but three different quantities in the proportion, and the last is said to be a *third proportional to the two others.* Thus, if we have,

$$A \ : \ B \ :: \ B \ : \ C,$$

B is a *mean* proportional between A and C, and C is a *third* proportional to A and B.

5. Quantities are in proportion by *alternation,* when antecedent is compared with antecedent, and consequent with consequent.

6. Quantities are in proportion by *inversion,* when antecedents are made consequents, and consequents, antecedents.

7. Quantities are in proportion by *composition,* when the sum of antecedent and consequent is compared with either antecedent or consequent.

8. Quantities are in proportion by *division,* when the difference of the antecedent and consequent is compared with either antecedent or consequent.

9. Four quantities are *reciprocally* proportional, when the first is to the second as the fourth is to the third. *Two varying* quantities are reciprocally proportional, when their product is a fixed quantity, as $xy = m$.

10. Equimultiples of two or more quantities, are the products obtained by multiplying each by the same quan tity. Thus, mA and mB, are equimultiples of A and B.

PROPOSITION I. THEOREM.

If four quantities are in proportion, the product of the means is equal to the product of the extremes.

Assume the proportion,

$$A : B :: C : D; \quad \text{whence} \quad \frac{B}{A} = \frac{D}{C};$$

clearing of fractions, we have,

$$BC = AD;$$

which was to be proved.

Cor. If B is equal to C, there are but three proportional quantities; in this case, *the square of the mean is equal to the product of the extremes.*

PROPOSITION II. THEOREM.

If the product of two factors is equal to the product of two other factors, either pair of factors may be made the extremes and the other pair the means of a proportion.

Assume
$$B \times C = A \times D;$$

dividing each member by $A \times C$, we have,

$$\frac{B}{A} = \frac{D}{C}, \quad \text{or} \quad A : B :: C : D;$$

in like manner, we have,

$$\frac{A}{B} = \frac{C}{D}, \quad \text{or} \quad B : A :: D : C;$$

which was to be proved.

PROPOSITION III. THEOREM.

If four quantities are in proportion, they are in proportion by alternation.

Assume the proportion,

$$A \ : \ B \ :: \ C \ : \ D; \quad \text{whence,} \quad \frac{B}{A} = \frac{D}{C}.$$

Multiplying each member by $\frac{C}{B}$, we have,

$$\frac{C}{A} = \frac{D}{B}; \quad \text{or} \quad A \ : \ C \ :: \ B \ : \ D;$$

which was to be proved.

PROPOSITION IV. THEOREM.

If one couplet in each of two proportions is the same, the other couplets form a proportion.

Assume the proportions,

$$A \ : \ B \ :: \ C \ : \ D; \quad \text{whence,} \quad \frac{B}{A} = \frac{D}{C};$$

and $\quad A \ : \ B \ :: \ F \ : \ G; \quad \text{whence,} \quad \frac{B}{A} = \frac{G}{F}.$

From Axiom 1, we have,

$$\frac{D}{C} = \frac{G}{F}; \quad \text{whence,} \quad C \ : \ D \ :: \ F \ : \ G;$$

which was to be proved.

Cor. If the antecedents, in two proportions, are the same, the consequents are proportional. For, the antecedents of the second couplets may be made the consequents of the first, by alternation (P. III.).

PROPOSITION V. THEOREM.

If four quantities are in proportion, they are in proportion by inversion.

Assume the proportion,

$$A : B :: C : D; \quad \text{whence,} \quad \frac{B}{A} = \frac{D}{C}.$$

If we take the reciprocals of each member (A. 7), we have,

$$\frac{A}{B} = \frac{C}{D}; \quad \text{whence,} \quad B : A :: D : C;$$

which was to be proved.

PROPOSITION VI. THEOREM.

If four quantities are in proportion, they are in proportion by composition or division.

Assume the proportion,

$$A : B :: C : D; \quad \text{whence,} \quad \frac{B}{A} = \frac{D}{C}.$$

If we add 1 to each member, and subtract 1 from each member, we have,

$$\frac{B}{A} + 1 = \frac{D}{C} + 1; \quad \text{and} \quad \frac{B}{A} - 1 = \frac{D}{C} - 1;$$

whence, by reducing to a common denominator, we have,

$$\frac{B + A}{A} = \frac{D + C}{C}, \quad \text{and} \quad \frac{B - A}{A} = \frac{D - C}{C}; \quad \text{whence,}$$

$$A : B + A :: C : D + C, \quad \text{and} \quad A : B - A :: C : D - C;$$

which was to be proved.

PROPOSITION VII. THEOREM.

Equimultiples of two quantities are proportional to the quantities themselves.

Let A and B be any two quantities; then $\dfrac{B}{A}$ will denote their ratio.

If we multiply each term of this fraction by m, its value will not be changed; and we shall have,

$$\frac{mB}{mA} = \frac{B}{A}; \quad \text{whence,} \quad mA : mB :: A : B;$$

which was to be proved.

PROPOSITION VIII. THEOREM.

If four quantities are in proportion, any equimultiples of the first couplet are proportional to any equimultiples of the second couplet.

Assume the proportion,

$$A : B :: C : D; \quad \text{whence,} \quad \frac{B}{A} = \frac{D}{C}.$$

If we multiply each term of the first member by m, and each term of the second member by n, we have,

$$\frac{mB}{mA} = \frac{nD}{nC}; \quad \text{whence,} \quad mA : mB :: nC : nD;$$

which was to be proved.

PROPOSITION IX.   THEOREM.

If two quantities are increased or diminished by like parts of each, the results are proportional to the quantities themselves.

We have, Prop. VII.,

$$A \ : \ B \ :: \ mA \ : \ mB.$$

If we make $m = 1 \pm \dfrac{p}{q}$, in which $\dfrac{p}{q}$ is any fraction, we have,

$$A \ : \ B \ :: \ A \pm \frac{p}{q}A \ : \ B \pm \frac{p}{q}B;$$

which was to be proved.

PROPOSITION X.   THEOREM.

If both terms of the first couplet of a proportion are increased or diminished by like parts of each; and if both terms of the second couplet are increased or diminished by any other like parts of each, the results are in proportion.

Since we have, Prop. VIII.,

$$mA \ : \ mB \ :: \ nC \ : \ nD;$$

if we make $m = 1 \pm \dfrac{p}{q}$, and $n = 1 \pm \dfrac{p'}{q'}$, we have,

$$A \pm \frac{p}{q}A \ : \ B \pm \frac{p}{q}B \ :: \ C \pm \frac{p'}{q'}C \ : \ D \pm \frac{p'}{q'}D;$$

which was to be proved.

PROPOSITION XI. THEOREM.

In any continued proportion, the sum of the antecedents is to the sum of the consequents, as any antecedent to its corresponding consequent.

From the definition of a continued proportion (D. 3),

$$A : B :: C : D :: E : F :: G : H, \&c.;$$

hence,

$$\frac{B}{A} = \frac{B}{A}; \qquad \text{whence,} \qquad BA = AB;$$

$$\frac{B}{A} = \frac{D}{C}; \qquad \text{whence,} \qquad BC = AD;$$

$$\frac{B}{A} = \frac{F}{E}; \qquad \text{whence,} \qquad BE = AF;$$

$$\frac{B}{A} = \frac{H}{G}; \qquad \text{whence,} \qquad BG = AH;$$

$$\&c., \qquad\qquad\qquad \&c.$$

Adding and factoring, we have,

$$B(A + C + E + G + \&c.) = A(B + D + F + H + \&c.):$$

hence, from Proposition II.,

$$A + C + E + G + \&c. : B + D + F + H + \&c. :: A : B;$$

which was to be proved.

PROPOSITION XII. THEOREM.

The products of the corresponding terms of two proportions are proportional.

Assume the two proportions,

$$A \,:\, B \,::\, C \,:\, D; \quad \text{whence,} \quad \frac{B}{A} = \frac{D}{C};$$

and $\quad E \,:\, F \,::\, G \,:\, H; \quad \text{whence,} \quad \frac{F}{E} = \frac{H}{G}.$

Multiplying the equations, member by member, we have,

$$\frac{BF}{AE} = \frac{DH}{CG}; \quad \text{whence,} \quad AE \,:\, BF \,::\, CG \,:\, DH;$$

which was to be proved.

Cor. 1. If the corresponding terms of two proportions are equal, each term of the resulting proportion is the square of the corresponding term in either of the given proportions: hence, *If four quantities are proportional, their squares are proportional.*

Cor. 2. If the principle of the proposition be extended to three or more proportions, and the corresponding terms of each be supposed equal, it will follow that, *like powers of proportional quantities are proportionals.*

BOOK III.

DEFINITIONS.

1. A Circle is a plane figure, bounded by a curved line, every point of which is equally distant from a point within, called the *centre*.

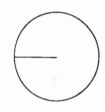

The bounding line is called the *circumference*.

2. A Radius is a straight line drawn from the centre to any point of the circumference.

3. A Diameter is a straight line drawn through the centre and terminating in the circumference.

All radii of the same circle are equal. All diameters are also equal, and each is double the radius.

4. An Arc is any part of a circumference.

5. A Chord is a straight line joining the extremities of an arc.

Any chord belongs to two arcs: the smaller one is meant, unless the contrary is expressed.

6. A Segment is a part of a circle included between an arc and its chord.

7. A Sector is a part of a circle included between an arc and the two radii drawn to its extremities.

8. An INSCRIBED ANGLE is an angle whose vertex is in the circumference, and whose sides are chords.

9. An INSCRIBED POLYGON is a polygon whose vertices are all in the circumference. The sides are chords.

10. A SECANT is a straight line which cuts the circumference in two points.

11. A TANGENT is a straight line which touches the circumference in one point only. This point is called, the *point of contact*, or the *point of tangency.*

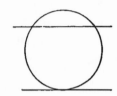

12. Two circles are *tangent to each other*, when they touch each other in one point only. This point is called, the *point of contact*, or the *point of tangency.*

13. A Polygon is *circumscribed about a circle*, when each of its sides is tangent to the circumference.

14. A Circle is *inscribed in a polygon*, when its circumference touches each of the sides of the polygon.

POSTULATE.

A circumference can be described from any point as a *centre*, and with any *radius*.

PROPOSITION I. THEOREM.

Any diameter divides the circle, and also its circumference, into two equal parts.

Let AEBF be a circle, and AB any diameter: then will it divide the circle and its circumference into two equal parts.

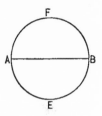

For, let AFB be applied to AEB, the diameter AB remaining common; then will they coincide; otherwise there would be some points in either one or the other of the curves unequally distant from the centre; which is impossible (D. 1): hence, AB divides the circle, and also its circumference, into two equal parts; *which was to be proved.*

PROPOSITION II. THEOREM.

A diameter is greater than any other chord.

Let AD be a chord, and AB a diameter through one extremity, as A: then will AB be greater than AD.

Draw the radius CD. In the triangle ACD, we have AD less than the sum of AC and CD (B. I., P. VII.). But this sum is equal to AB (D. 3): hence, AB is greater than AD; *which was to be proved.*

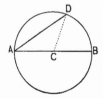

PROPOSITION III.   THEOREM.

A straight line can not meet a circumference in more than two points.

Let AEBF be a circumference, and AB a straight line: then AB can not meet the circumference in more than two points.

For, suppose that AB could meet the circumference in three points. By draw-

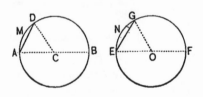

ing radii to these points, we should have three equal straight lines drawn from the same point to the same straight line; which is impossible (B. I.; P. XV., C. 2): hence, AB can not meet the circumference in more than two points; *which was to be proved.*

PROPOSITION IV.   THEOREM.

In equal circles, equal arcs are subtended by equal chords; and conversely, equal chords subtend equal arcs.

1°. In the equal circles ADB and EGF, let the arcs AMD and ENG be equal: then are the chords AD and EG equal.

Draw the diameters AB and EF. If the semicircle ADB be applied to the semicircle EGF, it will coincide with it, and the semi-circumference ADB will coincide with the semi-circumference EGF. But the part AMD is equal to the part ENG, by hypothesis: hence, the point D will fall on G; therefore, the chord AD will coincide with EG (A. 11), and is, therefore, equal to it; *which was to be proved.*

2°. Let the chords AD and EG be equal: then will the arcs AMD and ENG be equal.

Draw the radii CD and OG. The triangles ACD and EOG have all the sides of the one equal to the corresponding sides of the other; they are, therefore, equal in all respects: hence, the angle ACD is equal to EOG. If, now, the sector ACD be placed upon the sector EOG, so that the angle ACD shall coincide with the angle EOG, the sectors will coincide throughout; and, consequently, the arcs AMD and ENG will coincide: hence, they are equal; *which was to be proved.*

PROPOSITION V. THEOREM.

In equal circles, a greater arc is subtended by a greater chord; and conversely, a greater chord subtends a greater arc.

1°. In the equal circles ADL and EGK, let the arc EGP be greater than the arc AMD: then is the chord EP greater than the chord AD.

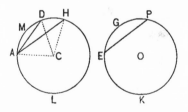

For, place the circle EGK upon AHL, so that the centre O shall fall upon the centre C, and the point E upon A; then, because the arc EGP is greater than AMD, the point P will fall at some point H, beyond D, and the chord EP will take the position AH.

Draw the radii CA, CD, and CH. Now, the sides AC, CH, of the triangle ACH, are equal to the sides AC, CD, of the triangle ACD, and the angle ACH is greater than ACD: hence, the side AH, or its equal EP, is greater than the side AD (B. I., P. IX.); *which was to be proved.*

2°. Let the chord EP, or its equal AH, be greater than AD: then is the arc EGP, or its equal ADH, greater than AMD.

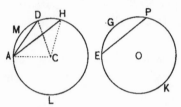

For, if ADH were equal to AMD, the chord AH would be equal to the chord AD (P. IV.); which contradicts the hypothesis. And, if the arc ADH were less than AMD, the chord AH would be less than AD; which also contradicts the hypothesis. Then, since the arc ADH, subtended by the greater chord, can neither be equal to, nor less than AMD, it must be greater than AMD; *which was to be proved.*

PROPOSITION VI.   THEOREM.

The radius which is perpendicular to a chord, bisects that chord, and also the arc subtended by it.

Let CG be the radius which is perpendicular to the chord AB: then this radius bisects the chord AB, and also the arc AGB.

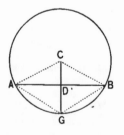

For, draw the radii CA and CB. Then, the right-angled triangles CDA and CDB have the hypothenuse CA equal to CB, and the side CD common; the triangles are, therefore, equal in all respects: hence, AD is equal to DB. Again, because CG is perpendicular to AB, at its middle point, the chords GA and GB are equal (B. I., P. XVI.); and consequently, the arcs GA and GB are also equal (P. IV.): hence, CG bisects the chord AB, and also the arc AGB; *which was to be proved.*

Cor. A straight line, perpendicular to a chord, at its middle point, passes through the centre of the circle.

Scholium. The centre C, the middle point D of the chord AB, and the middle point G of the subtended arc, are points of the radius perpendicular to the chord. But two points determine the position of a straight line (A. 11): hence, any straight line which passes through two of these points, passes through the third, and is per-pendicular to the chord.

PROPOSITION VII. THEOREM.

Through any three points, not in the same straight line, one circumference may be made to pass, and but one.

Let A, B, and C, be any three points, not in a straight line: then may one circumference be made to pass through them, and but one.

Join the points by the lines AB, BC, and bisect these lines by per-pendiculars DE and FG: then will these perpendiculars meet in some point O. For, if they do not meet, they are parallel. Draw DF. The sum of the angles EDF and GFD 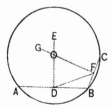 is less than the sum of the angles EDB and GFB, *i. e.,* less than two right angles: therefore, DE and FG are not parallel, and will meet at some point, as O (B. I., P. XXI.)

Now, O is on a perpendicular to AB at its middle point; it is, therefore, equally distant from A and B (B. I., P. XVI.). For a like reason, O is equally distant from B and C. If, therefore, a circumference be described from O as a centre, with a radius equal to the distance from O to A, it will pass through A, B, and C.

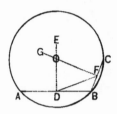

Again, O is the only point which is equally distant from A, B, and C: for, DE contains all of the points which are equally distant from A and B; and FG all of the points which are equally distant from B and C; and consequently, their point of intersection O, is the only point that is equally distant from A, B, and C: hence, one circumference may be made to pass through these points, and but one; *which was to be proved.*

Cor. Two circumferences can not intersect in more than two points; for, if they could intersect in three points, there would be two circumferences passing through the same three points; which is impossible.

PROPOSITION VIII.   THEOREM.

In equal circles, equal chords are equally distant from the centres; and of two unequal chords, the less is at the greater distance from the centre.

1°. In the equal circles ACH and KLG, let the chords AC and KL be equal; then are they equally distant from the centres.

For, let the circle KLG be placed upon ACH, so that the centre R shall fall upon the centre O, and the point K upon the point A: then will the chord KL coincide with AC (P. IV.); and consequently, they are equally distant from the centre; *which was to be proved.*

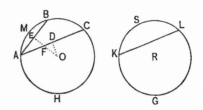

2°. Let AB be less than KL: then is it at a greater distance from the centre.

For, place the circle KLG upon ACH, so that R shall fall upon O, and K upon A. Then, because the chord KL is greater than AB, the arc KSL is greater than AMB; and consequently, the point L will fall at a point C, beyond B, and the chord KL will take the direction AC.

Draw OD and OE, respectively perpendicular to AC and AB; then OE is greater than OF (A. 8), and OF than OD (B. I., P. XV.): hence, OE is greater than OD. But, OE and OD are the distances of the two chords from the centre (B. I., P. XV., C. 1): hence, the less chord is at the greater distance from the centre; *which was to be proved.*

Scholium. All the propositions relating to chords and arcs of equal circles, are also true for chords and arcs of one and the same circle. For, any circle may be regarded as made up of two equal circles. so placed that they coincide in all their parts.

PROPOSITION IX.   THEOREM.

If a straight line is perpendicular to a radius at its outer extremity, it is tangent to the circle at that point; conversely, if a straight line is tangent to a circle at any point, it is perpendicular to the radius drawn to that point.

1°. Let BD be perpendicular to the radius CA, at A: then is it tangent to the circle at A.

For, take any other point of BD, as E, and draw CE: then CE is greater than CA (B. I., P. XV.); and consequently, the point E lies without the circle: hence, BD touches the circum-ference at the point A; it is,

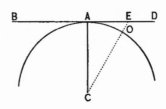

therefore, tangent to it at that point (D. 11); *which was to be proved.*

2°. Let BD be tangent to the circle at A: then is it perpendicular to CA.

For, let E be any point of the tangent, except the point of contact, and draw CE. Then, because BD is a tangent, E lies without the circle; and consequently, CE is greater than CA: hence, CA is shorter than any other line that can be drawn from C to BD; it is, therefore, perpendicular to BD (B. I., P. XV., C. 1); *which was to be proved.*

Cor. At a given point of a circumference, only one tangent can be drawn. For, if two tangents could be drawn, they would both be perpendicular to the same radius at the same point; which is impossible (B. I., P. XIV.).

PROPOSITION X. THEOREM.

Two, parallels intercept equal arcs of a circumference.

There may be three cases: both parallels may be secants; one may be a secant and the other a tangent; or, both may be tangents.

1°. Let the secants AB and DE be parallel: then the intercepted arcs MN and PQ are equal.

For, draw the radius CH per-pendicular to the chord MP; it is also perpendicular to NQ (B. I., P. XX., C. 1), and H is at the middle point of the arc MHP, and also of the arc NHQ: hence, MN, which is the differ-ence of HN and HM, is equal to

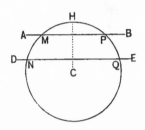

PQ, which is the difference of HQ and HP (A. 3); *which was to be proved.*

2°. Let the secant AB and tangent DE be parallel; then the intercepted arcs MH and PH are equal.

For, draw the radius CH to the point of contact H; it will be perpendicular to DE (P. IX.), and also to its parallel MP. But, because CH is perpendicu-lar to MP, H is the middle point of the arc MHP (P. VI.): hence, MH and PH are equal; *which was to be proved.*

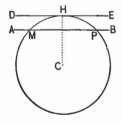

3°. Let the tangents DE and IL be parallel, and let H and K be their points of contact: then the intercepted arcs HMK and HPK are equal.

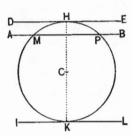

For, draw the secant AB parallel to DE; then, from what has just been shown, we have HM equal to HP, and MK equal to PK: hence, HMK, which is the sum of HM and MK, is equal to HPK, which is the sum of HP and PK; *which was to be proved.*

PROPOSITION XI. THEOREM.

If two circumferences intersect each other, the line joining their centres bisects at right angles the line joining the points of intersection.

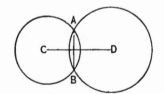

Let the circumferences, whose centres are C and D, intersect at the points A and B: then CD bisects AB at right angles. For the point C, being the centre of the circle, is equally distant from A and B; in like manner, D is equally distant from A and B: hence, CD bisects AB at right angles (B. I., P. XVI., C.); *which was to be proved.*

PROPOSITION XII. THEOREM.

If two circumferences intersect each other, the distance between their centres is less than the sum, and greater than· the difference, of their radii.

Let the circumferences, whose centres are C and D, intersect at A: then CD is less than the sum, and greater than the difference of the radii of the two circles.

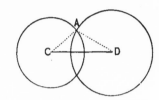

For, draw AC and AD, forming the triangle ACD. Then CD is less than the sum of AC and AD, and greater than their difference (B. I., P. VII.); *which was to be proved.*

PROPOSITION XIII. THEOREM.

If the distance between the centres of two circles is equal to the sum of their radii, the circles are tangent externally.

Let C and D be the centres of two circles, and let the distance between the centres be equal to the sum of the radii: then the circles are tangent externally.

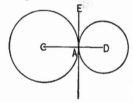

For, they have at least one point, A, on the line CD, common; for, if not, the distance between their centres would be greater than the sum of their radii, which contradicts the hypothesis, and is, therefore, impossible. Again, they have no other point in common; for, if they had two points in common, the distance between their centres would be less than the sum of their radii, which contradicts the hypothesis: hence, they have one and only one point in common, and are tangent externally; *which was to be proved.*

PROPOSITION XIV.   THEOREM.

*If the distance between the centres of two circles is **equal** to the difference of their radii, one circle is tangent to the other internally.*

Let C and D be the centres of two circles, and let the distance between these centres be equal to the difference of the radii : then one circle is tangent to the other internally.

For, the circles will have at least one point, A, on DC, common; for, if not, the distance between the centres would be less than the difference of their radii, which contradicts the hypothesis.   Again, they will have no other point in common; for, if they 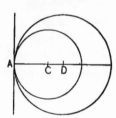 had two points in common, the distance between their centres would be greater than the difference of their radii, which contradicts the hypothesis : hence, they have one and only one point in common, and one is tangent to the other internally ; *which was to be proved.*

Cor. 1. If two circles are tangent, either externally or internally, the point of contact is on the straight line drawn through their centres.

Cor. 2. All circles whose centres are on the same straight line, and which pass through a common point of that line, are tangent to each other at that point.   And if a straight line be drawn tangent to one of the circles at their common point, it is tangent to them all at that point.

Scholium. From the preceding propositions, we infer that two circles may have any one of six positions with respect to each other, depending upon the distance between their centres :

1°. When the distance between their centres is greater

than the sum of their radii, *they are external, one to the other :*

2°. When this distance is equal to the sum of the radii, *they are tangent,* externally :

3°. When this distance is less than the sum, and greater than the difference of the radii, *they intersect each other :*

4°. When this distance is equal to the difference of their radii, *one is tangent to the other,* internally :

5°. When this distance is less than the difference of the radii, *one is wholly within the other :*

6°. When this distance is equal to zero, *they have a common centre ;* or, *they are concentric.*

PROPOSITION XV.   THEOREM.

In equal circles, radii making equal angles at the centre, intercept equal arcs of the circumference ; conversely, radii which intercept equal arcs, make equal angles at the centre.

1°. In the equal circles ADB and EGF, let the angles ACD and EOG be equal: then the arcs AMD and ENG are equal.

For, draw the chords AD and EG ; then the triangles ACD and EOG have two sides and their included angle, in the one, equal to two sides and their included angle, in 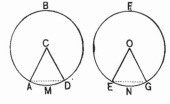 the other, each to each. They are, therefore, equal in all respects ; consequently, AD is equal to EG. But, since the chords AD and EG are equal, the arcs AMD and ENG are also equal (P. IV.) ; *which was to be proved.*

2°. Let the arcs AMD and ENG be equal: then the angles ACD and EOG are equal.

For, since the arcs AMD and ENG are equal, the chords AD and EG are equal (P. IV.); consequently, the triangles ACD and EOG have their sides equal, each to each; they are, therefore, equal in all respects:

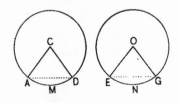

hence, the angle ACD is equal to the angle EOG; *which was to be proved.*

PROPOSITION XVI. THEOREM.

In equal circles, commensurable angles at the centre are proportional to their intercepted arcs.

In the equal circles, whose centres are C and O, let the angles ACB and DOE be commensurable; that is, be exactly measured by a common unit: then are they proportional to the intercepted arcs AB and DE.

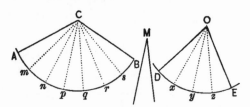

Let the angle M be a common unit; and suppose, for example, that this unit is contained 7 times in the angle ACB, and 4 times in the angle DOE. Then, suppose ACB be divided into 7 angles, by the radii Cm, Cn, Cp, &c.; and DOE into 4 angles, by the radii Ox, Oy, and Oz, each equal to the unit M.

From the last proposition, the arcs A*m*, *mn*, &c., D*x*, *xy*, &c., are equal to each other; and because there are 7 of these arcs in AB, and 4 in DE, we shall have,

arc AB : arc DE :: 7 : 4.

But, by hypothesis, we have,

angle ACB : angle DOE :: 7 : 4 ;

hence, from (B. II., P. IV.), we have,

angle ACB : angle DOE :: arc AB : arc DE.

If any other numbers than 7 and 4 had been used, the same proportion would have been found; *which was to be proved.*

Cor. If the intercepted arcs are commensurable, they are proportional to the corresponding angles at the centre, as may be shown by changing the order of the couplets in the above proportion.

PROPOSITION XVII. THEOREM.

In equal circles, incommensurable angles at the centre are proportional to their intercepted arcs.

In the equal circles, whose centres are C and O, let ACB and FOH be incommensurable: then are they proportional to the arcs AB and FH.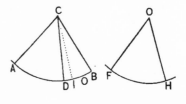

For, let the less angle FOH, be placed upon the greater angle ACB, so that it shall take the position ACD. Then, if the proposition is not true, let us suppose that the angle ACB is to the angle FOH, or its equal ACD, as the arc AB is to an arc AO, greater than FH, or its equal AD; whence,

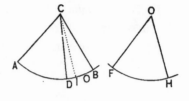

angle ACB : angle ACD :: arc AB : arc AO.

Conceive the arc AB to be divided into equal parts, each less than DO: there will be at least one point of division between D and O; let I be that point; and draw CI. Then the arcs AB, AI, will be commensurable, and we shall have (P. XVI.),

angle ACB : angle ACI :: arc AB : arc AI.

Comparing the two proportions, we see that the antecedents are the same in both: hence, the consequents are proportional (B. II., P. IV., C.); hence,

angle ACD : angle ACI :: arc AO : arc AI.

But, AO is greater than AI: hence, if this proportion is true, the angle ACD must be greater than the angle ACI. On the contrary, it is less: hence, the fourth term of the assumed proportion can not be greater than AD.

In a similar manner, it may be shown that the fourth term can not be less than AD: hence, it must be equal to AD; therefore, we have,

angle ACB : angle ACD :: arc AB : arc AD;

which was to be proved.

Cor. 1. The intercepted arcs are proportional to the corresponding angles at the centre, as may be shown by changing the order of the couplets in the preceding proportion.

Cor. 2. In equal circles, angles at the centre are proportional to their intercepted arcs, and the reverse, whether they are commensurable or incommensurable.

Cor. 3. In equal circles, sectors are proportional to their angles, and also to their arcs.

Scholium. Since the intercepted arcs are proportional to the corresponding angles at the centre, the arcs may be taken as the measures of the angles. That is, if a circumference be described from the vertex of any angle, as a centre, and with a fixed radius, the arc intercepted between the sides of the angle may be taken as the measure of the angle. In Geometry, the right angle, which is measured by a quarter of a circumference, or a *quadrant*, is taken as a unit. If, therefore, any angle is measured by one half or two thirds of a quadrant, it is equal to one half or two thirds of a right angle.

PROPOSITION XVIII. THEOREM.

An inscribed angle is measured by half of the arc included between its sides.

There may be three cases: the centre of the circle may lie on one of the sides of the angle; it may lie within the angle; or, it may lie without the angle.

1°. Let EAD be an inscribed angle, one of whose sides AE passes through the centre: then it is measured by half of the arc DE.

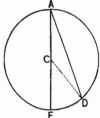

For, draw the radius CD. The external angle DCE, of
the triangle DCA, is equal to the sum of the opposite
interior angles CAD and CDA (B. I., P. XXV., C. 6). But,
the triangle DCA being isosceles, the
angles D and A are equal; therefore,
the angle DCE is double the angle DAE.
Because DCE is at the centre, it is
measured by the arc DE (P. XVII., S.):
hence, the angle DAE is measured by
half of the arc DE; *which was to be
proved.*

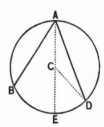

2°. Let DAB be an inscribed angle, and let the centre
lie within it: then the angle is measured by half of the
arc BED.

For, draw the diameter AE. Then, from what has just
been proved, the angle DAE is measured by half of DE,
and the angle EAB by half of EB: hence, BAD, which is
the sum of EAB and DAE, is measured by half of the sum
of DE and EB, or by half of BED; *which was to be
proved.*

3°. Let BAD be an inscribed angle, and let the centre
lie without it: then it is measured by half of the arc
BD.

For, draw the diameter AE. Then,
from what precedes, the angle DAE is
measured by half of DE, and the angle
BAE by half of BE: hence, BAD, which
is the difference of BAE and DAE, is
measured by half of the difference of
BE and DE, or by half of the arc BD;
which was to be proved.

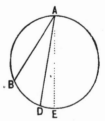

Cor. 1. All the angles BAC, BDC, BEC, inscribed in the same segment, are equal; because they are each measured by half of the same arc BOC.

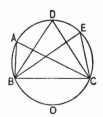

Cor. 2. Any angle BAD, inscribed in a semicircle, is a right angle; because it is measured by half the semi-circumference BOD, or by a quadrant (P. XVII., S.).

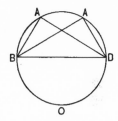

Cor. 3. Any angle BAC, inscribed in a segment greater than a semicircle, is acute; for it is measured by half the arc BOC, less than a semi-circumference.

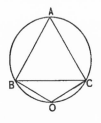

Any angle BOC, inscribed in a segment less than a semicircle, is obtuse; for it is measured by half the arc BAC, greater than a semi-circumference.

Cor. 4. The opposite angles A and C, of an inscribed quadrilateral ABCD, are together equal to two right angles; for the angle DAB is measured by half the arc DCB, the angle DCB by half the arc DAB: hence, the two angles, taken together, are measured by half the circumference: hence, their sum is equal to two right angles.

PROPOSITION XIX. THEOREM.

Any angle formed by two chords, which intersect, is meas-
ured by half the sum of the included arcs.

Let DEB be an angle formed by the intersection of the
chords AB and CD: then it is measured by half the sum
of the arcs AC and DB.

For, draw AD: then, the angle DEB,
being an exterior angle of the triangle
DEA, is equal to the sum of the angles
EDA and EAD (B. I., P. XXV., C. 6).
But, the angle EDA is measured by
half the arc AC, and EAD by half the
arc DB (P. XVIII.): hence, the angle

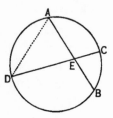

DEB is measured by half the sum of the arcs AC and DB;
which was to be proved.

PROPOSITION XX. THEOREM.

The angle formed by two secants, intersecting without the
circumference, is measured by half the difference of the
included arcs.

Let AB, AC, be two secants: then the angle BAC is
measured by half the difference of the
arcs BC and DF.

Draw DE parallel to AC: the arc EC
is equal to DF (P. X.), and the angle
BDE to the angle BAC (B. I., P. XX.,
C. 3). But BDE is measured by half
the arc BE (P. XVIII.): hence, BAC is
also measured by half the arc BE; that
is, by half the difference of BC and EC,
or by half the difference of BC and DF; *which was to be*
proved.

PROPOSITION XXI. THEOREM.

An angle formed by a tangent and a chord meeting it at the point of contact, is measured by half the included arc.

Let BE be tangent to the circle AMC, and let AC be a chord drawn from the point of contact A: then BAC is measured by half of the arc AMC.

For, draw the diameter AD. The angle BAD is a right angle (P. IX.), and is measured by half the semi-circumference AMD (P. XVII., S.); the angle DAC is measured by half of the arc DC (P. XVIII.): hence, the angle BAC, which is equal to the sum of the angles BAD and DAC, is measured by half the sum of the arcs AMD and DC, or by half of the arc AMC; *which was to be proved.*

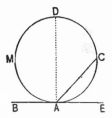

The angle CAE, which is the difference of DAE and DAC, is measured by half the difference of the arcs DCA and DC, or by half the arc CA.

PRACTICAL APPLICATIONS.

PROBLEM I.

To bisect a given straight line.

Let AB be a given straight line.

From A and B, as centres, with a radius greater than one half of AB, describe arcs intersecting at E and F: join E and F, by the straight line EF. Then EF bisects the given line AB. For, E and F are each equally distant from 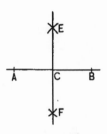 A and B; and consequently, the line EF bisects AB (B. I., P. XVI., C.).

PROBLEM II.

To erect a perpendicular to a given straight line, at a given point of that line.

Let EF be a given line, and let A be a given point of that line.

From A, lay off the equal distances AB and AC; from B and C, as centres, with a radius greater than one half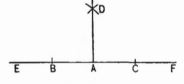

of BC, describe arcs intersecting at D; draw the line AD: then AD is the perpendicular required. For, D and A are each equally distant from B and C; consequently, DA is perpendicular to BC at the given point A (B. I., P. XVI., C.).

PROBLEM III.

To draw a perpendicular to a given straight line, from a given point without that line.

Let FG be the given line, and A the given point.

From A, as a centre, with a radius sufficiently great, describe an arc cutting FG in two points, B and D; with B and D as centres, and a radius greater than one half of BD, describe arcs intersecting at E; draw AE: then AE is the perpendicular 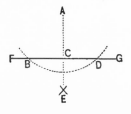 required. For, A and E are each equally distant from B and D: hence, AE is perpendicular to BD (B. I., P. XVI., C.).

PROBLEM IV.

At a point on a given straight line, to construct an angle equal to a given angle.

Let A be the given point, AB the given line, and IKL the given angle.

From the vertex K as a center, with any radius KI, describe the arc IL, terminating in the sides of the angle. From A as 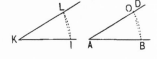 a centre, with a radius AB, equal to KI, describe the

indefinite arc BO; then, with a radius equal to the chord
LI, from B as a centre, describe an arc cutting the arc
BO in D; draw AD: then BAD is
equal to the angle K.

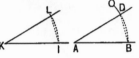

For the arcs BD, IL, have
equal radii and equal chords:
hence, they are equal (P. IV.);
therefore, the angles BAD, IKL, measured by them, are also
equal (P. XV.).

PROBLEM V.

To bisect a given arc or a given angle.

1°. Let AEB be a given arc, and C its centre.

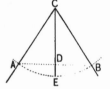

Draw the chord AB; through C,
draw CD perpendicular to AB (Prob.
III.): then CD bisects the arc AEB
(P. VI.).

2°. Let ACB be a given angle.

With C as a centre, and any radius CB, describe the
arc BA; bisect it by the line CD, as just explained: then
CD bisects the angle ACB.

For, the arcs AE and EB are equal, from what was just
shown; consequently, the angles ACE and ECB are also
equal (P. XV.).

Scholium. If each half of an arc or angle is bisected,
the original arc or angle is divided into four equal parts;
and if each of these is bisected, the original arc or angle
is divided into eight equal parts; and so on.

PROBLEM VI.

Through a given point, to draw a straight line parallel to a given straight line.

Let A be a given point, and BC a given line.

From the point A as a centre, with a radius AE, greater than the shortest distance from A to BC, describe an indefinite arc EO; from E as a centre, with the same radius, describe the arc AF; lay off ED equal to AF, and draw AD: then AD is the parallel required.

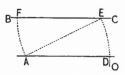

For, drawing AE, the angles AEF, EAD, are equal (P. XV.); therefore, the lines AD, EF are parallel (B. I., P. XIX., C. 1).

PROBLEM VII.

Given, two angles of a triangle, to construct the third angle.

Let A and B be given angles of a triangle.

Draw a line DF, and at some point of it, as E, construct the angle FEH equal to A, and HEC equal to B. Then, CED is equal to the required angle.

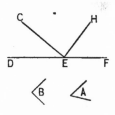

For, the sum of the three angles at E is equal to two right angles (B. I., P. I., C. 2), as is also the sum of the three angles of a triangle (B. I., P. XXV.). Consequently, the third angle CED must be equal to the third angle of the triangle.

PROBLEM VIII.

Given, two sides and the included angle of a triangle, to construct the triangle.

Let B and C denote the given sides, and A the given angle.

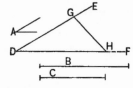

Draw the indefinite line DF, and at D construct an angle FDE, equal to the angle A; on DF, lay off DH equal to the side 🝆 and on DE, lay off DG equal to the side C; draw GH: then DGH is the required triangle (B. I., P. V.).

PROBLEM IX.

Given, one side and two angles of a triangle, to construct the triangle.

The two angles may be either both adjacent to the given side, or one may be adjacent and the other opposite to it. In the latter case, construct the third angle by Problem VII. We shall then have two angles and their included side.

Draw a straight line, and on it lay off DE equal to the given side; at D construct an angle equal to one of the adjacent angles, and at E construct an angle equal to the other adjacent angle; produce the sides DF and EG till they intersect at H: then DEH is the triangle required (B. I., P. VI.).

PROBLEM X.

Given, the three sides of a triangle, to construct the triangle.

Let A, B, and C, be ·the given sides.

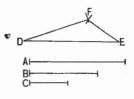

Draw DE, and make it equal to the side A; from D as a centre, with a radius equal to the side B, describe an arc; from E as a centre, with a radius equal to the side C, describe an arc intersecting the former at F; draw DF and EF: then DEF is the triangle required (B. I., P. X.).

Scholium. In order that the construction may be possible, any one of the given sides must be *less* than the sum of the two others, and *greater* than their difference (B. I., P. VII., S.).

PROBLEM XI.

Given, two sides of a triangle, and the angle opposite one of them, to construct the triangle.

Let A and B be the given sides, and C the given angle.

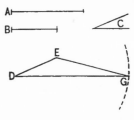

Draw an indefinite line DG, and at some point of it, as D, construct an angle GDE equal to the given angle; on one side of this angle lay off the distance DE equal to the side B adjacent to the given angle; from E as a centre, with a radius equal to the side opposite the given angle, describe an arc cutting the side DG at G: draw EG. Then DEG is the required triangle.

For, the sides DE and EG are equal to the given sides, and the angle D, opposite one of them, is equal to the given angle.

Scholium. If the side opposite the given angle is greater than the other given side, there is but one solution. If the given angle is acute, and the side opposite the given angle is less than the other given side, and greater than the shortest distance from E to DG, there are two solutions, DEG and DEF. If the side opposite the given angle is equal to the shortest distance from E to DG, the arc 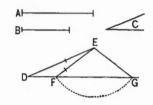 will be tangent to DG, the angle opposite DE is a right angle, and there is but one solution. If the side opposite the given angle is shorter than the distance from E to DG, there is no solution.

PROBLEM XII.

Given, two adjacent sides of a parallelogram and their included angle, to construct the parallelogram.

Let A and B be the given sides, and C the given angle.

Draw the line DH, and at some point as D, construct the angle HDF equal to the angle C. Lay off DE equal to the side A, and DF equal to the side B; draw FG parallel to DE, and EG parallel to DF; then DFGE is the parallelogram required.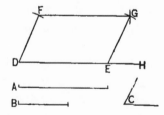

For, the opposite sides are parallel by construction; and consequently, the figure is a parallelogram (D. 28); it is also formed with the given sides and given angle.

PROBLEM XIII.

To find the centre of a given circumference or arc.

Take any three points A, B, and C, on the circumference or arc, and join them by the chords AB, BC; bisect these chords by the perpendiculars DE and FG: then their point of intersection, O, is the centre required (P. VII.).

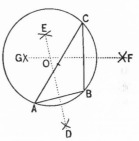

Scholium. The same construction enables us to pass a circumference through any three points not in a straight line. If the points are vertices of a triangle, the circle is circumscribed about it.

PROBLEM XIV.

Through a given point, to draw a tangent to a given circle.

There may be two cases: the given point may lie on the circumference of the given circle, or it may lie without the given circle.

1°. Let C be the centre of the given circle, and A a point on the circumference, through which the tangent is to be drawn.

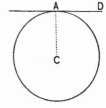

Draw the radius CA, and at A draw AD perpendicular to AC: then AD is the tangent required (P. IX.).

2°. Let C be the centre of the given circle, and A a point without the circle, through which the tangent is to be drawn.

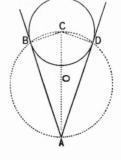

Draw the line AC; bisect it at O, and from O as a centre, with a radius OC, describe the circumference ABCD; join the point A with the points of intersection D and B: then both AD and AB are tangent to the given circle and there are two solutions.

For, the angles ABC and ADC are right angles (P. XVIII., C. 2): hence, each of the lines AB and AD is perpendicular to a radius at its extremity; and consequently, they are tangent to the given circle (P. IX.).

Corollary. The right-angled triangles ABC and ADC, have a common hypothenuse AC, and the side BC equal to DC; and consequently, they are equal in all respects (B. I., P. XVII.): hence, AB is equal to AD, and the angle CAB is equal to the angle CAD. The tangents are therefore equal, and the line AC bisects the angle between them.

PROBLEM XV.

To inscribe a circle in a given triangle.

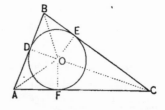

Let ABC be the given triangle.

Bisect the angles A and B, by the lines AO and BO, meeting in the point O (Prob. V.); from the point O let fall the perpendiculars OD, OE, OF, on the sides of the triangle: these perpendiculars are all equal.

For, in the triangles BOD and BOE, the angles OBE and OBD are equal, by construction; the angles ODB and OEB are equal, because each is a right angle; and consequently, the angles BOD and BOE are also equal (B. I., P. XXV., C. 2), and the side OB is common; and therefore, the triangles are equal in all respects (B. I., P. VI.): hence, OD is equal to OE. In like manner, it may be shown that OD is equal to OF.

From O as a centre, with a radius OD, describe a circle, and it will be the circle required. For, each side is perpendicular to a radius at its extremity, and is therefore tangent to the circle.

Corollary. The lines that bisect the three angles of a triangle all meet in one point.

PROBLEM XVI.

On a given straight line, to construct a segment that shall contain a given angle.

Let AB be the given line.

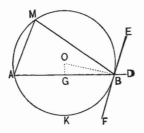 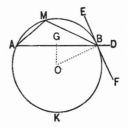

Produce AB towards D; at B construct the angle DBE equal to the given angle; draw BO perpendicular to BE,

and at the middle point G, of AB, draw GO perpendicular
to AB; from their point of intersection O, as a centre,
with a radius OB, describe the arc AMB: then the segment AMB is the segment required.

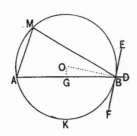 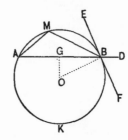

For, the angle ABF, equal to EBD, is measured by half
of the arc AKB (P. XXI.); and the inscribed angle AMB is
measured by half of the same arc: hence, the angle AMB
is equal to the angle EBD, and consequently, to the given
angle.

NOTE.—A *quadrant* or quarter of a circumference, as
CD, is, for convenience, divided into 90 equal parts, each
of which is called a *degree*. A degree
is denoted by the symbol °; thus, 25°
is read 25 degrees, etc. Since a quadrant contains 90°, the whole circumference contains 360°. A right angle, as
CAD, which is the unit of measure for
angles, being measured by a quadrant
(P. XVII., S.), is said to be an angle

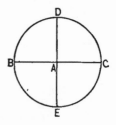

of 90°; an angle which is one third of a right angle is
an angle of 30°; an angle of 120° is $\frac{120}{90}$ or $\frac{4}{3}$ of a right
angle, etc.

EXERCISES.

1. Draw a circumference of given radius through two given points.

2. Construct an equilateral triangle, having given one of its sides.

3. At a point on a given straight line, construct an angle of 30°.

4. Through a given point without a given line, draw a line forming with the given line an angle of 30°.

5. A line 8 feet long is met at one extremity by a second line, making with it an angle of 30°; find the centre of the circle of which the first line is a chord and the second a tangent.

6. How many degrees in an angle inscribed in an arc of 135°?

7. How many degrees in the angle formed by two secants meeting without the circle and including arcs of 60° and 110°?

8. At one extremity of a chord, which divides the circumference into two arcs of 290° and 70° respectively, a tangent is drawn; how many degrees in each of the angles formed by the tangent and the chord?

9. Show that the sum of the alternate angles of an inscribed hexagon is equal to four right angles.

10. The sides of a triangle are 3, 5, and 7 feet; construct the triangle.

11. Show that the three perpendiculars erected at the middle points of the three sides of a triangle meet in a common point.

12. Construct an isosceles triangle with a given base and a given vertical angle.

13. At a point on a given straight line, construct an angle of 45°

14. Construct an isosceles triangle so that the base shall be a given line and the vertical angle a right angle.

15. Construct a triangle, having given one angle, one of its including sides, and the difference of the two other sides.

16. From a given point, A, without a circle, draw two tangents, AB and AC, and at any point, D, in the included arc, draw a third tangent and produce it to meet the two others; show that the three tangents form a triangle whose perimeter is constant.

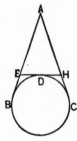

17. On a straight line 5 feet long, construct a circular segment that shall contain an angle of 30°.

18. Show that parallel tangents to a circle include semi-circumferences between their points of contact.

19. Show that four circles can be drawn tangent to three intersecting straight lines.

BOOK IV.

DEFINITIONS.

1. SIMILAR POLYGONS are polygons which are mutually equiangular, and which have the sides about the equal angles, taken in the same order, proportional.

2. In similar polygons, the parts which are similarly placed in each, are called *homologous*.

The corresponding angles are *homologous angles*, the corresponding sides are *homologous sides*, the corresponding diagonals are *homologous diagonals*, and so on.

3. SIMILAR ARCS, SECTORS, or SEGMENTS, in different circles, are those which correspond to equal angles at the centre.

Thus, if the angles A and O are equal, the arcs BFC and DGE are similar, the sectors BAC and DOE are similar, and the segments BFC and DGE are similar.

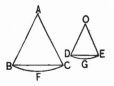

4. The ALTITUDE OF A TRIANGLE is the perpendicular distance from the vertex of any angle to the opposite side, or the opposite side produced.

The vertex of the angle from which the distance is measured, is called the *vertex of the triangle*, and the opposite side is called the *base of the triangle*.

5. The ALTITUDE OF A PARALLELOGRAM is the perpendicular distance between two opposite
sides.

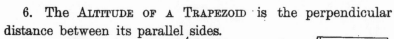

These sides are called *bases;* one the
upper, and the other, the *lower base.*

6. The ALTITUDE OF A TRAPEZOID is the perpendicular
distance between its parallel sides.

These sides are called *bases;* one the
upper, and the other, the *lower base.*

7. The AREA OF A SURFACE is its numerical value
expressed in terms of some other surface taken as a *unit.*
The unit adopted is a square described on the linear unit
as a side.

PROPOSITION I. THEOREM.

*Parallelograms which have equal bases and equal altitudes,
are equal.*

Let the parallelograms ABCD and EFGH have equal
bases and equal altitudes: then the parallelograms are
equal.

For, let them be so placed
that their lower bases shall
coincide; then, because they
have the same altitude, their
upper bases will be in the same line DG, parallel to AB.

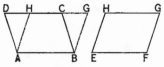

The triangles DAH and CBG, have the sides AD and BC
equal, because they are opposite sides of the parallelogram AC (B. I., P. XXVIII.); the sides AH and BG equal,
because they are opposite sides of the parallelogram AG;
the angles DAH and CBG equal, because their sides are

parallel and lie in the same direction (B. I., P. XXIV.): hence, the triangles are equal (B. I., P. V.).

If from the quadrilateral ABGD, we take away the triangle DAH, there will remain the parallelogram AG; if from the same quadrilateral ABGD, we take away the triangle CBG, there will remain the parallelogram AC: hence, the parallelogram AC is equal to the parallelogram EG (A. 3); *which was to be proved.*

PROPOSITION II. THEOREM.

A triangle is equal to one half of a parallelogram having an equal base and an equal altitude.

Let the triangle ABC, and the parallelogram ABFD, have equal bases and equal altitudes: then the triangle is equal to one half of the parallelogram.

For, let them be so placed that the base of the triangle shall coincide with the lower base of the parallelogram; then, be-

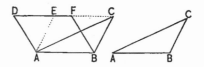

cause they have equal altitudes, the vertex of the triangle will lie in the upper base of the parallelogram, or in the prolongation of that base.

From A, draw AE parallel to BC, forming the parallelogram ABCE. This parallelogram is equal to the parallelogram ABFD, from Proposition I. But the triangle ABC is equal to half of the parallelogram ABCE (B. I., P. XXVIII., C. 1): hence, it is equal to half of the parallelogram ABFD (A. 7); *which was to be proved.*

Cor. Triangles having equal bases and equal altitudes are equal, for they are halves of equal parallelograms.

PROPOSITION III. THEOREM.

Rectangles having equal altitudes, are proportional to their bases.

There may be two cases: the bases may be commensurable, or they may be incommensurable.

1°. Let ABCD and HEFK, be two rectangles whose altitudes AD and HK are equal, and whose bases AB and HE are commensurable: then the areas of the rectangles are proportional to their bases.

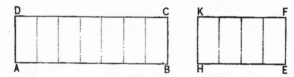

Suppose that AB is to HE, as 7 is to 4. Conceive AB to be divided into 7 equal parts, and HE into 4 equal parts, and at the points of division, let perpendiculars be drawn to AB and HE. Then will ABCD be divided into 7, and HEFK into 4 rectangles, all of which are equal, because they have equal bases and equal altitudes (P. I.): hence, we have,

ABCD : HEFK :: 7 : 4.

But we have, by hypothesis,

AB : HE :: 7 : 4.

From these proportions, we have (B. II., P. IV.),

ABCD : HEFK :: AB : HE.

Had any other numbers than 7 and 4 been used, the same proportion would have been found; *which was to be proved.*

2°. Let the bases of the rectangles be incommensura-ble : then the rectangles are proportional to their bases.

For, place the rectangle HEFK upon the rectangle ABCD, so that it shall take the position AEFD. Then, if the rectangles are not proportional to their bases, let us suppose that

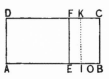

$$ABCD \ : \ AEFD \ :: \ AB \ : \ AO ;$$

in which AO is greater than AE. Divide AB into equal parts, each less than OE; at least one point of division, as I, will fall between E and O; at this point, draw IK perpendicular to AB. Then, because AB and AI are com-mensurable, we shall have, from what has just been shown,

$$ABCD \ : \ AIKD \ :: \ AB \ : \ AI.$$

The above proportions have their antecedents the same in each; hence (B. II., P. IV., C.),

$$AEFD \ : \ AIKD \ :: \ AO \ : \ AI.$$

The rectangle AEFD is less than AIKD; and if the above proportion were true, the line AO would be less than AI; whereas, it is greater. The fourth term of the proportion, therefore, cannot be greater than AE. In like manner, it may be shown that it cannot be less than AE; conse-quently, it must be equal to AE: hence,

$$ABCD \ : \ AEFD \ :: \ AB \ : \ AE ;$$

which was to be proved.

Cor. If rectangles have equal bases, they are to each other as their altitudes.

PROPOSITION IV.   THEOREM.

Any two rectangles are to each other as the products of their bases and altitudes.

Let ABCD and AEGF be two rectangles: then ABCD is to AEGF, as AB × AD is to AE × AF.

For, place the rectangles so that the angles DAB and EAF shall be opposite or vertical; then, produce the sides CD and GE till they meet in H.

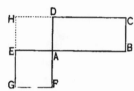

The rectangles ABCD and ADHE have the same altitude AD: hence (P. III.),

$$ABCD \ : \ ADHE \ :: \ AB \ : \ AE.$$

The rectangles ADHE and AEGF have the same altitude AE: hence,

$$ADHE \ : \ AEGF \ :: \ AD \ : \ AF.$$

Multiplying these proportions, term by term (B. II., P. XII.), and omitting the common factor ADHE (B. II., P. VII.), we have,

$$ABCD \ : \ AEGF \ :: \ AB \times AD \ : \ AE \times AF;$$

which was to be proved.

Cor. If we suppose AE and AF, each to be equal to the linear unit, the rectangle AEGF is the superficial unit, and we have,

$$ABCD \ : \ 1 \ :: \ AB \times AD \ : \ 1;$$

$$ABCD = AB \times AD:$$

hence, *the area of a rectangle is equal to the product of its base and altitude;* that is, the number of superficial units in the rectangle, is equal to the product of the number of linear units in its base by the number of linear units in its altitude.

The product of two lines is sometimes called the *rectangle* of the lines, because the product is equal to the area of a rectangle constructed with the lines as sides.

PROPOSITION V. THEOREM.

The area of a parallelogram is equal to the product of its base and altitude.

Let ABCD be a parallelogram, AB its base, and BE its altitude: then the area of ABCD is equal to AB × BE.

For, construct the rectangle ABEF, having the same base and altitude: then will the rectangle be equal to the parallelogram (P. I.); but the area of the rectangle is equal to AB × BE: hence, the area of the parallelogram is also equal to AB × BE; *which was to be proved.*

Cor. Parallelograms are to each other as the products of their bases and altitudes. If their altitudes are equal, they are to each other as their bases. If their bases are equal, they are to each other as their altitudes.

PROPOSITION VI.   THEOREM.

The area of a triangle is equal to half the product of its base and altitude.

Let ABC be a triangle, BC its base, and AD its altitude: then its area is equal to $\frac{1}{2}$BC × AD.

For, from C, draw CE parallel to
BA, and from A, draw AE parallel to
BC.   The area of the parallelogram
BCEA is BC × AD (P. V.); but the
triangle ABC is half of the parallel-

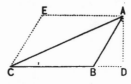

ogram BCEA: hence, its area is equal to $\frac{1}{2}$BC × AD; *which was to be proved.*

Cor. 1.   Triangles are to each other, as the products of their bases and altitudes (B. II., P. VII.).   If the altitudes are equal, they are to each other as their bases.   If the bases are equal, they are to each other as their altitudes.

Cor. 2.   The area of a triangle is equal to half the product of its perimeter and the radius of the inscribed circle.

For, let DEF be a circle inscribed in the triangle ABC.   Draw
OD, OE, and OF, to the points of
contact, and OA, OB, and OC, to
the vertices.

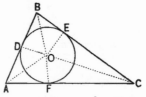

The area of OBC is equal to
$\frac{1}{2}$OE × BC; the area of OAC is equal to $\frac{1}{2}$OF × AC; and
the area of OAB is equal to $\frac{1}{2}$OD × AB; and since OD, OE,
and OF, are equal, the area of the triangle ABC (A. 9), is
equal to $\frac{1}{2}$OD (AB + BC + CA).

PROPOSITION VII. THEOREM.

The area of a trapezoid is equal to the product of its alti-
tude and half the sum of its parallel sides.

Let ABCD be a trapezoid, DE its altitude, and AB and
DC its parallel sides : then its area is
equal to DE × ½ (AB + DC).

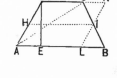

For, draw the diagonal AC, forming
the triangles ABC and ACD. The altitude of each of these triangles is equal
to DE. The area of ABC is equal to
½AB × DE (P. VI.) ; the area of ACD is equal to ½DC × DE :
hence, the area of the trapezoid, which is the sum of the
triangles, is equal to the sum of ½AB × DE and ½DC × DE,
or to DE × ½ (AB + DC) ; *which was to be proved.*

Scholium. Through I, the middle point of BC, draw IH
parallel to AB, and LI parallel to AD, meeting DC produced,
at K. Then, since AI and HK are parallelograms, we have
AL = HI = DK ; and therefore, HI = ½ (AL + DK). But since
the triangles LIB and CIK are equal in all respects,
LB = CK ; hence, AL + DK = AB + DC ; and we have HI =
½ (AB + DC) : hence,

The area of a trapezoid is equal to its altitude multi-
plied by the line which connects the middle points of its
inclined sides.

PROPOSITION VIII. THEOREM.

The square described on the sum of two lines is equal to
the sum of the squares described on the lines, increased
by twice the rectangle of the lines.

Let AB and BC be two lines, and AC their sum: then
$\overline{AC}^2 = \overline{AB}^2 + \overline{BC}^2 + 2AB \times BC$.

On AC, construct the square AD; from B, draw BH
parallel to AE; lay off AF equal to AB,
and from F, draw FG parallel to AC:
then IG and IH are each equal to BC;
and IB and IF, to AB.

The square ACDE is composed of four
parts. The part ABIF is a square described
on AB; the part IGDH is equal to a
square described on BC; the part BCGI is equal to the
rectangle of AB and BC; and the part FIHE is also equal
to the rectangle of AB and BC: hence, we have (A. 9),

$$\overline{AC}^2 = \overline{AB}^2 + \overline{BC}^2 + 2AB \times BC;$$

which was to be proved.

Cor. If the lines AB and BC are equal, the four parts
of the square on AC are also equal: hence, *the square
described on a line is equal to four times the square
described on half the line.*

PROPOSITION IX.   THEOREM.

*The square described on the difference of two lines is equal
to the sum of the squares described on the lines, diminished by twice the rectangle of the lines.*

Let AB and BC be two lines, and AC their difference;
then
$$\overline{AC}^2 = \overline{AB}^2 + \overline{BC}^2 - 2AB \times BC.$$

On AB construct the square ABIF; from C draw CG
parallel to BI; lay off CD equal to AC, and from D draw
DK parallel and equal to BA; complete the square EFLK;

then EK is equal to BC, and EFLK is equal to the square of BC.

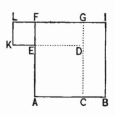

The whole figure ABILKE is equal to the sum of the squares described on AB ·and BC. The part CBIG is equal to the rectangle of AB and BG; the part DGLK is also equal to the rectangle of AB and BC. If from the whole figure ABILKE, the two parts CBIG and DGLK be taken, there will remain the part ACDE, which is equal to the square of AC: hence,

$$\overline{AC}^2 = \overline{AB}^2 + \overline{BC}^2 - 2AB \times BC;$$

which was to be proved.

PROPOSITION X. THEOREM.

The rectangle contained by the sum and difference of two lines, is equal to the difference of their squares.

Let AB and BC be two lines, of which AB is the greater: then

$$(AB + BC)(AB - BC) = \overline{AB}^2 - \overline{BC}^2.$$

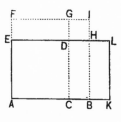

On AB, construct the square ABIF; prolong AB, and make BK equal to BC; then AK is equal to AB + BC; from K, draw KL parallel to BI, and make it equal to AC; draw LE parallel to KA, and CG parallel to BI: then DG is equal to BC, and the figure DHIG is equal to the square on BC, and EDGF is equal to BKLH.

If we add to the figure ABHE, the rectangle BKLH, we have the rectangle AKLE, which is equal to the rectangle of AB + BC and AB — BC. If to the same figure ABHE, we add the rectangle DGFE, equal to BKLH, we have the figure ABHDGF, which is equal to the difference of the squares of AB and BC. But the sums of equals are equal (A. 2), hence,

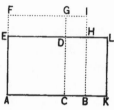

$$(AB + BC)(AB - BC) = \overline{AB}^2 - \overline{BC}^2;$$

which was to be proved.

PROPOSITION XI. THEOREM.

The square described on the hypothenuse of a right-angled triangle, is equal to the sum of the squares described on the two other sides.

Let ABC be a triangle, right-angled at A: then

$$\overline{BC}^2 = \overline{AB}^2 + \overline{AC}^2.$$

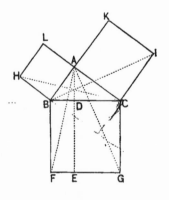

Construct the square BG on the side BC, the square AH on the side AB, and the square AI on the side AC; from A draw AD perpendicular to BC, and prolong it to E: then DE is parallel to BF; draw AF and HC.

In the triangles HBC and ABF, we have HB equal to AB, because they are sides of the same square; BC equal to BF, for the same reason, and the included angles HBC and ABF equal, because each is equal to the angle ABC plus a right angle: hence, the triangles are equal in all respects (B. I., P. V.).

The triangle ABF, and the rectangle BE, have the same base BF, and because DE is the prolongation of DA, their altitudes are equal: hence, the triangle ABF is equal to half the rectangle BE (P. II.). The triangle HBC, and the square BL, have the same base BH, and because AC is the prolongation of LA (B. I., P. IV.), their altitudes are equal: hence, the triangle HBC is equal to half the square of AH. But, the triangles ABF and HBC are equal: hence, the rectangle BE is equal to the square AH. In the same manner, it may be shown that the rectangle DG is equal to the square AI: hence, the sum of the rectangles BE and DG, or the square BG, is equal to the sum of the squares AH and AI; or, $\overline{BC}^2 = \overline{AB}^2 + \overline{AC}^2$; *which was to be proved.*

Cor. 1. The square of either side about the right angle is equal to the square of the hypothenuse diminished by the square of the other side: thus,

$$\overline{AB}^2 = \overline{BC}^2 - \overline{AC}^2; \quad \text{or,} \quad \overline{AC}^2 = \overline{BC}^2 - \overline{AB}^2.$$

Cor. 2. If from the vertex of the right angle, a perpendicular be drawn to the hypothenuse, dividing it into two *segments*, BD and DC, *the square of the hypothenuse is to the square of either of the other sides, as the hypothenuse is to the segment adjacent to that side.*

For, the square BG, is to the rectangle BE, as BC to BD (P. III.); but the rectangle BE is equal to the square AH: hence,

$$\overline{BC}^2 \ : \ \overline{AB}^2 \ :: \ BC \ : \ BD.$$

In like manner, we have,

$$\overline{BC}^2 \ : \ \overline{AC}^2 \ :: \ BC \ : \ DC.$$

Cor. 3. *The squares of the sides about the right angle are to each other as the adjacent segments of the hypothenuse.*

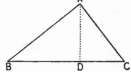

For, by combining the proportions of the preceding corollary (B. II., P. IV., C.), we have,

$$\overline{AB}^2 \ : \ \overline{AC}^2 \ :: \ BD \ : \ DC.$$

Cor. 4. *The square described on the diagonal of a square is double the given square.*

For, the square of the diagonal is equal to the sum of the squares of the two sides; but the square of each side is equal to the given square: hence,

$$\overline{AC}^2 \ = \ 2\overline{AB}^2 ; \quad \text{or,} \quad \overline{AC}^2 \ = \ 2\overline{BC}^2.$$

Cor. 5. From the last corollary, we have,

$$\overline{AC}^2 \ : \ \overline{AB}^2 \ :: \ 2 \ : \ 1 ;$$

hence, by extracting the square root of each term, we have,

$$AC \ : \ AB \ :: \ \sqrt{2} \ : \ 1 ;$$

that is, *the diagonal of a square is to the side, as the square root of two is to one;* consequently, *the diagonal and the side of a square are incommensurable.*

PROPOSITION XII. THEOREM.

In any triangle, the square of a side opposite an acute angle. is equal to the sum of the squares of the base and the other side, diminished by twice the rectangle of the base and the distance from the vertex of the acute angle to the foot of the perpendicular drawn from the vertex of the opposite angle to the base, or to the base produced.

Let ABC be a triangle, C one of its acute angles, BC its base, and AD the perpendicular drawn from A to BC, or BC produced ; then

$$\overline{AB}^2 = \overline{BC}^2 + \overline{AC}^2 - 2\,BC \times CD.$$

For, whether the perpendicular meets the base, or the base produced, we have BD equal to the difference of BC and CD : hence (P. IX.),

$$\overline{BD}^2 = \overline{BC}^2 + \overline{CD}^2 - 2\,BC \times CD.$$

Adding $\overline{AD}^2$ to both members, we have,

$$\overline{BD}^2 + \overline{AD}^2 = \overline{BC}^2 + \overline{CD}^2 + \overline{AD}^2 - 2\,BC \times CD.$$

But, $$\overline{BD}^2 + \overline{AD}^2 = \overline{AB}^2,$$

and $$\overline{CD}^2 + \overline{AD}^3 = \overline{AC}^2:$$

hence, $$\overline{AB}^2 = \overline{BC}^2 + \overline{AC}^2 - 2\,BC \times CD;$$

which was to be proved.

PROPOSITION XIII. THEOREM.

In any obtuse-angled triangle, the square of the side opposite the obtuse angle is equal to the sum of the squares of the base and the other side, increased by twice the rectangle of the base and the distance from the vertex of the obtuse angle to the foot of the perpendicular drawn from the vertex of the opposite angle to the base produced.

Let ABC be an obtuse-angled triangle, B its obtuse angle, BC its base, and AD the perpendicular drawn from A to BC produced; then

$$\overline{AC}^2 = \overline{BC}^2 + \overline{AB}^2 + 2BC \times BD$$

For, CD is the sum of BC and BD: hence (P. VIII.),

$$\overline{CD}^2 = \overline{BC}^2 + \overline{BD}^2 + 2BC \times BD.$$

Adding $\overline{AD}^2$ to both members, and reducing, we have,

$$\overline{AC}^2 = \overline{BC}^2 + \overline{AB}^2 + 2BC \times BD;$$

which was to be proved.

Scholium. The right-angled triangle is the only one in which the sum of the squares described on two sides is equal to the square described on the third side.

PROPOSITION XIV. THEOREM.

In any triangle, the sum of the squares described on two sides is equal to twice the square of half the third side, increased by twice the square of the line drawn from the middle point of that side to the vertex of the opposite angle.

Let ABC be any triangle, and EA a line drawn from the middle of the base BC to the vertex A: then

$$\overline{AB}^2 + \overline{AC}^2 = 2\overline{BE}^2 + 2\overline{EA}^2.$$

Draw AD perpendicular to BC; then, from Proposition XII., we have,

$$\overline{AC}^2 = \overline{EC}^2 + \overline{EA}^2 - 2EC \times ED.$$

From Proposition XIII., we have,

$$\overline{AB}^2 = \overline{BE}^2 + \overline{EA}^2 + 2BE \times ED.$$

Adding these equations, member to member (A. 2), recollecting that BE is equal to EC, we have,

$$\overline{AB}^2 + \overline{AC}^2 = 2\overline{BE}^2 + 2\overline{EA}^2;$$

which was to be proved.

Cor. Let ABCD be a parallelogram, and BD, AC, its diagonals. Then, since the diagonals mutually bisect each other (B. I., P. XXXI.), we have,

$$\overline{AB}^2 + \overline{BC}^2 = 2\overline{AE}^2 + 2\overline{BE}^2;$$

and, $\quad \overline{CD}^2 + \overline{DA}^2 = 2\overline{CE}^2 + 2\overline{DE}^2;$

whence, by addition, recollecting that AE is equal to CE, and BE to DE, we have,

$$\overline{AB}^2 + \overline{BC}^2 + \overline{CD}^2 + \overline{DA}^2 = 4\overline{CE}^2 + 4\overline{DE}^2;$$

but, $4\overline{CE}^2$ is equal to $\overline{AC}^2$, and $4\overline{DE}^2$ to $\overline{BD}^2$ (P. VIII., C.): hence,

$$\overline{AB}^2 + \overline{BC}^2 + \overline{CD}^2 + \overline{DA}^2 = \overline{AC}^2 + \overline{BD}^2.$$

That is, *the sum of the squares of the sides of a parallelogram, is equal to the sum of the squares of its diagonals.*

PROPOSITION XV.   THEOREM.

In any triangle, a line drawn parallel to the base divides the other sides proportionally.

Let ABC be a triangle, and DE a line parallel to the base BC : then

$$AD : DB :: AE : EC.$$

Draw EB and DC. Then, because the triangles AED and DEB have their bases in the same line AB, and their vertices at the same point E, they have a common altitude: hence (P. VI., C.),

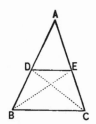

$$AED : DEB :: AD : DB.$$

The triangles AED and EDC, have their bases in the same line AC, and their vertices at the same point D; they have, therefore, a common altitude; hence,

$$AED : EDC :: AE : EC.$$

But the triangles DEB and EDC have a common base DE, and their vertices in the line BC, parallel to DE: they are, therefore, equal: hence, the two preceding proportions have a couplet in each equal; and consequently, the remaining terms are proportional (B. II., P. IV.), hence,

$$AD : DB :: AE : EC;$$

which was to be proved.

Cor. 1. We have, by composition (B. II., P. VI.),

$$AD + DB : AD :: AE + EC : AE;$$

or,   AB : AD :: AC : AE;

and, in like manner,

AB : DB :: AE : EC.

Cor. 2. If any number of parallels be drawn cutting two lines, they divide the lines proportionally.

For, let O be the point where AB and CD meet. In the triangle OEF, the line AC being parallel to the base EF, we have,

OE : AE :: OF : CF.

In the triangle OGH, we have,

OE : EG :: OF : FH;

hence (B. II., P. IV., C.),

AE : EG :: CF : FH.

In like manner,

EG : GB :: FH : HD;

and so on.

PROPOSITION XVI. THEOREM.

If a straight line divides two sides of a triangle proportionally, it is parallel to the third side.

Let ABC be a triangle, and let DE divide AB and AC, so that

AD : DB :: AE : EC;

then DE is parallel to BC.

Draw DC and EB. Then the triangles

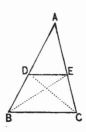

ADE and DEB have a common altitude; and consequently, we have,

ADE : DEB :: AD : DB.

The triangles ADE and EDC have also a common altitude; and consequently, we have,

ADE : EDC :: AE : EC;

but, by hypothesis,

AD : DB :: AE : EC;

hence (B. II., P. IV.),

ADE : DEB :: ADE : EDC.

The antecedents of this proportion being equal, the consequents are equal; that is, the triangles DEB and EDC are equal. But these triangles have a common base DE: hence, their altitudes are equal (P. VI., C.); that is, the points B and C, of the line BC, are equally distant from DE, or DE prolonged: hence, BC and DE are parallel (B. I., P. XXX., C.); *which was to be proved.*

PROPOSITION XVII. THEOREM.

In any triangle, the straight line which bisects the angle at the vertex, divides the base into two segments proportional to the adjacent sides.

Let AD bisect the vertical angle A of the triangle BAC: then the segments BD and DC are proportional to the adjacent sides BA and CA.

From C, draw CE parallel to DA, and produce it until

it meets BA prolonged, at E. Then, because CE and DA are parallel, the angles BAD and AEC are equal (B. I., P. XX., C. 3); the angles DAC and ACE are also equal (B. I., P. XX., C. 2). But, BAD and DAC are equal, by hypothesis; consequently, AEC and ACE are equal: hence, the triangle ACE is isosceles, AE being equal to AC.

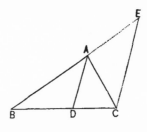

In the triangle BEC, the line AD is parallel to the base EC: hence (P. XV.),

$$BA : AE :: BD : DC;$$

or, substituting AC for its equal AE,

$$BA : AC :: BD : DC;$$

which was to be proved.

PROPOSITION XVIII. THEOREM.

Triangles which are mutually equiangular, are similar.

Let the triangles ABC and DEF have the angle A equal to the angle D, the angle B to the angle E, and the angle C to the angle F: then they are similar.

For, place the triangle DEF upon the triangle ABC, so that the angle E shall coincide with the angle B; then will the point F fall at some point H, of BC; the point D at some point G, of BA;

the side DF will take the position GH, and BGH will be equal to EDF.

Since the angle BHG is equal to BCA, GH will be parallel to AC (B. I., P. XIX., C. 2); and consequently, we have (P. XV.),

$$BA \ : \ BG \ :: \ BC \ : \ BH;$$

or, since BG is equal to ED, and BH to EF,

$$BA \ : \ ED \ :: \ BC \ : \ EF.$$

In like manner, it may be shown that

$$BC \ : \ EF \ :: \ CA \ : \ FD;$$

and also,

$$CA \ : \ FD \ :: \ AB \ : \ DE;$$

hence, the sides about the equal angles, taken in the same order, are proportional; and consequently, the triangles are similar (D. 1); *which was to be proved.*

Cor. If two triangles have two angles in one, equal to two angles in the other, each to each, they are similar (B. I., P. XXV., C. 2).

PROPOSITION XIX.   THEOREM.

Triangles which have their corresponding sides proportional, are similar.

In the triangles ABC and DEF, let the corresponding sides be proportional; that is, let

$$BA \ : \ ED \ :: \ BC \ : \ EF \ :: \ CA \ : \ FD;$$

then the triangles are similar.

For, on BA lay off BG equal to ED; on BC lay off BH equal to EF, and draw GH. Then, because BG is equal to ED, and BH to EF, we have,

$$BA \ : \ BG \ :: \ BC \ : \ BH;$$

hence, GH is parallel to AC (P. XVI.); and consequently, the triangles BAC and BGH are equiangular, and therefore similar; hence,

$$BC \ : \ BH \ :: \ CA \ : \ HG.$$

But, by hypothesis,

$$BC \ : \ EF \ :: \ CA \ : \ FD;$$

hence (B. II., P. IV., C.), we have,

$$BH \ : \ EF \ :: \ HG \ : \ FD.$$

But, BH is equal to EF; hence, HG is equal to FD. The triangles BHG and EFD have, therefore, their sides equal, each to each, and consequently, they are equal in all respects. Now, it has just been shown that BHG and BCA are similar; hence, EFD and BCA are also similar; *which was to be proved.*

Scholium. In order that polygons may be similar, they must fulfill two conditions: they must be *mutually equiangular*, and *the corresponding sides must be proportional.* In the case of triangles, either of these conditions involves the other, which is not true of any other species of polygons.

PROPOSITION XX.   THEOREM.

*Triangles which have an angle in each equal, and the in-
cluding sides proportional, are similar.*

In the triangles ABC and DEF, let the angle B be equal
to the angle E; and suppose that

$$BA \; : \; ED \; :: \; BC \; : \; EF;$$

then the triangles are similar.

For, place the angle E
upon its equal B; F will fall
at some point of BC, as H;
D will fall at some point of
BA, as G; DF will take the

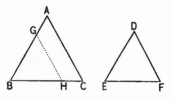

position GH, and the triangle DEF will coincide with GBH,
and consequently, is equal to it.

But, from the assumed proportion, and because BG is
equal to ED, and BH to EF, we have,

$$BA \; : \; BG \; :: \; BC \; : \; BH;$$

hence, GH is parallel to AC; and consequently, BAC and
BGH are mutually equiangular, and therefore similar. But,
EDF is equal to BGH: hence, it is also similar to BAC;
which was to be proved.

PROPOSITION XXI.   THEOREM.

*Triangles which have their sides parallel, each to each, or
perpendicular, each to each, are similar.*

1°. Let the triangles ABC and DEF have the side AB
parallel to DE, BC to EF, and CA to FD; then they are
similar.

For, since the side AB is parallel to DE, and BC to EF, the angle B is equal to the angle E (B. I., P. XXIV.); in like manner, the angle C is equal to the angle F, and the angle A to the angle D; the triangles are, therefore, mutually equiangular, and con-

sequently, are similar (P. XVIII.); *which was to be proved.*

2°. Let the triangles ABC and DEF have the side AB perpendicular to DE, BC to EF, and CA to FD: then they are similar.

For, prolong the sides of the triangle DEF till they meet the sides of the triangle ABC. The sum of the interior angles of the quadrilateral BIEG is equal to four right angles (B. I., P. XXVI.); but, the angles EIB and EGB are each right angles, by

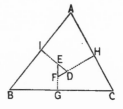

hypothesis; hence, the sum of the angles IEG, IBG is equal to two right angles; the sum of the angles IEG and DEF is equal to two right angles, because they are adjacent; and since things which are equal to the same thing are equal to each other, the sum of the angles IEG and IBG is equal to the sum of the angles IEG and DEF; or, taking away the common part IEG, we have the angle IBG equal to the angle DEF. In like manner, the angle GCH may be proved equal to the angle EFD, and the angle HAI to the angle EDF; the triangles ABC and DEF are, therefore, mutually equiangular, and consequently similar; *which was to be proved.*

Cor. 1. In the first case, the parallel sides are homologous; in the second case, the perpendicular sides are homologous.

Cor. 2. The homologous angles are those included by sides respectively parallel or perpendicular to each other.

Scholium. When two triangles have their sides perpendicular, each to each, they may have a different relative position from that shown in the figure. But we can always construct a triangle within the triangle ABC, whose sides shall be parallel to those of the other triangle, and then the demonstration will be the same as above.

PROPOSITION XXII.   THEOREM.

If a straight line is drawn parallel to the base of a triangle, and straight lines are drawn from the vertex of the triangle to points of the base, these lines divide the base and the parallel proportionally.

Let ABC be a triangle, BC its base, A its vertex, DE parallel to BC, and AF, AG, AH, lines drawn from A to points of the base : then

DI : BF :: IK : FG :: KL : GH :: LE : HC.

For, the triangles AID and AFB, being similar (P. XXI.), we have,

AI : AF :: DI : BF;

and, the triangles AIK and AFG, being similar, we have,

AI : AF :: IK : FG;

hence (B. II., P. IV.), we have,

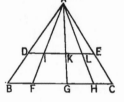

$$DI : BF :: IK : FG.$$

In like manner,

$$IK : FG :: KL : GH,$$

and,

$$KL : GH :: LE : CH;$$

hence (B. II., P. IV.),

$$DI : BF :: IK : FG :: KL : GH :: LE : HC;$$

which was to be proved.

Cor. If BC is divided into equal parts at F, G, and H, then DE is divided into equal parts, at I, K, and L.

PROPOSITION XXIII. THEOREM.

If, in a right-angled triangle, a perpendicular is drawn from the vertex of the right angle to the hypothenuse:

1°. *The triangles on each side of the perpendicular are similar to the given triangle, and to each other:*

2°. *Each side about the right angle is a mean proportional between the hypothenuse and the adjacent segment:*

3°. *The perpendicular is a mean proportional between the two segments of the hypothenuse.*

1°. Let ABC be a right-angled triangle, A the vertex of the right angle, BC the hypothenuse, and AD perpendicular to BC: then ADB and ADC are similar to ABC, and consequently, similar to each other.

The triangles ADB and ABC have the angle B common, and the angles ADB and BAC equal,

because each is a right angle; they are, therefore, similar (P. XVIII., C.). In like manner, it may be shown
that the triangles ADC and ABC are similar; and since
ADB and ADC are each similar to ABC, they are similar to
each other; *which was to be proved.*

2°. AB is a mean proportional
between BC and BD; and AC is a
mean proportional between CB and
CD.

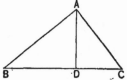

For, the triangles ADB and BAC
being similar, their homologous sides are proportional:
hence,

$$BC \;:\; AB \;::\; AB \;:\; BD.$$

In like manner,

$$BC \;:\; AC \;::\; AC \;:\; DC;$$

which was to be proved.

3°. AD is a mean proportional between BD and DC.
For, the triangles ADB and ADC being similar, their homologous sides are proportional; hence,

$$BD \;:\; AD \;::\; AD \;:\; DC;$$

which was to be proved.

Cor. 1. From the proportions,

$$BC \;:\; AB \;::\; AB \;:\; BD,$$

and,

$$BC \;:\; AC \;::\; AC \;:\; DC,$$

we have (B. II., P. I.),

$$\overline{AB}^2 = BC \times BD,$$

and,

$$\overline{AC}^2 = BC \times DC;$$

whence, by addition,

$$\overline{AB}^2 + \overline{AC}^2 = BC\,(BD + DC);$$

or,

$$\overline{AB}^2 + \overline{AC}^2 = \overline{BC}^2;$$

as was shown in Proposition XI.

Cor. 2. If from any point A, in a semi-circumference BAC, chords are drawn to the extremities B and C of the diameter BC, and a perpendicular AD is drawn to the diameter: then ABC is a right-angled triangle, right-angled at A; and from what was proved above, *each chord is* 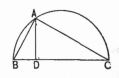 *a mean proportional between the diameter and the adjacent segment;* and, *the perpendicular is a mean proportional between the segments of the diameter.*

PROPOSITION XXIV. THEOREM.

Triangles which have an angle in each equal, are to each other as the rectangles of the including sides.

Let the triangles GHK and ABC have the angles G and A equal: then are they to each other as the rectangles of the sides about these angles.

For, lay off AD equal to GH, AE to GK, and draw DE; then the triangles ADE and GHK are equal in all respects. Draw EB.

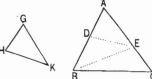

The triangles ADE and ABE have their bases in the same line AB, and a common vertex E; therefore, they have the same altitude, and consequently, are to each other as their bases; that is,

ADE : ABE :: AD : AB.

The triangles ABE and ABC, have their bases in the same line AC, and a common vertex B: hence,

ABE : ABC :: AE : AC;

multiplying these proportions, term by term, and omitting the common factor ABE (B. II., P. VII.), we have,

ADE : ABC :: AD × AE : AB × AC;

substituting for ADE, its equal, GHK, and for AD × AE, its equal, GH × GK, we have,

GHK : ABC :: GH × GK : AB × AC,

which was to be proved.

Cor. If ADE and ABC are similar, the angles D and B being homologous, DE is parallel to BC, and we have,

AD : AB :: AE : AC;

hence (B. II., P. IV.), we have,

ADE : ABE :: ABE : ABC;

that is, ABE is a mean proportional between ADE and ABC.

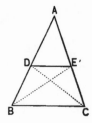

PROPOSITION XXV. THEOREM.

Similar triangles are to each other as the squares of their homologous sides.

Let the triangles ABC and DEF be similar, the angle A being equal to the angle D, B to E, and C to F: then the triangles are to each other as the squares of any two homologous sides.

Because the angles A and D are equal, we have (P. XXIV.),

$$ABC \ : \ DEF \ :: \ AB \times AC \ : \ DE \times DF;$$

and, because the triangles are similar, we have,

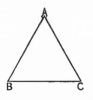

$$AB \ : \ DE \ :: \ AC \ : \ DF;$$

multiplying the terms of this proportion by the corresponding terms of the proportion,

$$AC \ : \ DF \ :: \ AC \ : \ DF,$$

we have (B. II., P. XII.),

$$AB \times AC \ : \ DE \times DF \ :: \ \overline{AC}^2 \ : \ \overline{DF}^2;$$

combining this with the first proportion (B. II., P. IV.), we have,

$$ABC \ : \ DEF \ :: \ \overline{AC}^2 \ : \ \overline{DF}^2.$$

In like manner, it may be shown that the triangles are to each other as the squares of AB and DE, or of BC and EF; *which was to be proved.*

PROPOSITION XXVI.   THEOREM.

Similar polygons may be divided into the same number of triangles, similar, each to each, and similarly placed.

Let ABCDE and FGHIK be two similar polygons, the angle A being equal to the angle F, B to G, C to H, and so on : then can they be divided into the same number of similar triangles, similarly placed.

For, from A draw the diagonals AC, AD, and from F, homologous with A, draw the diagonals FH, FI, to the vertices H and I, homologous with C and D.

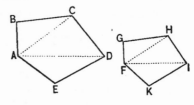

Because the polygons are similar, the triangles ABC and FGH have the angles B and G equal, and the sides about these angles proportional; they are, therefore, similar (P. XX.). Since these triangles are similar, we have the angle ACB equal to FHG, and the sides AC and FH, proportional to BC and GH, or to CD and HI.   The angle BCD being equal to the angle GHI, if we take from the first the angle ACB, and from the second the .equal angle FHG, we have the angle ACD equal to the angle FHI : hence, the triangles ACD and FHI have an angle in each equal, and the including sides proportional; they are therefore similar.

In like manner, it may be shown that ADE and FIK are similar; *which was to be proved.*

Cor. 1. The corresponding triangles in the two polygons are *homologous triangles*, and the corresponding diagonals are *homologous diagonals*.

Any two homologous triangles are *like parts* of the polygons to which they belong.

For, the homologous triangles being similar, we have,

$$ABC : FGH :: \overline{AC}^2 : \overline{FH}^2;$$

and, $$ACD : FHI :: \overline{AC}^2 : \overline{FH}^2;$$

whence, $$ABC : FGH :: ACD : FHI.$$

In like manner, $$ACD : FHI :: ADE : FIK;$$

hence, $$ABC : FGH :: ACD : FHI :: ADE : FIK.$$

Whence, by composition (B. II., P. X.),

$$ABC : FGH :: ACD + ABC + ADE : FHI + FGH + FI\blacktriangleright$$

that is, $$ABC : FGH :: ABCDE : FGHIK.$$

Cor. 2. If two polygons are made up of similar triangles, similarly placed, the polygons themselves are similar.

PROPOSITION XXVII. THEOREM.

The perimeters of similar polygons are to each other as any two homologous sides; and the polygons are to each other as the squares of any two homologous sides.

1°. Let ABCDE and FGHIK be similar polygons: then their perimeters are to each other as any two homologous sides.

For, any two homologous sides, as AB and FG, are like parts of the perimeters to which they belong: hence (B. II., P. IX.), the perimeters of the polygons are 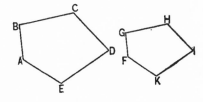 to each other as AB to FG, or as any other two homologous sides; *which was to be proved.*

2°. The polygons are to each other as the squares of any two homologous sides.

For, let the polygons be divided into homologous triangles (P. XXVI., C. 1); then, because the homologous triangles ABC and FGH are like parts of the polygons to

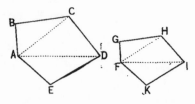

which they belong, the polygons are to each other as these triangles; but these triangles, being similar, are to each other as the squares of AB and FG: hence, the polygons are to each other as the squares of AB and FG, or as the squares of any other two homologous sides; *which was to be proved.*

Cor. 1. Perimeters of similar polygons are to each other as their homologous diagonals, or as any other homologous lines; and the polygons are to each other as the squares of their homologous diagonals, or as the squares of any other homologous lines.

Cor. 2. If the three sides of a right-angled triangle are made homologous sides of three similar polygons, these polygons are to each other as the squares of the sides of the triangle. But the square of the hypothenuse is equal to the sum of the squares of the other sides, and consequently, *the polygon on the hypothenuse will be equal to the sum of the polygons on the other sides.*

PROPOSITION XXVIII. THEOREM.

If two chords intersect in a circle, their segments are reciprocally proportional.

Let the chords AB and CD intersect at O: then are

their segments reciprocally proportional; that is, one segment of the first will be to one segment of the second, as the remaining segment of the second is to the remaining segment of the first.

For, draw CA and BD. Then the angles ODB and OAC are equal, because each is measured by half of the arc CB (B. III., P. XVIII.). The angles OBD and OCA are also equal, because each is measured by half of the arc AD: hence, the triangles OBD and OCA are similar (P. XVIII., C.), and consequently, their homologous sides are proportional: hence,

$$DO \ : \ AO \ :: \ OB \ : \ OC;$$

which was to be proved.

Cor. From the above proportion, we have,

$$DO \times OC \ = \ AO \times OB;$$

that is, *the rectangle of the segments of one chord is equal to the rectangle of the segments of the other.*

PROPOSITION XXIX. THEOREM.

If from a point without a circle, two secants are drawn terminating in the concave arc, they are reciprocally proportional to their external segments.

Let OB and OC be two secants terminating in the concave arc of the circle BCD: then

$$OB \ : \ OC \ :: \ OD \ : \ OA.$$

For, draw AC and DB. The triangles ODB and OAC
have the angle O common, and the angles OBD and OCA
equal, because each is measured by half of
the arc AD: hence, they are similar, and
consequently, their homologous sides are
proportional; whence,

$$OB : OC :: OD : OA;$$

which was to be proved.

Cor. From the above proportion, we have,

$$OB \times OA = OC \times OD;$$

that is, *the rectangles of each secant and its external segment are equal.*

PROPOSITION XXX. THEOREM.

*If from a point without a circle, a tangent and a secant
are drawn, the secant terminating in the concave arc,
the tangent is a mean proportional between the secant
and its external segment.*

Let ADC be a circle, OC a secant, and OA a tangent:
then

$$OC : OA :: OA : OD.$$

For, draw AD and AC. The triangles
OAD and OAC have the angle O common,
and the angles OAD and ACD equal, because each is measured by half of the arc
AD (B. III., P. XVIII., P. XXI.); the triangles are therefore similar, and consequently, their homologous sides are proportional: hence,

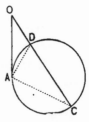

$$OC \; : \; OA \; :: \; OA \; : \; OD;$$

which was to be proved.

Cor. From the above proportion, we have,

$$\overline{AO}^2 = OC \times OD;$$

that is, *the square of the tangent is equal to the rectangle of the secant and its external segment.*

PRACTICAL APPLICATIONS.

PROBLEM I.

To divide a given straight line into parts proportional to given straight lines: also into equal parts.

1°. Let AB be a given straight line, and let it be required to divide it into parts proportional to the lines P, Q, and R.

From one extremity A, draw the indefinite line AG, making any angle with AB; lay off AC equal to P, CD equal to Q, and DE equal to R; draw EB, and from the points C and D, draw CI and DF parallel to EB: then

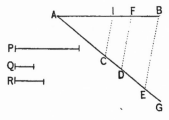

AI, IF, and FB, are proportional to P, Q, and R (P. XV., C. 2).

2°. Let AH be a given straight line, and let it be required to divide it into any number of equal parts, say five.

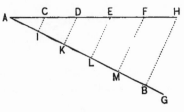

From one extremity A, draw the indefinite line AG; take AI equal to any convenient line, and lay off IK, KL, LM, and MB, each equal to AI. Draw BH, and from I, K, L, and M, draw the lines IC, KD, LE, and MF, parallel to BH: then AH is divided into equal parts at C, D, E, and F (P. XV., C. 2).

PROBLEM II.

To construct a fourth proportional to three given straight lines.

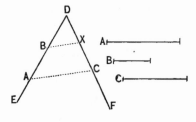

Let A, B, and C, be the given lines. Draw DE and DF, making any convenient angle with each other. Lay off DA equal to A, DB equal to B, and DC equal to C; draw AC, and from B draw BX parallel to AC: then DX is the fourth proportional required.

For (P. XV., C.), we have,

$$DA \; : \; DB \; :: \; DC \; : \; DX;$$

or,

$$A \; : \; B \; :: \; C \; : \; DX.$$

Cor. If DC is made equal to DB, DX is a third proportional to DA and DB, or to A and B.

PROBLEM III.

To construct a mean proportional between two given straight lines.

Let A and B be the given lines. On an indefinite line, lay off DE equal to A, and EF equal to B; on DF as a diameter describe the semicircle DGF, and draw EG perpendicular to DF: then EG is the mean proportional required.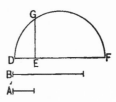

For (P. XXIII., C. 2), we have,

$$DE \ : \ EG \ :: \ EG \ : \ EF;$$

$$A \ : \ EG \ :: \ EG \ : \ B.$$

PROBLEM IV.

To divide a given straight line into two such parts, that the greater part shall be a mean proportional between the whole line and the other part.

Let AB be the given line.

At the extremity B, draw BC perpendicular to AB, and make it equal to half of AB. With C as a centre, and CB as a radius, describe the arc DBE; draw AC, and produce 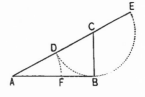 it till it terminates in the concave arc at E; with A as centre and AD as radius, describe the arc DF: then AF is the greater part required.

For, **AB** being perpendicular to **CB** at **B**, is tangent to the arc **DBE**: hence (P. XXX.),

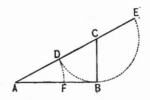

$$AE \ : \ AB \ :: \ AB \ : \ AD \ ;$$

and, by division (B. II., P. VI.),

$$AE - AB \ : \ AB \ :: \ AB - AD \ : \ AD.$$

But, **DE** is equal to twice **CB**, or to **AB**: hence, **AE** — **AB** is equal to **AD**, or to **AF**; and **AB** — **AD** is equal to **AB** — **AF**, or to **FB**: hence, by substitution,

$$AF \ : \ AB \ :: \ FB \ : \ AF \ ;$$

and, by inversion (B. II., P. V.),

$$AB \ : \ AF \ :: \ AF \ : \ FB.$$

Scholium. When a straight line is divided so that the greater segment is a mean proportional between the whole line and the less segment, it is said to be divided *in extreme and mean ratio.*

Since **AB** and **DE** are equal, the line **AE** is divided in extreme and mean ratio at **D**; for we have, from the first of the above proportions, by substitution,

$$AE \ : \ DE \ :: \ DE \ : \ AD.$$

PROBLEM V.

Through a given point, in a given angle, to draw a straight line so that the segments between the point and the sides of the angle shall be equal.

Let BCD be the given angle, and A the given point.

Through A, draw AE parallel to DC; lay off EF equal to CE, and draw FAD: then AF and AD are the segments required.

For (P. XV.), we have,

FA : AD :: FE : EC;

but, FE is equal to EC; hence, FA is equal to AD.

PROBLEM VI.

To construct a triangle equal to a given polygon.

Let ABCDE be the given polygon.

Draw CA; produce EA, and draw BG parallel to CA; draw the line CG. Then the triangles BAC and GAC have the common base AC, and because their vertices B and G lie in the same line BG parallel to the base, their altitudes are equal, and consequently, the triangles are equal: hence, the polygon GCDE is equal to the polygon ABCDE.

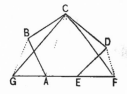

Again, draw CE; produce AE and draw DF parallel to CE; draw also CF; then will the triangles FCE and DCE be equal: hence, the triangle GCF is equal to the polygon GCDE, and consequently, to the given polygon. In like manner, a triangle may be constructed equal to any other given polygon.

PROBLEM VII.

To construct a square equal to a given triangle.

Let ABC be the given triangle, AD its altitude, and BC its base.

Construct a mean proportional between AD and half of BC (Prob. III.). Let XY be that mean proportional, and on it, as a side, con-

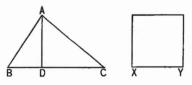

struct a square: then this is the square required. For, from the construction,

$$\overline{XY}^2 = \tfrac{1}{2}BC \times AD = \text{area } ABC.$$

Scholium. By means of Problems VI. and VII., a square may be constructed equal to any given polygon.

PROBLEM VIII.

On a given straight line, to construct a polygon similar to a given polygon.

Let FG be the given line, and ABCDE the given polygon. Draw AC and AD.

At F, construct the angle GFH equal to BAC, and at G the angle FGH equal to ABC; then FGH is similar to ABC (P. XVIII. C.). In like manner, construct the

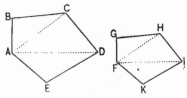

triangle FHI similar to ACD, and FIK similar to ADE; then the polygon FGHIK is similar to the polygon ABCDE (P. XXVI., C. 2).

PROBLEM IX.

To construct a square equal to the sum of two given squares; also a square equal to the difference of two given squares.

1°. Let A and B be the sides of the given squares, and let A be the greater.

Construct a right angle CDE; make DE equal to A, and DC equal to B; draw CE, and on it construct a square : this square 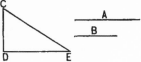 will be equal to the sum of the given squares (P. XI.).

2°. Construct a right angle CDE.

Lay off DC equal to B; with C as a centre, and CE, equal to A, as a radius, describe an arc cutting DE at E; draw CE, and on DE construct a square : this square will be equal to the difference of the given squares (P. XI, C. 1).

Scholium. A polygon may be constructed similar to either of two given polygons, and equal to their sum or difference.

For, let A and B be homologous sides of the given polygons. Find a square equal to the sum or difference of the squares on A and B; and let X be a side of that square. On X as a side, homologous to A or B, construct a polygon similar to the given polygons, and it will be equal to their sum or difference (P. XXVII., C. 2).

EXERCISES.

1. The altitude of an isosceles triangle is 3 feet, each of the equal sides is 5 feet; find the area.

2. The parallel sides of a trapezoid are 8 and 10 feet, and the altitude is 6 feet; what is the area?

3. The sides of a triangle are 60, 80, and 100 feet, the diameter of the inscribed circle is 40 feet; find the area.

4. Construct a square equal to the sum of the squares whose sides are respectively 16, 12, 8, 4, and 2 units in length.

5. Show that the sum of the three perpendiculars drawn from any point within an equilateral triangle to the three sides is equal to the altitude of the triangle.

6. Show that the sum of the squares of two lines, drawn from any point in the circumference of a circle to two points on the diameter of the circle equidistant from the centre, will be always the same.

7. The distance of a chord, 8 feet long, from the centre of a circle is 3 feet; what is the diameter of the circle?

8. Construct a triangle, having given the vertical angle, the line bisecting the base, and the angle which the bisecting line makes with the base.

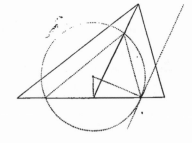

9. Show that if a line bisecting the exterior vertical angle of a triangle is not parallel to the base, the distances of the point in which it meets the base produced, from the extremities of the base, are proportional to the other two sides of the triangle.

10. The segments made by a perpendicular, drawn from a point on the circumference of a circle to a diameter, are 16 feet and 4 feet; find the length of the perpendicular.

11. Two similar triangles, ABC and DEF, have the homologous sides AC and DF equal respectively to 4 feet and 6 feet, and the area of DEF is 9 square feet; find the area of ABC.

12. Two chords of a circle intersect; the segments of one are respectively 6 feet and 8 feet, and one segment of the other is 12 feet; find the remaining segment.

13. Two circles, whose radii are 6 feet and 10 feet, intersect, and the line joining their points of intersection is 8 feet; find the distance between their centres.

14. Find the area of a triangle whose sides are respectively 31, 28, and 20 feet.

15. Show that the area of an equilateral triangle is equal to one fourth the square of one side multiplied by $\sqrt{3}$; or to the square of one side multiplied by .433.

16. From a point, O, in an equilateral triangle, ABC, the distances to the vertices were measured and found to be: OB = 20, OA = 28, OC = 31; find the area of the triangle and the length of each side.

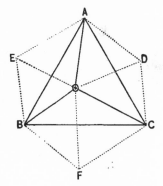

[AD is made equal to OA, CD to OB, CF to OC, BF to OA, BE to OB, AE to OC.]

BOOK V.

DEFINITION.

1. A REGULAR POLYGON is a polygon which is both equilateral and equiangular.

PROPOSITION I. THEOREM.

Regular polygons of the same number of sides are similar.

Let ABCDEF and *abcdef* be regular polygons of the same number of sides: then they are similar.

For, the corresponding angles in each are equal, because any angle in either polygon is equal to twice as many right angles as the polygon has sides, less four right angles, divided 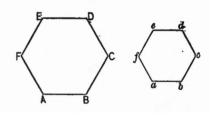 by the number of angles (B. I., P. XXVI., C. 4); and further, the corresponding sides are proportional, because all the sides of either polygon are equal (D. 1): hence, the polygons are similar (B. IV., D. 1); *which was to be proved.*

PROPOSITION II. THEOREM.

The circumference of a circle may be circumscribed about any regular polygon; a circle may also be inscribed in it.

1°. Let ABCF be a regular polygon: then can the circumference of a circle be circumscribed about it.

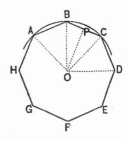

For, through three consecutive vertices A, B, C, describe the circumference of a circle (B. III., Problem XIII., S.). Its centre O lies on PO, drawn perpendicular to BC, at its middle point P; draw OA and OD.

Let the quadrilateral OPCD be turned about the line OP, until PC falls on PB; then, because the angle C is equal to B, the side CD will take the direction BA: and because CD is equal to BA, the vertex D, will fall upon the vertex A; and consequently, the line OD will coincide with OA, and is, therefore, equal to it: hence, the circumference which passes through A, B, and C, passes through D. In like manner, it may be shown that it passes through each of the other vertices: hence, it is circumscribed about the polygon; *which was to be proved.*

2°. A circle may be inscribed in the polygon.

For, the sides AB, BC, &c., being equal chords of the circumscribed circle, are equidistant from the centre O; hence, a circle described from O as a centre, with OP as a radius, is tangent to each of the sides of the polygon, and consequently, is inscribed in it; *which was to be proved.*

Scholium. If the circumference of a circle is divided into equal arcs, the chords of these arcs are sides of a regular inscribed polygon.

For, the sides are equal, because they are chords of equal arcs, and the angles are equal, because they are measured by halves of equal arcs.

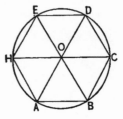

If the vertices A, B, C, &c., of a regular inscribed polygon be joined with the centre O, the triangles thus formed will be equal, because their sides are equal, each to each: hence, all of the angles about the point O are equal to each other.

DEFINITIONS.

1. The CENTRE OF A REGULAR POLYGON is the common centre of the circumscribed and inscribed circles.

2. The ANGLE AT THE CENTRE is the angle formed by drawing lines from the centre to the extremities of any side.

The angle at the centre is equal to four right angles divided by the number of sides of the polygon.

3. The APOTHEM is the shortest distance from the centre to any side.

The apothem is equal to the radius of the inscribed circle.

PROPOSITION III. PROBLEM.

To inscribe a square in a given circle.

Let ABCD be the given circle. Draw any two diameters AC and BD perpendicular to each other; they divide the circumference into four equal arcs (B. III., P. XVII., S.). Draw the chords AB, BC, CD, and DA: then the figure ABCD is the square required (P. II., S.).

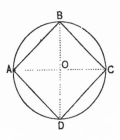

Scholium. The radius is to the side of the inscribed square as 1 is to $\sqrt{2}$.

PROPOSITION IV. THEOREM.

If a regular hexagon is inscribed in a circle, any side is equal to the radius of the circle.

Let ABD be a circle, and ABCDEH a regular inscribed hexagon: then any side, as AB, is equal to the radius of the circle.

Draw the radii OA and OB. Then the angle AOB is equal to one sixth of four right angles, or to two thirds of one right angle, because it is an angle at the centre (P. II., D. 2). The sum of the two angles OAB and OBA is, consequently, equal to four

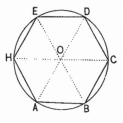

thirds of a right angle (B. I., P. XXV., C. 1); but, the angles OAB and OBA are equal, because the opposite sides OB and OA are equal: hence, each is equal to two thirds of a right angle. The three angles of the triangle AOB are therefore equal, and consequently, the triangle is equilateral: hence, AB is equal to OA; *which was to be proved.*

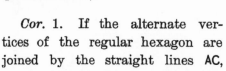

PROPOSITION V. PROBLEM.

To inscribe a regular hexagon in a given circle.

Let ABE be a circle, and O its centre.

Beginning at any point of the circumference, as A, apply the radius OA six times as a chord; then ABCDEF is the hexagon required (P. IV.).

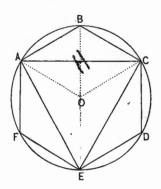

Cor. 1. If the alternate vertices of the regular hexagon are joined by the straight lines AC, CE, and EA, the inscribed triangle ACE is equilateral (P. II., S.).

Cor. 2. If we draw the radii OA and OC, the figure AOCB is a rhombus, because its sides are equal: hence (B. IV., P. XIV., C.), we have,

$$\overline{AB}^2 + \overline{BC}^2 + \overline{OA}^2 + \overline{OC}^2 = \overline{AC}^2 + \overline{OB}^2;$$

or, taking away from the first member the quantity $\overline{OA}^2$, and from the second its equal $\overline{OB}^2$, and reducing, we have,

$$3\overline{OA}^2 = \overline{AC}^2;$$

whence (B. II., P. II.),

$$\overline{AC}^2 : \overline{OA}^2 :: 3 : 1;$$

or (B. II., P. XII., C. 2),

$$AC \; : \; OA \; :: \; \sqrt{3} \; : \; 1;$$

that is, *the side of an inscribed equilateral triangle is to the radius, as the square root of 3 is to 1.*

PROPOSITION VI. THEOREM.

If the radius of a circle is divided in extreme and mean ratio, the greater segment is equal to one side of a regular inscribed decagon.

Let ACG be a circle, OA its radius, and AB, equal to OM, the greater segment of OA when divided in extreme and mean ratio: then AB is equal to the side of a regular inscribed decagon.

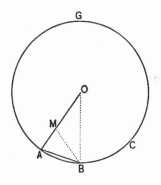

Draw OB and BM. We have, by hypothesis,

$$AO \; : \; OM \; :: \; OM \; : \; AM;$$

or, since AB is equal to OM, we have,

$$AO \; : \; AB \; :: \; AB \; : \; AM;$$

hence, the triangles OAB and BAM have the sides about their common angle BAM, proportional; they are, therefore, similar (B. IV., P. XX.). But, the triangle OAB is isosceles; hence, BAM is also isosceles, and consequently, the side BM is equal to AB. But, AB is equal to OM, by hypothesis: hence, BM is equal to OM, and consequently, the angles MOB and MBO are equal. The angle

AMB being an exterior angle of the triangle OMB, is equal
to the sum of the angles MOB and MBO, or to twice the
angle MOB; and because AMB is
equal to OAB, and also to OBA,
the sum of the angles OAB and
OBA is equal to four times the
angle AOB: hence, AOB is equal
to one fifth of two right angles,
or to one tenth of four right
angles; and consequently, the arc
AB is equal to one tenth of the
circumference: hence, the chord
AB is equal to the side of a
regular inscribed decagon; *which was to be proved.*

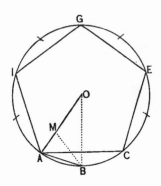

Cor. 1. If AB is applied ten times as a chord, the resulting polygon is a regular inscribed decagon.

Cor. 2. If the vertices A, C, E, G, and I, of the alternate angles of the decagon are joined by straight lines,
the resulting figure is a regular inscribed pentagon.

Scholium 1. If the arcs subtended by the sides of any
regular inscribed polygon are bisected, and chords of the
semi-arcs drawn, the resulting figure is a regular inscribed
polygon of double the number of sides.

Scholium 2. The area of any regular inscribed polygon
is less than that of a regular inscribed polygon of double
the number of sides, because a part is less than the whole.

PROPOSITION VII. PROBLEM.

To circumscribe, about a circle, a polygon which shall be similar to a given regular inscribed polygon.

Let TNQ be a circle, O its centre, and ABCDEF a regular inscribed polygon.

At the middle points T, N, P, &c., of the arcs subtended by the sides of the inscribed polygon, draw tangents to the circle, and prolong them till they intersect; then the resulting figure is the polygon required.

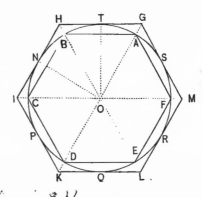

1°. The side HG being parallel to BA, and HI to BC, the angle H is equal to the angle B. In like manner, it may be shown that any other angle of the circumscribed polygon is equal to the corresponding angle of the inscribed polygon: hence, the circumscribed polygon is *equiangular*.

2°. Draw the straight lines OG, OT, OH, ON, and OI. Then, because the lines HT and HN are tangent to the circle, OH bisects the angle NHT, and also the angle NOT (B. III., Prob. XIV., C.); consequently, it passes through the middle point B of the arc NBT. In like manner, it may be shown that the straight line drawn from the centre to the vertex of any other angle of the circumscibed polygon, passes through the corresponding vertex of the inscribed polygon.

The triangles OHG and OHI have the angles OHG and

OHI equal, from what has just been shown; the **angles**
GOH and HOI equal, because they are measured by the
equal arcs AB and BC, and
the side OH common; they
are, therefore, equal in all
respects; hence, GH is equal
to HI. In like manner, it
may be shown that HI is
equal to IK, IK to KL, and so
on: hence, the circumscribed
polygon is *equilateral*.

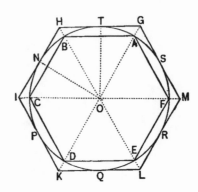

The circumscribed poly-
gon being both equiangular
and equilateral, is *regular ;*
and since it has the same number of sides as the inscribed polygon, it is similar to it.

Cor. 1. If straight lines are drawn from the centre of
a regular circumscribed polygon to its vertices, and the
consecutive points in which they intersect the circumference joined by chords, the resulting figure is a regular
inscribed polygon similar to the given polygon.

Cor. 2. The sum of the lines HT and HN is equal to
the sum of HT and TG, or to HG; that is, to one of the
sides of the circumscribed polygon.

Cor. 3. If at the vertices A, B, C, &c., of the inscribed
polygon, tangents are drawn to the circle and prolonged
till they meet the sides of the circumscribed polygon, the
resulting figure is a circumscribed polygon of double the
number of sides.

Sch. 1. The area of any regular circumscribed polygon

is greater than that of a regular circumscribed polygon of double the number of sides, because the whole is greater than any of its parts.

Sch. 2. By means of a circumscribed and inscribed square, we may construct, in succession, regular circumscribed and inscribed polygons of 8, 16, 32, &c., sides. By means of the regular hexagon we may, in like manner, construct regular polygons of 12, 24, 48, &c., sides. By means of the decagon, we may construct regular polygons of 20, 40, 80, &c., sides.

PROPOSITION VIII. THEOREM.

The area of a regular polygon is equal to half the product of its perimeter and apothem.

Let GHIK be a regular polygon, O its centre, and OT its apothem, or the radius of the inscribed circle: then the area of the polygon is equal to half the product of the perimeter and the apothem.

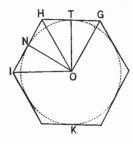

For, draw lines from the centre to the vertices of the polygon. These lines divide the polygon into triangles whose bases are the sides of the polygon, and whose altitudes are equal to the apothem. Now, the area of any triangle, as OHG, is equal to half the product of the side HG and the apothem: hence, the area of the polygon is equal to half the product of the perimeter and the apothem; *which was to be proved.*

PROPOSITION IX. THEOREM.

*The perimeters of similar regular polygons are to each other
as the radii of their circumscribed or inscribed circles;
and their areas are to each other as the squares of those
radii.*

1°. Let ABC and KLM be similar regular polygons. Let
OA and QK be the radii of their circumscribed, OD and
QR be the radii of their inscribed circles: then the perimeters of the polygons are to each other as OA is to QK,
or as OD is to QR.

For, the lines OA
and QK are homolo-
gous lines of the
polygons to which
they belong, as are
also the lines OD
and QR: hence, the
perimeter of ABC is
to the perimeter of

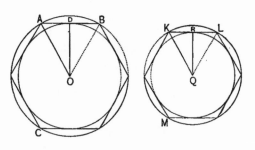

KLM, as OA is to QK, or as OD is to QR (B. IV., P.
XXVII., C. 1); *which was to be proved.*

2°. The areas of the polygons are to each other as
$\overline{OA}^2$ is to $\overline{QK}^2$, or as $\overline{OD}^2$ is to $\overline{QR}^2$.

For, OA being homologous with QK, and OD with QR,
we have, the area of ABC is to the area of KLM as
$\overline{OA}^2$ is to $\overline{QK}^2$, or as $\overline{OD}^2$ is to $\overline{QR}^2$ (B. IV., P. XXVII.,
C. 1); *which was to be proved.*

PROPOSITION X. THEOREM.

Two regular polygons of the same number of sides can be constructed, the one circumscribed about a circle and the other inscribed in it, which shall differ from each other by less than any given surface.

Let ABCE be a circle, O its centre, and Q the side of a square equal to or less than the given surface; then can two similar regular polygons be constructed, the one circumscribed about, and the other inscribed in the given circle, which shall differ from each other by less than the square of Q, and consequently, by less than the given surface.

Inscribe a square in the given circle (P. III.), and by means of it, inscribe, in succession, regular polygons of 8, 16, 32, &c., sides (P. VII., S. 2), until one is found whose side is less than Q; let AB be the side of such a polygon.

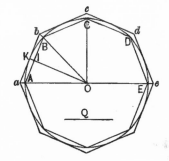

Construct a similar circumscribed polygon *abcde*: then these polygons differ from each other by less than the square of Q.

For, from *a* and *b*, draw the lines *a*O and *b*O; they pass through the points A and B. Draw also OK to the point of contact K; it bisects AB at I and is perpendicular to it. Prolong AO to E.

Let P denote the circumscribed, and *p* the inscribed polygon; then, because they are regular and similar, we have (P. IX.),

$$P : p :: \overline{OK}^2 \text{ or } \overline{OA}^2 : \overline{OI}^2 :$$

hence, by division (B. II., P. VI.), we have,

$$P : P - p :: \overline{OA}^2 : \overline{OA}^2 - \overline{OI}^2;$$

or,

$$P : P - p :: \overline{OA}^2 : \overline{AI}^2.$$

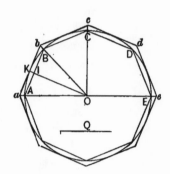

Multiplying the terms of the second couplet by 4 (B. II., P. VII), we have

$$P : P - p :: 4\overline{OA}^2 : 4\overline{AI}^2;$$

whence (B. IV., P. VIII., C.),

$$P : P - p :: \overline{AE}^2 : \overline{AB}^2$$

But P is less than the square of AE (P. VII., S. 1); hence, P − p is less than the square of AB, and consequently, less than the square of Q, or than the given surface; *which was to be proved.*

DEFINITION.—The *limit* of a variable quantity is a quantity to which it may be made to approach nearer than any given quantity, and which it reaches under a particular supposition.

LEMMA.—*Two variable quantities which constantly approach to equality, and of which the difference becomes less than any finite magnitude, are ultimately equal.*

For if they are not ultimately equal, let D be their ultimate difference. Now, by hypothesis, the quantities have approached nearer to equality than any given quantity, as D; hence D denotes their difference and a quantity greater than their difference, at the same time, which is impossible; therefore, the two quantities are ultimately equal.*

* Newton's Principia, Book I., Lemma I.

Cor. If we take any two similar regular polygons, the one circumscribed about, and the other inscribed in the circle, and bisect the arcs, and then circumscribe and inscribe two regular polygons having double the number of sides, it is plain that by continuing the operation, two new polygons may be found which shall differ from each other by less than any given surface ; hence, by the lemma, the two polygons will become ultimately equal. But this equality can not take place for any finite number of sides ; hence, the number of sides in each will be infinite, and each will coincide with the circle, which is their common limit. Under this hypothesis, the perimeter of each polygon will coincide with the circumference of the circle.

Scholium. The circle may be regarded as a regular polygon having an infinite number of sides. The circumference may be regarded as the *perimeter*, and the radius as the *apothem*.

PROPOSITION XI.   PROBLEM.

The area of a regular inscribed polygon, and that of a similar circumscribed polygon being given, to find the areas of the regular inscribed and circumscribed polygons having double the number of sides.

Let AB be the side of the given inscribed, and EF that of the given circumscribed polygon. Let C be their common centre, AMB a portion of the circumference of the circle, and M the middle point of the arc AMB.

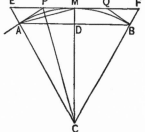

Draw the chord AM, and at A and B draw the tangents AP and BQ ; then AM is the side of the inscribed polygon, and PQ the side of the circumscribed polygon of double the number of sides (P. VII.). Draw CE, CP, CM, and CF.

Denote the area of the given inscribed polygon by p, the area of the given circumscribed polygon by P, and the areas of the inscribed and circumscribed polygons having double the number of sides, respectively by p' and P'.

1°. The triangles CAD, CAM, and CEM, are like parts of the polygons to which they belong: hence, they are proportional to the polygons themselves. But CAM is a mean proportional between CAD and CEM (B. IV., P. XXIV., C.); consequently, p' is a mean proportional between p and P: hence,

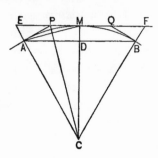

$$p' = \sqrt{p \times P}. \quad \ldots \ldots \ldots \quad (1.)$$

2°. Because the triangles CPM and CPE have the common altitude CM, they are to each other as their bases: hence,

$$\text{CPM} \;:\; \text{CPE} \;::\; \text{PM} \;:\; \text{PE};$$

and because CP bisects the angle ACM, we have (B. IV., P. XVII.),

$$\text{PM} \;:\; \text{PE} \;::\; \text{CM} \;:\; \text{CE} \;::\; \text{CD} \;:\; \text{CA};$$

hence (B. II., P. IV.),

$$\text{CPM} \;:\; \text{CPE} \;::\; \text{CD} \;:\; \text{CA or CM.}$$

But, the triangles CAD and CAM have the common altitude AD; they are, therefore, to each other as their bases: hence,

$$\text{CAD} \;:\; \text{CAM} \;::\; \text{CD} \;:\; \text{CM};$$

or, because CAD and CAM are to each other as the poly-gons to which they belong,

$$p \; : \; p' \; :: \; CD \; : \; CM;$$

hence (B. II., P. IV.), we have,

$$CPM \; : \; CPE \; :: \; p \; : \; p';$$

and, by composition,

$$CPM \; : \; CPM + CPE \text{ or } CME \; :: \; p \; : \; p + p';$$

hence (B. II., P. VII.),

$$2CPM \text{ or } CMPA \; : \; CME \; :: \; 2p \; : \; p + p'.$$

But, CMPA and CME are like parts of P' and P; hence,

$$P' \; : \; P \; :: \; 2p \; : \; p + p';$$

or,

$$P' = \frac{2p \times P}{p + p'}. \quad \cdots \cdots \cdots \text{ (2.)}$$

Scholium. By means of Equation (1), we can find p and then, by means of Equation (2), we can find P'.

PROPOSITION XII. PROBLEM.

To find the approximate area of a circle whose radius is 1.

The area of an inscribed square is equal to twice the square described on the radius (P. III., S.); the area of a circumscribed square is equal to the square described on the *diameter*. If the radius be taken as the unit of linear measure, and the square described on it as the unit of area, the area of the inscribed square will be 2, and that of the circumscribed square will be 4. Making p equal to 2, and P equal to 4, we have, from Equations (1) and (2) of Proposition XI.,

$$p' = \sqrt{8} = 2.8284271 \; \cdots \text{ inscribed octagon,}$$

$$P' = \frac{16}{2 + \sqrt{8}} = 3.3137085 \; \cdots \text{ circumscribed octagon.}$$

Making p equal to 2.8284271, and P equal to 3.3137085, we have, from the same equations,

$p' = 3.0614674$ . . . inscribed polygon of 16 sides.

P$' = 3.1825979$ . . . circumscribed polygon of 16 sides.

By a continued application of these equations, we find the areas indicated below:

NUMBER OF SIDES.	INSCRIBED POLYGONS.	CIRCUMSCRIBED POLYGONS.
4 . .	2.0000000 . .	4.0000000
8 . .	2.8284271 . .	3.3137085
16 . .	3.0614674 . .	3.1825979
32 . .	3.1214451 . .	3.1517249
64 . .	3.1365485 . .	3.1441184
128 . .	3.1403311 . .	3.1422236
256 . ⁓	3.1412772 . .	3.1417504
512 . .	3.1415138 . .	3.1416321
1024 . .	3.1415729 . .	3.1416025
2048 .	3.1415877 . .	3.1415951
4096 . .	3.1415914 . .	3.1415933
8192 . .	3.1415923 . .	3.1415928
16384 . .	3.1415925 . .	3.1415927

Now, the figures which express the areas of the last two polygons are the same for six decimal places; hence, those areas differ from each other by less than one millionth part of the measuring unit. But the circle differs from either of the polygons by less than they differ from each other. Hence, for all ordinary computation, it is sufficiently accurate to consider the area of a circle, whose radius is 1, equal to 3.141592; the unit of measure being, as shown above, the square described on the radius. This value, 3.141592, is represented by the Greek letter π.

Sch. For ordinary accuracy, π is taken equal to 3.1416.

PROPOSITION XIII. THEOREM.

The circumferences of circles are to each other as their radii, and the areas are to each other as the squares of their radii.

Let C and O be the centres of two circles whose radii are CA and OB: then the circumferences are to each other as their radii, and the areas are to each other as the squares of their radii.

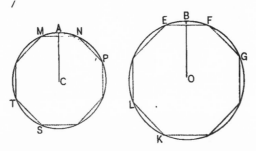

For, let similar regular polygons MNPST and EFGKL be inscribed in the circles: then the perimeters of these polygons are to each other as their apothems, and the areas are to each other as the squares of their apothems, whatever may be the number of their sides (P. IX.).

If the number · of sides is made infinite (P. X., Sch.), the polygons coincide with the circles, the perimeters with the circumferences, and the apothems with the radii: hence, the circumferences of the circles are to each other as their radii, and the areas are to each other as the squares of the radii; *which was to be proved.*

Cor. 1. Diameters of circles are proportional to their radii: hence, *the circumferences of circles are proportional to their diameters, and the areas are proportional to the squares of the diameters.*

Cor. 2. Similar arcs, as AB and DE, are like parts of the circumferences to which they belong, and similar sectors, as ACB and DOE, are like parts of the circles to which they belong: hence, *similar arcs are to each other as their radii,* and similar sectors are to each other as the squares of their radii.

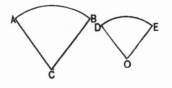

Scholium. The term *infinite*, used in the proposition, is to be understood in its *technical sense.* When it is proposed to make the number of sides of the polygons *infinite*, by the method indicated in the scholium of Proposition X., it is simply meant to express the condition of things, when the inscribed polygons reach their limits; in which case, the difference between the area of either circle and its inscribed polygon, is less than any appreciable quantity. We have seen (P. XII.), that when the number of sides is 16384, the areas differ by less than the millionth part of the measuring unit. By increasing the number of sides, we approximate still nearer.

PROPOSITION XIV. THEOREM.

The area of a circle is equal to half the product of its circumference and radius.

Let O be the centre of a circle, OC its radius, and ACDE its circumference: then the area of the circle is equal to half the product of the circumference and radius.

For, inscribe in it a regular polygon ACDE. Then the area of this polygon is equal to half the product

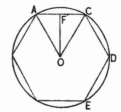

of its perimeter and apothem, whatever may be the number of its sides (P. VIII.).

If the number of sides is made infinite, the polygon coincides with the circle, the perimeter with the circumference, and the apothem with the radius: hence, the area of the circle is equal to half the product of its circumference and radius; *which was to be proved.*

Cor. 1. The area of a sector is equal to half the product of its arc and radius.

Cor. 2. The area of a sector is to the area of the circle, as the arc of the sector to the circumference.

PROPOSITION XV. PROBLEM.

To find an expression for the area of any circle in terms of its radius.

Let C be the centre of a circle, and CA its radius. Denote its area by *area* CA, its radius by R, and the area of a circle whose radius is 1, by π (P. XII., S.).

Then, because the areas of circles are to each other as the squares of their radii (P. XIII.), we have,

$$\text{area } \mathsf{CA} \ : \ \pi \ :: \ \mathsf{R}^2 \ : \ 1;$$

whence, $\qquad\qquad$ *area* $\mathsf{CA} = \pi\mathsf{R}^2.$

That is, *the area of any circle is 3.1416 times the square of its radius.*

PROPOSITION XVI. PROBLEM.

To find an expression for the circumference of a circle, in terms of its radius, or diameter.

Let C be the centre of a circle, and CA its radius.

Denote its circumference by *circ.* CA, its radius by R, and its diameter by D.   From the last Proposition, we have,

$$area \ \text{CA} = \pi \text{R}^2;$$

and, from Proposition XIV., we have,

$$area \ \text{CA} = \tfrac{1}{2}circ. \ \text{CA} \times \text{R};$$

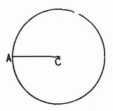

hence, $\tfrac{1}{2}circ.$ CA$\times$R $= \pi$R^2;

whence, by reduction,

$$circ. \ \text{CA} = 2\pi\text{R}, \quad \text{or}, \quad circ. \ \text{CA} = \pi\text{D}.$$

That is, *the circumference of any circle is equal to* 3.1416 *times its diameter.*

Scholium 1.   The abstract number π, equal to 3.1416, denotes the number of times that the diameter of a circle is contained in the circumference, and also the number of times that the square constructed on the radius is contained in the area of the circle (P. XV.). Now, it has been proved by the methods of higher mathematics, that the value of π is incommensurable with 1; hence, it is impossible to express, by means of numbers, the *exact* length of a circumference in terms of the radius, or the *exact* area in terms of the square described on the radius. It is not possible, therefore, to *square the circle*—that is, to construct a square whose area shall be *exactly* equal to that of the circle.

Scholium 2.   Besides the approximate value of π, 3.1416, usually employed, the fractions $\frac{22}{7}$ and $\frac{355}{113}$ are also sometimes used to express the ratio of the diameter to the circumference.

EXERCISES.

1. The side of an equilateral triangle inscribed in a circle is 6 feet; find the radius of the circle.

2. The radius of a circle is 10 feet; find the apothem of a regular inscribed hexagon.

3. Find the side of a square inscribed in a circle whose radius is 5 feet.

4. Draw a line whose length shall be $\sqrt{3}$.

5. The radius of a circle is 4 feet; find the area of an inscribed equilateral triangle.

6. Show that the sums of the alternate angles of an octagon inscribed in a circle are equal to each other.

7. The area of a regular hexagon, whose side is 20 feet, is 1039.23 square feet; find the apothem.

8. One side of a regular decagon is 20 feet, and its apothem 15.4 feet; find the perimeter and the area of a similar decagon whose apothem is 8 feet.

9. The area of a regular hexagon inscribed in a circle is 9 square feet, and the area of a similar circumscribed hexagon is 12 square feet; find the areas of regular inscribed and circumscribed polygons of 12 sides.

10. Given two diagonals of a regular pentagon that intersect; show that the greater segments will be equal to each other and to a side of the pentagon, and that the diagonals cut each other in extreme and mean ratio.

11. Show how to inscribe in a given circle a regular polygon of 15 sides.

12. Find the side and the altitude of an equilateral triangle in terms of the radius of the inscribed circle.

13. Given an equilateral triangle inscribed in a circle, and a similar circumscribed triangle; determine the ratio of the two triangles to each other.

14. The diameter of a circle is 20 feet; find the area of a sector whose arc is 120°.

15. The circumference of a circle is 200 feet; find its area.

16. The area of a circle is 78.54 square yards; find its diameter.

17. The radius of a circle is 10 feet, and the area of a circular sector 100 square feet; find the arc of the sector in degrees.

18. Show that the area of an equilateral triangle circumscribed about a circle is greater than that of a square circumscribed about the same circle.

19. Let AC and BD be diameters perpendicular to each other; from P, the middle point of the radius OA, as a centre, and a radius equal to PB, describe an arc cutting OC in Q; show that the radius OC is divided in extreme and mean ratio at Q.

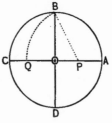

20. Show that the square of the side of a regular inscribed pentagon is equal to the square of the side of a regular inscribed decagon increased by the square of the radius of the circumscribing circle.

21. Show how, from 19 and 20, to inscribe a regular pentagon in a given circle.

22. The side of a regular pentagon, inscribed in a circle, is 5 feet, and that of a regular inscribed decagon is 2.65 feet; find the side and the area of a regular hexagon inscribed in the same circle.

BOOK VI.

DEFINITIONS.

1. A straight line is PERPENDICULAR TO A PLANE, when it is perpendicular to every straight line of the plane which passes through its FOOT; that is, through the *point* in which it meets the plane.

In this case, the plane is also perpendicular to the line.

2. A straight line is PARALLEL TO A PLANE, when it can not meet the plane, how far soever both may be produced.

In this case, the plane is also parallel to the line.

3. Two PLANES ARE PARALLEL, when they can not meet, how far soever both may be produced.

4. A DIEDRAL ANGLE is the amount of divergence of two planes.

The line in which the planes meet is called the *edge of the angle,* and the planes themselves are called *faces of the angle.*

The measure of a diedral angle is the same as that of a plane angle formed by two straight lines, one in each face, and both perpendicular to the edge at the same point. A diedral angle may be *acute*, *obtuse*, or a *right angle*. In the latter case, the faces are *perpendicular* to each other.

5. A POLYEDRAL ANGLE is the amount of divergence of several planes meeting at a common point.

This point is called the *vertex of the angle;* the lines in which the planes meet are called *edges of the angle,* and the portions of the planes lying between the edges are called *faces of the angle.* Thus, S is the vertex of the polyedral angle, whose edges are SA, SB, SC, SD, and whose faces are ASB, BSC, CSD, DSA.

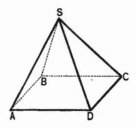

A polyedral angle which has but three faces, is called a *triedral angle.*

POSTULATE.

A straight line may be drawn perpendicular to a plane from any point of the plane, or from any point without the plane.

PROPOSITION I. THEOREM.

If a straight line has two of its points in a plane, it lies wholly in that plane.

For, by definition, a plane is a surface such, that if any two of its points are joined by a straight line, that line lies wholly in the surface (B. I., D. 8).

Cor. Through any point of a plane, an infinite number of straight lines may be drawn which lie in the plane. For, if a straight line is drawn from the given point to any other point of the plane, that line lies wholly in the plane.

Scholium. If any two points of a plane are joined by a straight line, the plane may be turned about that line as

an axis, so as to take an infinite number of positions.
Hence, we infer that an infinite number of planes may be
passed through a given straight line.

PROPOSITION II. THEOREM.

*Through three points, not in the same straight line, one
plane can be passed, and only one.*

Let A, B, and C be the three points: then can one
plane be passed through them, and only one.

Join two of the points, as A and B,
by the line AB. Through AB let a plane
be passed, and let this plane be turned
around AB until it contains the point C;
in this position it will pass through the

three points A, B, and C. If now, the plane be turned
about AB, in either direction, it will no longer contain
the point C: hence, one plane can always be passed
through three points, and only one; *which was to be proved.*

Cor. 1. Three points, not in a straight line, determine
the position of a plane, because only one plane can be
passed through them.

Cor. 2. A straight line and a point without that line
determine the position of a plane, because only one plane
can be passed through them.

Cor. 3. Two straight lines which intersect determine
the position of a plane. For, let AB and AC intersect at
A: then either line, as AB, and one point of the other,
as C, determine the position of a plane.

Cor. 4. Two parallel straight lines determine the position

of a plane. For, let AB and CD be parallel. By definition
(B. I., D. 16) two parallel lines always lie
in the same plane. But either line, as
AB, and any point of the other, as F, determine the position of a plane: hence,
two parallels determine the position of a plane.

A————————B

C————F————D

PROPOSITION III.   THEOREM.

The intersection of two planes is a straight line.

Let AB and CD be two planes: then is their intersection a straight line.

For, let E and F be any two points common to the planes; draw the straight line
EF. This line having two points in the
plane AB, lies wholly in that plane; and
having two points in the plane CD, lies
wholly in that plane: hence, every point of
EF is common to both planes. Furthermore,

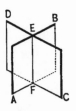

the planes can have no common point lying without EF,
otherwise there would be two planes passing through a
straight line and a point lying without it, which is impossible (P. II., C. 2); hence, the intersection of the two
planes is a straight line; *which was to be proved.*

PROPOSITION IV.   THEOREM.

*If a straight line is perpendicular to two straight lines at
their point of intersection, it is perpendicular to the
plane of those lines.*

Let MN be the plane of the two lines BB, CC, and let
AP be perpendicular to these lines at P: then is AP per-

pendicular to every straight line of the plane which passes through P, and consequently, to the plane itself.

For, through P, draw in the plane MN, any line PQ; through any point of this line, as Q, draw the line BC, so that BQ shall be equal to QC (B. IV., Prob. V.); draw AB, AQ, and AC.

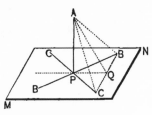

The base BC, of the triangle BPC, being bisected at Q, we have (B. IV., P. XIV.),

$$\overline{PC}^2 + \overline{PB}^2 = 2\overline{PQ}^2 + 2\overline{QC}^2.$$

In like manner, we have, from the triangle ABC,

$$\overline{AC}^2 + \overline{AB}^2 = 2\overline{AQ}^2 + 2\overline{QC}^2.$$

Subtracting the first of these equations from the second, member from member, we have,

$$\overline{AC}^2 - \overline{PC}^2 + \overline{AB}^2 - \overline{PB}^2 = 2\overline{AQ}^2 - 2\overline{PQ}^2.$$

But, from Proposition XI., C. 1, Book IV., we have,

$$\overline{AC}^2 - \overline{PC}^2 = \overline{AP}^2, \qquad \text{and} \qquad \overline{AB}^2 - \overline{PB}^2 = \overline{AP}^2;$$

hence, by substitution,

$$2\overline{AP}^2 = 2\overline{AQ}^2 - 2\overline{PQ}^2;$$

whence,

$$\overline{AP}^2 = \overline{AQ}^2 - \overline{PQ}^2; \qquad \text{or,} \qquad \overline{AP}^2 + \overline{PQ}^2 = \overline{AQ}^2.$$

The triangle APQ is, therefore, right-angled at P (B. IV., P. XIII., S.), and consequently, AP is perpendicular to PQ: hence, AP is perpendicular to every line of the plane MN passing through P, and consequently, to the plane itself; *which was to be proved.*

Cor. 1. Only one perpendicular can be drawn to a plane from a point without the plane.

For, suppose two perpendiculars, as AP and AQ, could be drawn from the point A to the plane MN. Draw PQ; then the triangle APQ would have two right angles, APQ and AQP; which is impossible (B. I., P. XXV., C. 3).

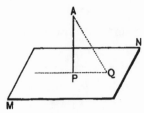

Cor. 2. Only one perpendicular can be drawn to a plane from a point of that plane. For, suppose that two perpendiculars could be drawn to the plane MN, from the point P. Pass a plane through the perpendiculars, and let PQ be its intersection with MN; then we should have two perpendiculars drawn to the same straight line from a point of that line; which is impossible (B. I., P. XIV.).

PROPOSITION V.   THEOREM.

If from a point without a plane, a perpendicular is drawn to the plane, and oblique lines drawn to different points of the plane:

1°. *The perpendicular is shorter than any oblique line:*

2°. *Oblique lines which meet the plane at equal distances from the foot of the perpendicular, are equal:*

3°. *Of two oblique lines which meet the plane at unequal distances from the foot of the perpendicular, the one which meets it at the greater distance is the longer.*

Let A be a point without the plane MN; let AP be perpendicular to the plane; let AC, AD, be any two oblique lines meeting the plane at equal distances from the foot of the perpendicular; and let AC and AE be any

two oblique lines meeting the plane at unequal distances
from the foot of the perpendicular:

1°. AP is shorter than any
oblique line AC.

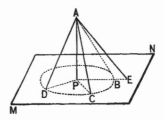

For, draw PC; then is AP
less than AC (B. I., P. XV.);
which was to be proved.

2°. AC and AD are equal.

For, draw PD; then the right-angled triangles APC,
APD, have the side AP common, and the sides PC, PD,
equal: hence, the triangles are equal in all respects,
and consequently, AC and AD are equal; *which was to be
proved.*

3°. AE is greater than AC.

For, draw PE, and take PB equal to PC; draw AB:
then is AE greater than AB (B. I., P. XV.); but AB and AC
are equal: hence, AE is greater than AC; *which was to be
proved.*

Cor. The equal oblique lines AB, AC, AD, meet the plane
MN in the circumference of a circle whose centre is P,
and whose radius is PB: hence, to draw a perpendicular
to a given plane MN, from a point A, without that plane,
find three points B, C, D, of the plane equally distant
from A, and then find the centre, P, of the circle whose
circumference passes through these points: then AP is the
perpendicular required.

Scholium. The angle ABP is called *the inclination of
the oblique line* AB to the plane MN. The equal oblique
lines AB, AC, AD, are all equally inclined to the plane MN.
The inclination of AE is less than the inclination of any
shorter line AB.

PROPOSITION VI. THEOREM.

*If from the foot of a perpendicular to a plane, a straight
line is drawn at right angles to any straight line of
that plane, and the point of intersection joined with any
point of the perpendicular, the last line is perpendicular
to the line of the plane.*

Let AP be perpendicular to the plane MN, P its foot,
BC the given line, and A any point of the perpendicular;
draw PD at right angles to BC, and join the point D with
A : then is AD perpendicular to BC.

For, lay off DB equal to DC,
and draw PB, PC, AB, and AC.
Because PD is perpendicular to
BC, and DB equal to DC, we have,
PB equal to PC (B. I., P. XV.);
and because AP is perpendicular
to the plane MN, and PB equal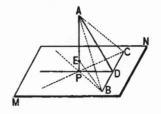
to PC, we have AB equal to AC (P. V.). The line AD has,
therefore, two of its points A and D, each equally distant
from B and C : hence, it is perpendicular to BC (B. I.,
P. XVI., C.); *which was to be proved.*

Cor. 1. The line BC is perpendicular to the plane of
the triangle APD; because it is perpendicular to AD and
PD, at D (P. IV.).

Cor. 2. The shortest distance between AP and BC is
measured on PD, perpendicular to both. For, draw BE
between any other points of the lines : then BE is greater
than PB, and PB greater than PD : hence, PD is less than
BE.

Scholium. The lines AP and BC, though not in the same plane, are considered perpendicular to each other. In general, any two straight lines not in the same plane are considered as making an angle with each other, which angle is equal to that formed by drawing, through a given point, two lines respectively parallel to the given lines.

PROPOSITION VII. THEOREM.

If one of two parallels is perpendicular to a plane, the other one is also perpendicular to the same plane.

Let AP and ED be two parallels, and let AP be perpendicular to the plane MN: then is ED also perpendicular to the plane MN.

For, pass a plane through the parallels; its intersection with MN is PD; draw AD, and in the plane MN draw BC perpendicular to PD at D. Now, BD is perpendicular to the plane APDE (P. VI., C. 1); the angle BDE is consequently a right angle; but the angle EDP is a

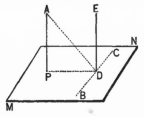

right angle; but the angle EDP is a right angle, because ED is parallel to AP (B. I., P. XX., C. 1): hence, ED is perpendicular to BD and PD, at their point of intersection, and consequently, to their plane MN (P. IV.); *which was to be proved.*

Cor. 1. If the lines AP and ED are perpendicular to the plane MN, they are parallel to each other. For, if not, conceive a line drawn through D parallel to PA; it would be perpendicular to the plane MN, from what has just been proved; we would, therefore, have two perpendiculars to the plane MN, at the same point; which is impossible (P. IV., C. 2).

Cor. 2. If two straight lines, A and B, are parallel to a third line C, they are parallel to each other. For, pass a plane perpendicular to C; it will be perpendicular to both A and B: hence, A and B are parallel.

PROPOSITION VIII.   THEOREM.

If a straight line is parallel to a line of a plane, it is parallel to that plane.

Let the line AB be parallel to the line CD of the plane MN; then is AB parallel to the plane MN.

For, through AB and CD pass a plane (P. II., C. 4); CD is its intersection with the plane MN. Now, since AB lies in this plane, if it can meet the plane MN, it will meet it at some point of CD; but 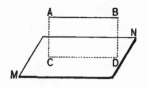 this is impossible, because AB and CD are parallel: hence, AB can not meet the plane MN, and consequently, it is parallel to it; *which was to be proved.*

PROPOSITION IX.   THEOREM.

If two planes are perpendicular to the same straight line, they are parallel to each other.

Let the planes MN and PQ be perpendicular to the line AB, at the points A and B: then are they parallel to each other.

For, if they are not parallel, they will meet; and let O be a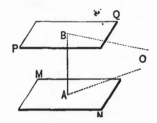

point common to both. From O draw the lines OA and OB: then, since OA lies in the plane MN, it is perpendicular to BA at A (D. 1). For a like reason, OB is perpendicular to AB at B: hence, the triangle OAB has two right angles, which is impossible; consequently, the planes can not meet, and are therefore parallel; *which was to be proved.*

PROPOSITION X. THEOREM.

If a plane intersects two parallel planes, the lines of intersection are parallel.

Let the plane EH intersect the parallel planes MN and PQ, in the lines EF and GH: then are EF and GH parallel.

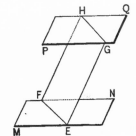

For, if they are not parallel, they will meet if sufficiently prolonged, because they lie in the same plane; but if the lines meet, the planes MN and PQ, in which they lie, also meet; but this is impossible, because these planes are parallel: hence, the lines EF and GH can not meet; they are, therefore, parallel; *which was to be proved.*

PROPOSITION XI. THEOREM.

If a straight line is perpendicular to one of two parallel planes, it is also perpendicular to the other.

Let MN and PQ be two parallel planes, and let the line AB be perpendicular to PQ: then is it also perpendicular to MN.

For, through AB pass any plane; its intersections with MN and PQ are parallel (P. X.); but, its intersection with PQ is perpendicular to AB at B (D. 1); hence, its inter-section with MN is also perpendicular to AB at A (B. I., P. XX., C. 1): hence, AB is perpendicular to every line of the plane MN through A, and is, therefore, perpendicular to that plane; *which was to be proved.*

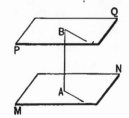

PROPOSITION XII.   THEOREM.

Parallel straight lines included between parallel planes, are equal.

Let EG and FH be any two parallel lines included be-tween the parallel planes MN and PQ: then are they equal.

Through the parallels conceive a plane to be passed; it will intersect the plane MN in the line EF, and PQ in the line GH; and these lines are parallel (Prop. X.). The figure EFHG is, therefore, a parallelogram: hence, GE and HF are equal (B. I., P. XXVIII.); *which was to be proved.*

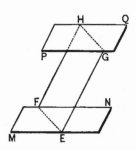

Cor. 1. The distance between two parallel planes is measured on a perpendicular to both; but any two per-pendiculars between the planes are equal: hence, parallel planes are every-where equally distant.

Cor. 2. If a straight line GH is parallel to any plane MN, then can a plane be passed through GH parallel to MN: hence, if a straight line is parallel to a plane, all of its points are equally distant from that plane.

PROPOSITION XIII. THEOREM.

If two angles, not situated in the same plane, have their sides parallel, and lying in the same direction, the angles are equal and their planes parallel.

Let CAE and DBF be two angles lying in the planes MN and PQ, and let the sides AC and AE be respectively parallel to BD and BF, and lying in the same direction: then are the angles CAE and DBF equal, and the planes MN and PQ parallel.

Take any two points of AC and AE, as C and E, and make BD equal to AC, and BF to AE; draw CE, DF, AB, CD, and EF

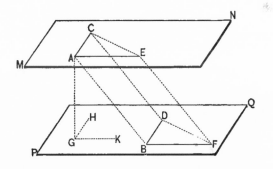

1°. The angles CAE and DBF are equal.

For, AE and BF being parallel and equal, the figure ABFE is a parallelogram (B. I., P. XXX.); hence, EF is parallel and equal to AB. For a like reason, CD is parallel and equal to AB: hence, CD and EF are parallel and equal to each other, and consequently, CE and DF are also parallel and equal to each other. The triangles CAE and DBF have, therefore, their corresponding sides equal, and consequently, the corresponding angles CAE and DBF are equal; *which was to be proved.*

2°. The planes of the angles, MN and PQ, are parallel. For, from A draw AG perpendicular to the plane PQ; at the point G, where it meets the plane, draw in the plane PQ, GH and GK parallel, respectively, to BD and BF; then is AC parallel to GH, and AE to GK (P. VII., C. 2). AG, being perpendicular to GH and GK (D. 1), is perpendicular to their parallels, AC and AE (B. I., P. XX., C. 1), and is, therefore, perpendicular to the plane MN (P. IV.). The planes MN and PQ, being perpendicular to the same straight line, AG, are parallel to each other (P. IX.); *which was to be proved.*

Cor. If two parallel planes, MN and PQ, are met by two other planes, AD and AF, the angles CAE and DBF, formed by their intersections, are equal.

PROPOSITION XIV.   THEOREM.

If three straight lines, not situated in the same plane, are equal and parallel, the triangles formed by joining the extremities of these lines are equal, and their planes parallel.

Let AB, CD, and EF be equal parallel lines not in the same plane: then are the triangles ACE and BDF equal, and their planes parallel.

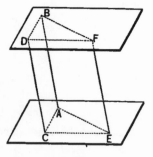

For, AB being equal and parallel to EF, the figure ABFE is a parallelogram, and consequently, AE is equal and parallel to BF. For a like reason, AC is equal and parallel to BD: hence, the included angles CAE and DBF are equal and their planes parallel (P. XIII.). Now, the triangles CAE and DBF have two sides and their included angles equal, each to each: hence, they are equal in all respects. The triangles are, therefore, equal and their planes parallel; *which was to be proved.*

PROPOSITION XV. THEOREM.

If two straight lines are cut by three parallel planes, they are divided proportionally.

Let the lines AB and CD be cut by the parallel planes MN, PQ, and RS, in the points A, E, B, and C, F, D; then

$$AE : EB :: CF : FD.$$

For, draw the line AD, and suppose it to pierce the plane PQ in G; draw AC, BD, EG, and GF.

The plane ABD intersects the parallel planes RS and PQ in the lines BD and EG; consequently, these lines are parallel (P. X.): hence (B. IV., P. XV.),

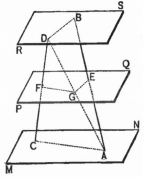

$$AE : EB :: AG : GD.$$

The plane ACD intersects the parallel planes MN and PQ, in the parallel lines AC and GF: hence,

$$AG : GD :: CF : FD.$$

Combining these proportions (B. II., P. IV.), we have,

$$AE : EB :: CF : FD;$$

which was to be proved.

Cor. 1. If two straight lines are cut by any number of parallel planes, they are divided proportionally.

Cor. 2. If any number of straight lines are cut by three parallel planes, they are divided proportionally.

PROPOSITION XVI. THEOREM.

If a straight line is perpendicular to a plane, every plane passed through the line is also perpendicular to that plane.

Let AP be perpendicular to the plane MN, and let BF be a plane passed through AP: then is BF perpendicular to MN.

In the plane MN, draw PD perpendicular to BC, the intersection of BF
and MN. Since AP is perpendicular
to MN, it is perpendicular to BC and
DP (D. 1); and since AP and DP, in
the planes BF and MN, are perpendicular to the intersection of these planes

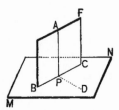

at the same point, the angle which they form is equal to the angle formed by the planes (D. 4); but this angle is a right angle: hence, BF is perpendicular to MN; *which was to be proved.*

Cor. If three lines AP, BP, and DP, are perpendicular to each other at a common point P, each line is perpendicular to the plane of the two others, and the three planes are perpendicular to each other.

PROPOSITION XVII. THEOREM.

If two planes are perpendicular to each other, a straight line drawn in one of them, perpendicular to their inter-section, is perpendicular to the other.

Let the planes BF and MN be perpendicular to each other, and let the line AP, drawn in the plane BF, be per-pendicular to the intersection BC; then is AP perpendicu-lar to the plane MN.

For, in the plane MN, draw PD perpendicular to BC at
P. Then because the planes BF and
MN are perpendicular to each other,
the angle APD is a right angle:
hence, AP is perpendicular to the two
lines PD and BC, at their intersection,
and consequently, is perpendicular to
their plane MN; *which was to be proved.*

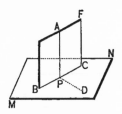

Cor. If the plane BF is perpendicular to the plane MN,
and if at a point P of their intersection, a perpendicular
is erected to the plane MN, that perpendicular is in the
plane BF. For, if not, draw in the plane BF, PA perpendicular to PC, the common intersection; AP is perpendicular to the plane MN, by the theorem; therefore, at the
same point P, there are two perpendiculars to the plane
MN; which is impossible (P. IV., C. 2).

PROPOSITION XVIII. THEOREM.

If two planes cut each other, and are perpendicular to a
third plane, their intersection is also perpendicular to
that plane.

Let the planes BF, DH, be perpendicular to MN: then
is their intersection AP perpendicular
to MN.

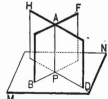

For, at the point P, erect a perpendicular to the plane MN; that perpendicular must be in the plane BF,
and also in the plane DH (P. XVII.,
C.); therefore, it is their common intersection AP; *which was to be proved.*

PROPOSITION XIX.   THEOREM.

The sum of any two of the plane angles formed by the edges of a triedral angle, is greater than the third.

Let SA, SB, and SC, be the edges of a triedral angle: then is the sum of any two of the plane angles formed by them, as ASC and CSB, greater than the third ASB.

If the plane angle ASB is equal to, or less than, either of the other two, the truth of the proposition is evident. Let us suppose, then, that ASB is greater than either.

In the plane ASB, construct the angle BSD equal to BSC; draw AB in that plane, at pleasure; lay off SC equal to SD, and draw AC and CB. The triangles BSD and BSC have the side SC equal to SD, by construction, the side SB common, and the included angles BSD and BSC equal, by construction; the triangles are therefore equal in all

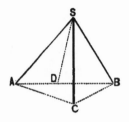

respects: hence, BD is equal to BC.   But, from Proposition VII., Book I., we have,

$$BC + CA > BD + DA.$$

Taking away the equal parts BC and BD, we have,

$$CA > DA;$$

hence (B. I., P. IX.), we have,

$$\text{angle ASC} > \text{angle ASD};$$

and, adding the equal angles BSC and BSD,

angle ASC + angle CSB > angle ASD + angle DSB;

or,   angle ASC + angle CSB > angle ASB;

which was to be proved.

PROPOSITION XX. THEOREM.

The sum of the plane angles formed by the edges of any polyedral angle, is less than four right angles.

Let S be the vertex of any polyedral angle whose edges are SA, SB, SC, SD, and SE; then is the sum of the angles about S less than four right angles.

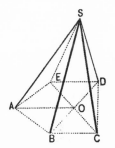

For, pass a plane cutting the edges in the points A, B, C, D, and E, and the faces in the lines AB, BC, CD, DE, and EA. From any point within the polygon thus formed, as O, draw the straight lines OA, OB, OC, OD, and OE.

We then have two sets of triangles, one set having a common vertex S, the other having a common vertex O, and both having common bases AB, BC, CD, DE, EA. Now, in the set which has the common vertex S, the sum of all the angles is equal to the sum of all the plane angles formed by the edges of the polyedral angle whose vertex is S, together with the sum of all the angles at the bases: viz., SAB, SBA, SBC, &c.; and the entire sum is equal to twice as many right angles as there are triangles. In the set whose common vertex is O, the sum of all the angles is equal to the four right angles about O, together with the interior angles of the polygon, and this sum is equal to twice as many right angles as there are triangles. Since

the number of triangles, in each set, is the same, it follows that these sums are equal. But in the triedral
angle whose vertex is B, we have (P. XIX.),

ABS + SBC > ABC;

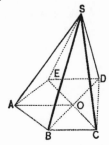

and the like may be shown at each of
the other vertices, C, D, E, A: hence,
the sum of the angles at the bases, in
the triangles whose common vertex is
S, is greater than the sum of the
angles at the bases, in the set whose
common vertex is O: therefore, the sum
of the vertical angles about S, is less than the sum of
the angles about O: that is, less than four right angles;
which was to be proved.

Scholium. The above demonstration is made on the
supposition that the polyedral angle is convex, that is,
that the diedral angles of the consecutive faces are each
less than two right angles.

PROPOSITION XXI.   THEOREM.

*If the plane angles formed by the edges of two triedral
angles are equal, each to each, the planes of the equal
angles are equally inclined to each other.*

Let S and T be the vertices of two triedral angles, and
let the angle ASC be equal to DTF, ASB to DTE, and BSC
to ETF: then the planes of the equal angles are· equally
inclined to each other.

For, take any point of SB, as B, and from it draw in
the two faces ASB and CSB, the lines BA and BC, respectively perpendicular to SB: then the angle ABC measures
the inclination of these faces.   Lay off TE equal to SB

and from E draw in the faces DTE and FTE, the lines ED and EF, respectively perpendicular to TE: then the angle DEF measures the inclination of these faces. Draw AC and DF

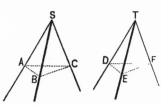

The right-angled triangles SBA and TED, have the side SB equal to TE, and the angle ASB equal to DTE; hence, AB is equal to DE, and AS to DT In like manner, it may be shown that BC is equal to EF, and CS to FT. The triangles ASC and DTF, have the angle ASC equal to DTF, by hypothesis, the side AS equal to DT, and the side CS to FT, from what has just been shown; hence, the triangles are equal in all respects, and consequently, AC is equal to DF. Now, the triangles ABC and DEF have their sides equal, each to each, and consequently, the corresponding angles are also equal; that is, the angle ABC is equal to DEF: hence, the inclination of the planes ASB and CSB, is equal to the inclination of the planes DTE and FTE. In like manner, it may be shown that the planes of the other equal angles are equally inclined; *which was to be proved.*

Cor. If the plane angles ASB and BSC are equal, respectively, to the plane angles DTE and ETF, and the inclination of the faces ASB and BSC is equal to that of the faces DTE and ETF, then are the remaining plane angles, ASC and DTF, equal to each other.

Scholium 1. If the planes of the equal plane angles are like placed, the triedral angles are equal in all respects, for they may be placed so as to coincide. If the planes of the equal angles are not similarly placed, the triedral angles are said to be angles *equal by symmetry*, or symmetrical triedral angles. In this case, they may be placed so that two of the homologous faces shall coincide, the triedral angles lying on opposite sides of the plane, which is then called a *plane of symmetry*. In this position, for every point on one side of the plane of symmetry, there is a corresponding point on the other side.

Scholium 2. If the plane angles ASB and DTE are equal to each other, and the inclination of the face ASB to each of the faces BSC and ASC is equal, respectively, to the inclination of DTE to each of the faces ETF and DTF, then are the plane angles BSC and CSA equal, respectively, to the plane angles ETF and FTD. For, place the plane angle ASB upon its equal DTE, so that the point S shall coincide with T, the edge SA with TD, and the edge SB with TE, then will the face BSC take the direction of the face ETF, and the edge SC will lie somewhere in the plane ETF; the face ASC will take the direction of the face DTF, and the edge SC will lie somewhere in the plane DTF. Since SC is at the same time in both the planes ETF and DTF, it must be on their intersection (P. III.): hence, the plane angles BSC and CSA coincide with and are equal, respectively, to ETF and FTD.

If the triedral angle whose vertex is S can not be made to coincide with the triedral angle whose vertex is T, it may be made to coincide with its symmetrical triedral angle, and the corresponding plane angles would be equal, as before.

NOTE 1.—The projection of a point on a plane is the foot of a perpendicular drawn from the point to the plane.

NOTE 2.—The projection of a line on a plane is that line of the plane which joins the projection of the two extreme points of the given line on the plane.

EXERCISES.

1. Find a point in a plane equidistant from two given points without and on the same side of the plane.

2. From two given points on the same side of a given plane, draw two lines that shall meet the plane in the same point and make equal angles with it.

[The angle made by a line with a plane is the angle which the line makes with its projection on the plane.]

3. What is the greatest number of equilateral triangles that can be grouped about a point so as to form a convex polyedral angle?

4. Show that if from any two points in the edge of a diedral angle straight lines are drawn in each of its faces perpendicular to the edge, these lines contain equal angles.

5. From any point within a diedral angle, draw a perpendicular to each of its two faces, and show that the angle contained by the perpendiculars is the supplement of the diedral angle.

6. Show that if a plane meets another plane, the sum of the adjacent diedral angles is equal to two right angles.

7. Show that if two planes intersect each other, the opposite or vertical diedral angles are equal to each other.

8. Show that if a plane intersects two parallel planes, the sum of the interior diedral angles on the same side is equal to two right angles.

9. Show that if two diedral angles have their faces parallel and lying in the same or in opposite directions, they are equal.

10. Show that every point of a plane bisecting a diedral angle is equidistant from the faces of the angle.

11. Show that the inclination of a line to a plane—that is, the angle which the line makes with its own projection on the plane—is the least angle made by the line with any line of the plane.

12. Show that if three lines are perpendicular to a fourth at the same point, the first three are in the same plane.

13. Show that when a plane is perpendicular to a given line at its middle point, every point of the plane is equally distant from the extremities of the line, and that every point out of the plane is unequally distant from the extremities of the line.

14. Show that through a line parallel to a given plane, but one plane can be passed perpendicular to the given plane.

15. Show that if two planes which intersect contain two lines parallel to each other, the intersection of the planes is parallel to the lines.

16. Show that when a line is parallel to one plane and perpendicular to another, the two planes are perpendicular to each other.

17. Draw a perpendicular to two lines not in the same plane.

18. Show that the three planes which bisect the diedral angles formed by the consecutive faces of a triedral angle, meet in the same line.

BOOK VII.

DEFINITIONS.

1. A POLYEDRON is a volume bounded by polygons.

The bounding polygons are called *faces* of the polyedron; the lines in which the faces meet, are called *edges* of the polyedron; the points in which the edges meet, are called *vertices* of the polyedron.

2. A PRISM is a polyedron in which two of the faces are polygons equal in all respects, and having their homologous sides parallel. The other faces are parallelograms (B. I., P. XXX.).

The equal polygons are called *bases* of the prism; one the *upper*, and the other the *lower base;* the parallelograms taken together make up the *lateral* or *convex surface* of the prism; the lines in which the lateral faces meet, are called *lateral edges*, and the lines in which the lateral faces meet either base are called *basal edges* of the prism.

3. The ALTITUDE of a prism is the perpendicular distance between the planes of its bases.

4. A RIGHT PRISM is one whose lateral edges are perpendicular to the planes of the bases.

In this case, any lateral edge is equal to the altitude.

5. An OBLIQUE PRISM is one whose lateral edges are oblique to the planes of the bases.

In this case, any lateral edge is greater than the altitude.

6. Prisms are named from the number of sides of their bases; a *triangular prism* is one whose bases are triangles; a *pentagonal* prism is one whose bases are pentagons, &c.

7. A PARALLELOPIPEDON is a prism whose bases are parallelograms.

A *Right Parallelopipedon* is one whose lateral edges are perpendicular to the planes of the bases.

A *Rectangular Parallelopipedon* is one whose faces are all rectangles.

A *Cube* is a rectangular parallelopipedon whose faces are squares.

8. A PYRAMID is a polyedron bounded by a polygon called the *base*, and by triangles meeting at a common point, called the *vertex* of the pyramid.

The triangles taken together make up the *lateral* or *convex surface* of the pyramid; the lines in which the lateral faces meet, are called the *lateral edges*, and the lines in which the lateral faces meet the base are called *basal edges* of the pyramid.

9. Pyramids are named from the number of sides of their bases; a *triangular pyramid* is one whose base is a triangle; a *quadrangular* pyramid is one whose base is a quadrilateral, and so on.

10. The ALTITUDE of a pyramid is the perpendicular distance from the vertex to the plane of its base.

11. A RIGHT PYRAMID is one whose base is a regular polygon, and in which the perpendicular, drawn from the vertex to the plane of the base, passes through the centre of the base.

This perpendicular is called the axis of the pyramid.

12. The SLANT HEIGHT of a right pyramid, is the perpendicular distance from the vertex to any side of the base.

13. A TRUNCATED PYRAMID is that portion of a pyramid included between the base and any plane which cuts the pyramid.

When the cutting plane is parallel to the base, the truncated pyramid is called a FRUSTUM OF A PYRAMID, and the intersection of the cutting plane with the pyramid, is called the *upper base* of the frustum; the base of the pyramid is called the *lower* base of the frustum.

14. The ALTITUDE of a frustum of a pyramid, is the perpendicular distance between the planes of its bases.

15. The SLANT HEIGHT of a frustum of a right pyramid, is that portion of the slant height of the pyramid which lies between the planes of its upper and lower bases.

16. SIMILAR POLYEDRONS are those which are bounded by the same number of similar polygons, similarly placed.

Parts which are similarly placed, whether faces, edges, or angles, are called *homologous*.

17. A DIAGONAL of a polyedron, is a straight line joining the vertices of two polyedral angles not in the same face.

`18. The VOLUME OF A POLYEDRON is its numerical value expressed in terms of some other polyedron taken as a unit.

The unit generally employed is a cube constructed on the linear unit as an edge.

PROPOSITION I.   THEOREM.

The convex surface of a right prism is equal to the perimeter of either base multiplied by the altitude.

Let ABCDE–K be a right prism: then is its convex surface equal to,

$$(AB + BC + CD + DE + EA) \times AF.$$

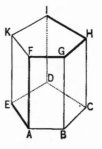

For, the convex surface is equal to the sum of all the rectangles AG, BH, CI, DK, EF, which compose it. Now, the altitude of each of the rectangles AF, BG, CH, &c., is equal to the altitude of the prism, and the area of each rectangle is equal to its base multiplied by its altitude (B. IV., P. V.): hence, the sum of these rectangles, or the convex surface of the prism, is equal to,

$$(AB + BC + CD + DE + EA) \times AF ;$$

that is, to the perimeter of the base multiplied by the altitude; *which was to be proved.*

Cor. If two right prisms have the same altitude, their convex surfaces are to each other as the perimeters of their bases.

PROPOSITION II. THEOREM.

In any prism, the sections made by parallel planes are polygons equal in all respects.

Let the prism AH be intersected by the parallel planes NP, SV: then are the sections NOPQR, STVXY, equal polygons.

For, the sides NO, ST, are parallel, being the intersections of parallel planes with a third plane ABGF; these sides, NO, ST, are included between the parallels NS, OT: hence, NO is equal to ST (B. I., P. XXVIII., C. 2). For like reasons, the sides OP, PQ, QR, &c., of NOPQR, are equal to the sides TV, VX, &c., of STVXY, each to each; and since the equal sides are parallel, each to each, it follows that the angles NOP, OPQ, &c., of the first section, are equal to the angles STV, TVX, &c., of the second section, each to each (B. VI., P. XIII.): hence, the two sections NOPQR, STVXY, are equal in all respects; *which was to be proved.*

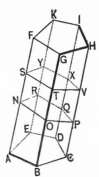

Cor. The bases of a prism and any section of a prism parallel to the bases, are equal in all respects.

PROPOSITION III. THEOREM.

If a pyramid is cut by a plane parallel to the base:
1°. *The edges and the altitude are divided proportionally:*
2°. *The section is a polygon similar to the base.*

Let the pyramid S–ABCDE, whose altitude is SO, be cut by the plane *abcde*, parallel to the base ABCDE.

1°. The edges and altitude are divided proportionally. For, let a plane be passed through the vertex S, parallel to the base AC; then the edges and the altitude are cut by three parallel planes, and are consequently divided proportionally (B. VI., P. XV., C. 2); *which was to be proved.*

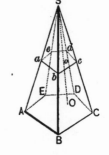

2°. The section *abcde* is similar to the base ABCDE.

For, each side of the section is parallel to the corresponding side of the base (B. VI., P. X.); hence, the corresponding angles of the section and of the base are equal (B. VI., P. XIII.); the two polygons are therefore mutually equiangular. Again, because *ab* is parallel to AB, and *bc* to BC, the triangle S*ba* is similar to SBA, and S*bc* to SBC; hence,

$$ab : AB :: Sb : SB, \quad \text{and} \quad bc : BC :: Sb : SB,$$

whence (B. II., P. IV.), $ab : AB :: bc : BC.$

In like manner, it may be shown that the remaining sides of *abcde* are proportional to the corresponding sides of ABCDE; hence (B. IV., D. 1), the polygons are similar; *which was to be proved.*

Cor. 1. If two pyramids S–ABCD and S–XYZ, having a common vertex S and their bases in the same plane, are cut by a plane *aoz* parallel to the plane of their bases, the sections are to each other as the bases.

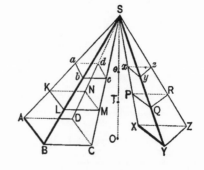

For the polygons *abcd* and ABCD, being similar, are to each other as the squares of any homologous sides (B. IV., P. XXVII.); but

$$\overline{ab}^2 \ : \ \overline{AB}^2 \ :: \ \overline{Sa}^2 \ : \ \overline{SA}^2 \ :: \ \overline{So}^2 \ :: \ \overline{SO}^2;$$

hence (B. II., P. IV.), we have, *abcd* : ABCD :: $\overline{So}^2 : \overline{SO}^2$.

In like manner, we have, *xyz* : XYZ :: $\overline{So}^2 : \overline{SO}^2;$

hence, *abcd* : ABCD :: *xyz* : XYZ.

Cor. 2. If the bases are equal, any sections at equal distances from the vertex, or from the bases, are equal.

Cor. 3. The area of any section parallel to the base is proportional to the square of its distance from the vertex.

Cor. 4. If the two pyramids are cut by a plane KTR, so that ST is a mean proportional between S*o* and SO, that is, so that $\overline{ST}^2$ is a mean proportional between $\overline{So}^2$ and $\overline{SO}^2$, the section KLMN is a mean proportional between *abcd* and ABCD, and also PQR is a mean proportional between *xyz* and XYZ.

PROPOSITION IV. THEOREM.

The convex surface of a right pyramid is equal to the perimeter of its base multiplied by half the slant height.

Let S be the vertex, ABCDE the base, and SF, perpendicular to EA, the slant height of a right pyramid: then is the convex surface equal to,

(AB + BC + CD + DE + EA) × ½SF.

Draw SO perpendicular to the plane of the base.

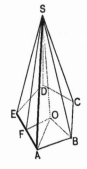

From the definition of a right pyramid, the point O is the centre of the base (D. 11): hence, the lateral edges, SA, SB, &c., are all equal (B. VI., P. V.); but the sides of the base are all equal, being sides of a regular polygon: hence, the lateral faces are all equal, and consequently their altitudes are all equal, each being equal to the slant height of the pyramid.

Now, the area of any lateral face, as SEA, is equal to its base EA, multiplied by half its altitude SF: hence, the sum of the areas of the lateral faces, or the convex surface of the pyramid, is equal to,

$$(AB + BC + CD + DE + EA) \times \tfrac{1}{2}SF;$$

which was to be proved.

Scholium. The convex surface of a frustum of a right pyramid is equal to half the sum of the perimeters of its upper and lower bases, multiplied by the slant height.

Let ABCDE-*e* be a frustum of a right pyramid, whose vertex is S: then the section *abcde* is similar to the base ABCDE, and their homologous sides are parallel (P. III.). Any lateral face of the frustum, as AE*ea*, is a trapezoid, whose altitude is equal to F*f*, the slant height of the frustum; hence, its area is equal to $\tfrac{1}{2}(EA + ea) \times Ff$ (B. IV., P. VII.). But 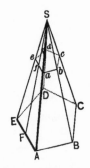 the area of the convex surface of the frustum is equal to the sum of the areas of its lateral faces; it is, therefore, equal to the half sum of the perimeters of its upper and lower bases, multiplied by the slant height.

PROPOSITION V. THEOREM.

If the three faces which include a triedral angle of a prism are equal in all respects to the three faces which include a triedral angle of a second prism, each to each, and are like placed, the two prisms are equal in all respects.

Let B and b be the vertices of two triedral angles, included by faces respectively equal to each other, and similarly placed: then the prism ABCDE–K is equal to the prism *abcde–k* in all respects.

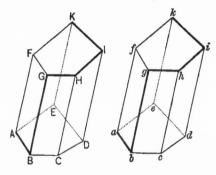

For, place the base *abcde* upon the equal base ABCDE, so that they shall coincide; then because the triedral angles whose vertices are *b* and B, are equal, the parallelogram *bh* will coincide with BH, and the parallelogram *bf* with BF: hence, the two sides *fg* and *gh*, of one upper base, will coincide with the homologous sides FG and GH, of the other upper base; and because the upper bases are equal in all respects, and have been shown to coincide in part, they must coincide throughout; consequently, each of the lateral faces of one prism will coincide with the corresponding lateral face of the other prism; the prisms, therefore, coincide throughout, and are therefore equal in all respects; *which was to be proved.*

Cor. If two right prisms have their bases equal in all respects, and have also equal altitudes, the prisms themselves are equal in all respects. For, the faces which include any triedral angle of the one, are equal in all respects to the faces which include the corresponding triedral angle of the other, each to each, and they are similarly placed.

PROPOSITION VI. THEOREM.

In any parallelopipedon, the opposite faces are equal in all respects, each to each, and their planes are parallel.

Let ABCD–H be a parallelopipedon: then its opposite faces are equal and their planes are parallel.

For, the bases, ABCD and EFGH are equal, and their planes parallel by definition (D. 7). The opposite faces AEHD and BFGC, have the sides AE and BF parallel, because they are opposite sides of the parallelogram BE; and the sides EH and FG parallel, because they 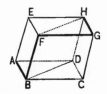 are opposite sides of the parallelogram EG; and consequently, the angles AEH and BFG are equal (B. VI., P. XIII.). But the side AE is equal to BF, and the side EH to FG; hence, the faces AEHD and BFGC are equal; and because AE is parallel to BF, and EH to FG, the planes of the faces are parallel (B. VI., P. XIII.). In like manner, it may be shown that the parallelograms ABFE and DCGH, are equal and their planes parallel: hence, the opposite faces are equal, each to each, and their planes are parallel; *which was to be proved.*

Cor. 1. Any two opposite faces of a parallelopipedon may be taken as bases.

Cor. 2. In a rectangular parallelopipedon, the square of any of the diagonals is equal to the sum of the squares of the three edges which meet at the same vertex.

For, let FD be one of the diagonals, and draw FH.

Then, in the right-angled triangle FHD, we have,

$$\overline{FD}^2 = \overline{DH}^2 + \overline{FH}^2.$$

But DH is equal to FB, and $\overline{FH}^2$ is equal to $\overline{FA}^2$ plus $\overline{AH}^2$ or $\overline{FC}^2$: hence,

$$\overline{FD}^2 = \overline{FB}^2 + \overline{FA}^2 + \overline{FC}^2.$$

Cor. 3. A parallelopipedon may be constructed on three straight lines AB, AD, and AE, intersecting in a common point A, and not lying in the same plane. For, pass through the extremity of each line, a plane parallel to the plane of the two others; then will these planes, together with the planes of the given lines, be the faces of a parallelopipedon.

PROPOSITION VII.   THEOREM.

If a plane is passed through the diagonally opposite edges of a parallelopipedon, it divides the parallelopipedon into two equal triangular prisms.

Let ABCD–H be a parallelopipedon, and let a plane be passed through the edges BF and DH; then are the prisms ABD–H and BCD–H equal in volume.

For, through the vertices F and B let planes be passed perpendicular to FB, the former cutting the other lateral edges in the points e, h, g, and the latter cutting those edges produced, in the points a, d, and c. The sections Fehg and Badc are parallelograms, because their opposite sides are parallel,

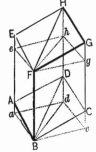

each to each (B. VI., P. X.); they are also equal (P. II.): hence, the polyedron B*adc-g* is a right prism (D. 2, 4), as are also the polyedrons B*ad–h* and B*cd–h*.

Place the triangle F*eh* upon B*ad*, so that F shall coincide with B, *e* with *a*, and *h* with *d*; then, because *e*E, *h*H, are perpendicular to the plane F*eh*, and *a*A, *d*D, to the plane B*ad*, the line *e*E takes the direction *a*A, and the line *h*H the direction *d*D. The lines AE and *ae* are equal, because each is equal to BF (B. I., P. XXVIII.). If we take away from the line *a*E the part *ae*, there remains the part *e*E; and if from the same line, we take away the part AE, there remains the part A*a*: hence, *e*E and *a*A are equal (A. 3); for a like reason *h*H is equal to *d*D: hence, the point E coincides with A, and the point H with D, and consequently, the polyedrons F*eh*-H and B*ad*-D coincide throughout, and are therefore equal.

If from the polyedron B*ad*-H, we take away the part B*ad*-D, there remains the prism BAD-H; and if from the same polyedron we take away the part F*eh*-H, there remains the prism B*ad–h*: hence, these prisms are equal in volume. In like manner, it may be shown that the prisms BCD-H and B*cd–h* are equal in volume.

The prisms B*ad–h*, and B*cd–h*, have equal bases, because these bases are halves of equal parallelograms (B. I., P. XXVIII., C. 1); they have also equal altitudes; they are therefore equal (P. V., C.): hence, the prisms BAD-H and BCD-H are equal (A. 1); *which was to be proved.*

Cor. Any triangular prism ABD-H, is equal to half of the parallelopipedon AG, which has the same triedral angle A, and the same edges AB, AD, and AE.

PROPOSITION VIII. THEOREM.

If two parallelopipedons have a common lower base, and their upper bases between the same parallels, they are equal in volume.

Let the parallelopipedons AG and AL have the common lower base ABCD, and their upper bases EFGH and IKLM, between the same parallels EK and HL: then are they equal in volume.

For, in the triangular prisms AEI–M and BFK–L, the faces AEI and BKF are equal, having their sides respectively equal; the faces AEHD and BFGC are equal (P. VI.);

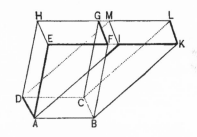

the faces EHMI and FGLK are equal, as they consist, respectively, of the common part FGMI and the equal parts EHGF and IMLK: hence, the triangular prisms AEI–M and BFK–L are equal (P. V.).

If from the polyedron ABKE–H, we take away the prism BFK–L, there remains the parallelopipedon AG; and if from the same polyedron we take away the prism AEI-M, there remains the parallelopipedon AL: hence, these parallelopipedons are equal in volume (A. 3); *which was to be proved.*

PROPOSITION IX.   THEOREM.

If two parallelopipedons have a common lower base and the same altitude, they are equal in volume.

Let the parallelopipedons AG and AL have the common lower base ABCD and the same altitude: then are they equal in volume.

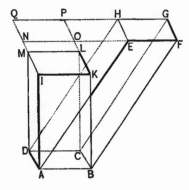

Because they have the same altitude, their upper bases lie in the same plane. Let the sides IM and KL be prolonged, and also the sides FE and GH; these prolongations form a parallelogram OQ, which is equal to the common base of the given parallelopipedons, because its sides are respectively parallel and equal to the corresponding sides of that base.

Now, if a third parallelopipedon be constructed, having for its lower base the parallelogram ABCD, and for its upper base NOPQ, this third parallelopipedon will be equal in volume to the parallelopipedon AG, since they will have the same lower base, and their upper bases between the same parallels, QG, NF (P. VIII.). For a like reason, this third parallelopipedon will also be equal in volume to the parallelopipedon AL: hence, the two parallelopipedons AG, AL, are equal in volume; *which was to be proved.*   .

Cor. Any oblique parallelopipedon is equal in volume to a right parallelopipedon having the same base and the same altitude.

PROPOSITION X. PROBLEM.

To construct a rectangular parallelopipedon equal in volume to a right parallelopipedon whose base is any parallelogram.

Let ABCD–M be a right parallelopipedon, having for its base the parallelogram ABCD.

Through the edges Al and BK pass the planes AQ and BP, respectively perpendicular to the plane AK, the former meeting the face DL in OQ, and the latter meeting that face produced in NP: then the polyedron AP is a rectangular parallelopipedon equal to the given parallelopipedon. It is a rectangular parallelopipedon, because all of its faces are rectangles, and 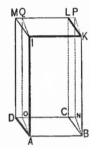 it is equal to the given parallelopipedon, because the two may be regarded as having the common base AK (P. VI., C. 1), and an equal altitude AO (P. IX.).

Cor. 1. Since any oblique parallelopipedon is equal in volume to a right parallelopipedon, having the same base and altitude (P. IX., Cor.); and since any right parallelopipedon is equal in volume to a rectangular parallelopipedon having an equal base and altitude; it follows, that any oblique parallelopipedon is equal in volume to a rectangular parallelopipedon, having an equal base and an equal altitude.

Cor. 2. Any two parallelopipedons are equal in volume when they have equal bases and equal altitudes.

PROPOSITION XI.   THEOREM.

Two rectangular parallelopipedons having a common lower base, are to each other as their altitudes.

Let the parallelopipedons AG and AL have the common lower base ABCD : then are they to each other as their altitudes AE and AI.

1°. Let the altitudes be commensurable, and suppose, for example, that AE is to AI, as 15 is to 8.

Conceive AE to be divided into 15 equal parts, of which AI contains 8 ; through the points of division let planes be passed parallel to ABCD.   These planes divide the parallelopipedon AG into 15 parallelopipedons, which have equal bases (P. II., C.) and equal altitudes ; hence, they are equal (P. X., Cor. 👌).

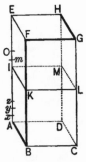

Now, AG contains 15, and AL 8 of these equal parallelopipedons ; hence, AG is to AL, as 15 is to 8, or as AE is to AI.   In like manner, it may be shown that AG is to AL, as AE is to AI, when the altitudes are to each other as any other whole numbers.

2°. Let the altitudes be incommen- surable.

Now, if AG is not to AL, as AE is to AI, let us suppose that

$$AG \ : \ AL \ :: \ AE \ : \ AO,$$

in which AO is greater than AI.

Divide AE into equal parts, such that each is less than OI ; there is at least one point of division m, between O and I. Let P denote the parallelopipedon, whose base is ABCD, and altitude A*m*; since the altitudes AE, A*m*, are to each other as two whole numbers, we have,

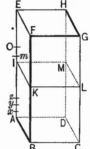

AG : P :: AE : A*m*.

But, by hypothesis, we have,

AG : AL :: AE : AO;

therefore (B. II., P. IV., C.),

AL : P :: AO : A*m*.

But AO is greater than A*m*; hence, if the proportion is true, AL must be greater than P. On the contrary, it is less; consequently, the fourth term of the proportion can not be greater than Al. In like manner, it may be shown that the fourth term can not be less than Al; it is, therefore, equal to Al. In this case, therefore, AG is to AL as AE is to Al.

Hence, in all cases, the given parallelopipedons are to each other as their altitudes; *which was to be proved.*

Sch. Any two rectangular parallelopipedons whose bases are equal in all respects, are to each other as their altitudes.

PROPOSITION XII. THEOREM.

Two rectangular parallelopipedons having equal altitudes, are to each other as their bases.

Let the rectangular parallelopipedons AG and AK have the same altitude AE: then are they to each other as their bases.

For, place them so that the plane angle EAO shall be common, and produce the plane of the face NL, until it intersects the plane of the face HC, in PQ; we thus form a third rectangular parallelopipedon AQ.

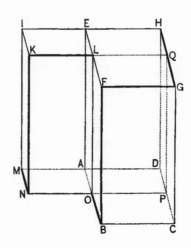

The parallelopipedons AG and AQ have a common base AH; they are therefore to each other as their altitudes AB and AO (P. XI.): hence, we have the proportion,

$$vol.\ AG\ :\ vol.\ AQ\ ::\ AB\ :\ AO.$$

The parallelopipedons AQ and AK have the common base AL; they are therefore to each other as their altitudes AD and AM: hence,

$$vol.\ AQ\ :\ vol.\ AK\ ::\ AD\ :\ AM.$$

Multiplying these proportions, term by term (B. II., P. XII.), and omitting the common factor, vol. AQ, we have,

$$vol.\ AG\ :\ vol.\ AK\ ::\ AB \times AD\ :\ AO \times AM.$$

But AB × AD is equal to the area of the base ABCD, and AO × AM is equal to the area of the base AMNO: hence, two rectangular parallelopipedons having equal altitudes, are to each other as their bases; *which was to be proved.*

PROPOSITION XIII. THEOREM.

Any two rectangular parallelopipedons are to each other as the products of their bases and altitudes; that is, as the products of their three dimensions.

Let AZ and AG be any two rectangular parallelopipedons: then are they to each other as the products of their three dimensions.

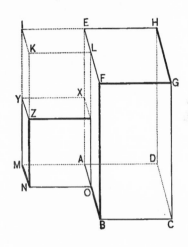

For, place them so that the plane angle EAO shall be common, and produce the faces necessary to complete the rectangular parallelopipedon AK. The parallelopipedons AZ and AK have a common base AN; hence (P. XI.),

$$vol.\ AZ\ :\ vol.\ AK\ ::\ AX\ :\ AE.$$

The parallelopipedons AK and AG have a common altitude AE; hence (P. XII.),

$$vol.\ AK\ :\ vol.\ AG\ ::\ AMNO\ :\ ABCD.$$

Multiplying these proportions, term by term, and omitting the common factor, *vol.* AK, we have,

$$vol.\ AZ\ :\ vol.\ AG\ ::\ AMNO \times AX\ :\ ABCD \times AE;$$

or, since AMNO is equal to AM × AO, and ABCD to AB × AD,

$$vol.\ AZ\ :\ vol.\ AG\ ::\ AM \times AO \times AX\ :\ AB \times AD \times AE;$$

which was to be proved.

Cor. 1. If we make the three edges AM, AO, and AX, each equal to the linear unit, the parallelopipedon AZ becomes a cube constructed on that unit, as an edge; and consequently, it is the unit of volume. Under this supposition, the last proportion becomes,

$$1 \; : \; vol. \; AG \; :: \; 1 \; : \; AB \times AD \times AE;$$

whence, $\qquad\qquad vol. \; AG = AB \times AD \times AE.$

Hence, *the volume of any rectangular parallelopipedon is equal to the product of its three dimensions;* that is, the number of times which it contains the unit of volume, is equal to the continued product of the number of linear units in its length, the number of linear units in its breadth, and the number of linear units in its height.

Cor. 2. *The volume of a rectangular parallelopipedon is equal to the product of its base and altitude;* that is, the number of times which it contains the unit of volume, is equal to the number of superficial units in its base, multiplied by the number of linear units in its altitude.

Cor. 3. The volume of any parallelopipedon is equal to the product of its base and altitude (P. X., C. 1).

PROPOSITION XIV. THEOREM.

The volume of any prism is equal to the product of its base and altitude.

Let ABCDE–K be any prism: then is its volume equal to the product of its base and altitude.

For, through any lateral edge, as AF, and the other lateral edges not in the same faces, pass the planes AH, AI, dividing the prism into triangular prisms. These prisms all have a common altitude equal to that of the given prism.

Now, the volume of any one of the triangular prisms, as ABC–H, is equal to half that of a parallelopipedon constructed on the edges BA, BC, BG (P. VII., C.) ; but the volume of this parallelopipedon is equal to the product of its base and altitude (P. XIII., C. 3) ; and because the base of the prism is half that of the parallelopipedon, the volume of the prism is also equal to the product of its base and altitude : hence, the sum of the triangular prisms, which make up the given prism, is

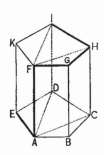

equal to the sum of their bases, which make up the base of the given prism, into their common altitude ; *which was to be proved.*

Cor. Any two prisms are to each other as the products of their bases and altitudes. Prisms having equal bases are to each other as their altitudes. Prisms having equal altitudes are to each other as their bases.

PROPOSITION XV. THEOREM.

Two triangular pyramids having equal bases and equal altitudes are equal in volume.

Let S–ABC, and S–abc, be two pyramids having their equal bases ABC and abc in the same plane, and let AT be their common altitude : then are they equal in volume.

For, if they are not equal in volume, suppose one of them, as S–ABC, to be the greater, and let their difference be equal to a prism whose base is ABC, and whose altitude is A*a.*

Divide the altitude AT into equal parts, Ax, xy, &c., each of which is less than Aa, and let k denote one of these parts; through the points of division pass planes parallel to the plane of the bases; the sections of the two pyramids, by each of these planes, are equal, namely, DEF to *def*, GHI to *ghi*, &c. (P. III., C. 2).

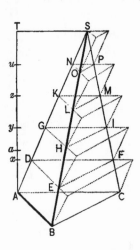
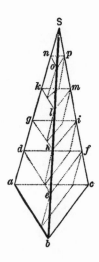

On the triangles ABC, DEF, &c., as lower bases, construct exterior prisms whose lateral edges are parallel to AS, and whose altitudes are equal to k: and on the triangles *def*, *ghi*, &c., taken as upper bases, construct interior prisms, whose lateral edges are parallel to aS, and whose altitudes are equal to k. It is evident that the sum of the exterior prisms is greater than the pyramid S–ABC, and also that the sum of the interior prisms· is less than the pyramid S–*abc*: hence, the difference between the sum of the exterior and the sum of the interior prisms, is greater than the difference between the two pyramids.

Now, beginning at the bases, the second exterior prism EFD–G, is equal to the first interior prism *efd–a*, because they have the same altitude k, and their bases EFD, *efd*, are equal: for a like reason, the third exterior prism HIG–K, and the second interior prism *hig–d*, are equal, and so on to the last in each set: hence, each of the exterior prisms, excepting the first BCA–D, has an equal corresponding ing interior prism; the prism BCA–D, is, therefore, the difference between the sum of all the exterior prisms, and the sum of all the interior prisms. But the difference between these two sets of prisms is greater than that between the two pyramids, which latter difference was supposed to be equal to a prism whose base is BCA, and whose altitude is equal to Aa, greater than k; consequently, the prism BCA–D is greater than a prism having the same base and a greater altitude, which is impossible: hence, the supposed inequality between the two pyramids can not exist; they are, therefore, equal in volume; *which was to be proved.*

PROPOSITION XVI. THEOREM.

Any triangular prism may be divided into three triangular pyramids, equal to each other in volume.

Let ABC–D be a triangular prism: then can it be divided into three equal triangular pyramids.

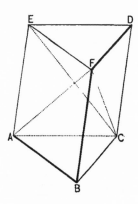

For, through the edge AC, pass the plane ACF, and through the edge EF pass the plane EFC. The pyramids ACE–F and ECD–F, have their bases ACE and ECD equal, because they are halves of the same parallelogram ACDE; and they have a common altitude, because their bases are in the same plane AD, and their vertices at the same point F; hence, they are equal in volume (P. XV.). The pyramids ABC–F and DEF–C, have their bases ABC and DEF, equal, because they are the bases of the given prism, and their altitudes are equal because each is equal to the altitude of the prism; they are, therefore, equal in volume: hence, the three pyramids into which the prism is divided, are all equal in volume; *which was to be proved.*

Cor. 1. A triangular pyramid is one third of a prism having an equal base and an equal altitude.

Cor. 2. The volume of a triangular pyramid is equal to one third of the product of its base and altitude.

PROPOSITION XVII. THEOREM.

The volume of any pyramid is equal to one third of the product of its base and altitude.

Let S–ABCDE, be any pyramid: then is its volume equal to one third of the product of its base and altitude.

For, through any lateral edge, as SE, pass the planes SEB, SEC, dividing the pyramid into triangular pyramids. The altitudes of these pyramids are equal to each other, because each is equal to that of the given pyramid. Now, the volume of each triangular pyramid is equal to one third of the product of its base and alti- 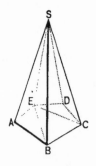 tude (P. XVI., C. 2); hence, the sum of the volumes of the triangular pyramids, is equal to one third of the product of the sum of their bases by their common altitude. But the sum of the triangular pyramids is equal to the given pyramid, and the sum of their bases is equal to the base of the given pyramid: hence, the volume of the given pyramid is equal to' one third of the product of its base and altitude; *which was to be proved.*

Cor. 1. The volume of a pyramid is equal to one third of the volume of a prism having an equal base and an equal altitude.

Cor. 2. Any two pyramids are to each other as the products of their bases and altitudes. Pyramids having equal bases are to each other as their altitudes. Pyramids having equal altitudes are to each other as their bases.

Scholium. The volume of a polyedron may be found by dividing it into triangular pyramids, and computing their volumes separately. The sum of these volumes is equal to the volume of the polyedron.

PROPOSITION XVIII. THEOREM.

The volume of a frustum of any triangular pyramid is equal to the sum of the volumes of three pyramids whose common altitude is that of the frustum, and whose bases are the lower base of the frustum, the upper base of the frustum, and a mean proportional between the two bases.

Let FGH–*h* be a frustum of any triangular pyramid: then is its volume equal to that of three pyramids whose common altitude is that of the frustum, and whose bases are the lower base FGH, the upper base *fgh*, and a mean proportional between these bases.

For, through the edge FH, pass the plane FH*g*, and

through the edge *fg*, pass the plane *fg*H, dividing the frustum into three pyramids. The pyramid *g*–FGH, has for its base the lower base FGH of the frustum, and its altitude is equal to that of the frustum, because its vertex *g* is in the plane of the upper base. The pyramid H–*fgh*, has for its base the upper base *fgh* of the frustum, and its altitude is equal to that of the frustum, because its vertex lies in the plane of the lower base.

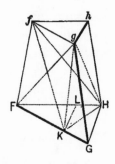

The remaining pyramid may be regarded as having the triangle F*f*H for its base, and the point *g* for its vertex. From *g*, draw *g*K parallel to *f*F, and draw also KH and K*f*. Then the pyramids K–F*f*H and *g*–F*f*H, are equal; for they have a common base, and their altitudes are equal, because their vertices K and *g* are in a line parallel to the base (B. VI., P. XII., C. 2).

Now, the pyramid K–F*f*H may be regarded as having FKH for its base and *f* for its vertex. From K, draw KL parallel to GH; it is parallel to *gh*: then the triangle FKL is equal to *fgh*, for the side FK is equal to *fg*, the angle F to the angle *f*, and the angle K to the angle *g*. But, FKH is a mean proportional between FKL and FGH (B. IV., P. XXIV., C.), or between *fgh* and FGH. The pyramid *f*–FKH, has, therefore, for its base a mean proportional between the upper and lower bases of the frustum, and its altitude is equal to that of the frustum; but the pyramid *f*–FKH is equal in volume to the pyramid *g*–F*f*H: hence, the volume of the given frustum is equal to that of three pyramids whose common altitude is equal to that of the frustum, and whose bases are the upper base, the lower base, and a mean proportional between them; *which was to be proved.*

Cor. The volume of the frustum of any pyramid is equal to the sum of the volumes of three pyramids whose common altitude is that of the frustum, and whose bases are the lower base of the frustum, the upper base of the frustum, and a mean proportional between them.

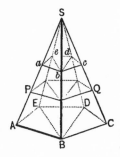

For, let ABCDE–e be a frustum of a pyramid whose vertex is S, and let PQ be a section parallel to the bases, such that distance from S is a mean proportional between the distances from S to the two bases of the frustum. Let planes be passed through SB, and SE, SD, dividing the frustum into triangular frustums; the section of each of the triangular frustums is a mean proportional between its bases (P. III., C. 4). Now the sum of the triangular frustums is equal to the sum of three sets of pyramids, whose altitude is that of the given frustum. The sum of the bases of the first set is the lower base of the frustum, the sum of the bases of the second set is the upper base of the frustum, and the sum of the bases of the third set is a mean proportional between these bases. Hence, the sum of the partial frustums, that is, the given frustum, is equal to the sum of three pyramids having the same altitude as the given frustum, and whose bases are the two bases of the frustum and a mean proportional between them.

PROPOSITION XIX. THEOREM.

Similar triangular prisms are to each other as the cubes of their homologous edges.

Let CBD–P, *cbd–p*, be two similar triangular prisms, and let BC, *bc*, be any two homologous edges: then is the prism CBD–P to the prism *cbd–p*, as $\overline{BC}^3$ to $\overline{bc}^3$.

For, the homologous angles B and b are equal, and the faces which bound them are similar (D. 16): hence, these triedral angles may be applied, one to the other, so that the angle *cbd* will coincide with CBD, the edge *ba* with BA. In this case, the prism *cbd–p* will take the position B*cd–p*. From A draw AH perpendicular to the common base of the prisms: 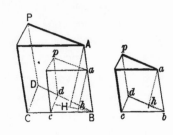 then the plane BAH is perpendicular to the plane of the common base (B. VI., P. XVI.). From a, in the plane BAH, draw *ah* perpendicular to BH: then *ah* is also perpendicular to the base BDC (B. VI., P. XVII.); and AH, *ah*, are the altitudes of the two prisms.

Since the bases CBD, *cbd*, are similar, we have (B. IV., P. XXV.),

$$base \text{ CBD} \quad : \quad base \text{ } cbd \quad :: \quad \overline{CB}^2 \quad : \quad \overline{cb}^2.$$

Now, because of the similar triangles ABH, *aBh*, and of the similar parallelograms AC, *ac*, we have,

$$AH \quad : \quad ah \quad :: \quad CB \quad : \quad cb\,;$$

hence, multiplying these proportions term by term, we have,

$$base \text{ CBD} \times AH \quad : \quad base \text{ } cbd \times ah \quad : \quad \overline{CB}^3 \quad : \quad \overline{cb}^3.$$

But, *base* CBD × AH is equal to the volume of the prism CDB–A, and *base cbd* × *ah* is equal to the volume of the prism *cbd–p*: hence,

$$prism \text{ CDB–P} \quad : \quad prism \text{ } cbd–p \quad :: \quad \overline{CB}^3 \quad : \quad \overline{cb}^3\,;$$

which was to be proved.

Cor. 1. *Any two similar prisms are to each other as the cubes of their homologous edges.*

For, since the prisms are similar, their bases are similar polygons (D. 16); and these similar polygons may each be divided into the same number of similar triangles, similarly placed (B. IV., P. XXVI.); therefore, each prism may be divided into the same number of triangular prisms, having their faces similar and like placed; consequently, the triangular prisms are similar (D. 16). But these triangular prisms are to each other as the cubes of their homologous edges, and being like parts of the polygonal prisms, the polygonal prisms themselves are to each other as the cubes of their homologous edges.

Cor. 2. Similar prisms are to each other as the cubes of their altitudes, or as the cubes of any other homologous lines.

PROPOSITION XX. THEOREM.

Similar pyramids are to each other as the cubes of their homologous edges.

Let S–ABCDE, and S–*abcde*, be two similar pyramids, so placed that their homologous angles at the vertex shall coincide, and let AB and *ab* be any two homologous edges: then are the pyramids to each other as the cubes of AB and *ab*.

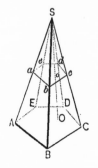

For, the face SAB, being similar to S*ab*, the edge AB is parallel to the edge *ab*, and the face SBC being similar to S*bc*, the edge BC is parallel to *bc*; hence, the planes of the bases are parallel (B. VI., P. XIII.).

Draw SO perpendicular to the base ABCDE; it will also be perpendicular to the base *abcde*. Let it pierce that plane at the point *o*; then SO is to So, as SA is to S*a* (P. III.), or as AB is to *ab*; hence,

$$\tfrac{1}{3}SO \;\; : \;\; \tfrac{1}{3}So \;\; :: \;\; AB \;\; : \;\; ab.$$

But the bases being similar polygons, we have (B. IV., P. XXVII.),

$$base\; ABCDE \;\; : \;\; base\; abcde \;\; :: \;\; \overline{AB}^2 \;\; : \;\; \overline{ab}^2.$$

Multiplying these proportions, term by term, we have,

$$base\; ABCDE \times \tfrac{1}{3}SO \;\; : \;\; base\; abcde \times \tfrac{1}{3}So \;\; :: \;\; \overline{AB}^3 \;\; : \;\; \overline{ab}^3.$$

But, *base* ABCDE $\times \tfrac{1}{3}$SO is equal to the volume of the pyramid S–ABCDE, and *base abcde* $\times \tfrac{1}{3}$So is equal to the volume of the pyramid S–*abcde*; hence,

$$pyramid\; S\text{–}ABCDE \;\; : \;\; pyramid\; S\text{-}abcde \;\; :: \;\; \overline{AB}^3 \;\; : \;\; \overline{ab}^3;$$

which was to be proved.

Cor. Similar pyramids are to each other as the cubes of their altitudes, or as the cubes of any other homologous lines.

GENERAL FORMULAS.

If we denote the volume of any prism by V, its base by B, and its altitude by H, we shall have (P. XIV.),

$$V = B \times H \quad \cdots \quad \cdots \quad (1.)$$

If we denote the volume of any pyramid by V, its base by B, and its altitude by H, we have (P. XVII.),

$$V = B \times \tfrac{1}{3}H \quad \cdots \quad \cdots \quad (2.)$$

If we denote the volume of the frustum of any pyramid by V, its lower base by B, its upper base by b, and its altitude by H, we shall have (P. XVIII., C.),

$$V = (B + b + \sqrt{B \times b}) \times \tfrac{1}{3}H \quad \cdots \quad (3.)$$

REGULAR POLYEDRONS.

A REGULAR POLYEDRON is one whose faces are all equal regular polygons, and whose polyedral angles are equal, each to each.

There are five regular polyedrons, namely:

1. The TETRAEDRON, or *regular pyramid*—a polyedron bounded by four equal equilateral triangles.

2. The HEXAEDRON, or *cube*—a polyedron bounded by six equal squares.

3. The OCTAEDRON—a polyedron bounded by eight equal equilateral triangles.

4. The DODECAEDRON—a polyedron bounded by twelve equal and regular pentagons.

5. The Icosaedron—a polyedron bounded by twenty equal equilateral triangles.

In the Tetraedron, the triangles are grouped about the polyedral angles in sets of three, in the Octaedron they are grouped in sets of four, and in the Icosaedron they are grouped in sets of five. Now, a greater number of equilateral triangles can not be grouped so as to form a salient polyedral angle; for, if they could, the sum of the plane angles formed by the edges would be equal to, or greater than, four right angles, which is impossible (B. VI., P. XX.).

In the Hexaedron, the squares are grouped about the polyedral angles in sets of three. Now, a greater number of squares can not be grouped so as to form a salient polyedral angle; for the same reason as before.

In the Dodecaedron, the regular pentagons are grouped about the polyedral angles in sets of three, and for the same reason as before, they can not be grouped in any greater number so as to form a salient polyedral angle.

Furthermore, no other regular polygons can be grouped so as to form a salient polyedral angle; therefore,

Only five regular polyedrons can be formed.

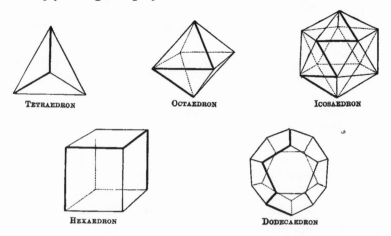

TETRAEDRON   OCTAEDRON   ICOSAEDRON

HEXAEDRON   DODECAEDRON

EXERCISES.

1. What is the convex surface of a right prism whose altitude is 20 feet and whose base is a pentagon each side of which is 15 feet?

2. The altitude of a pyramid is 10 feet and the area of its base 25 square feet; find the area of a section made by a plane 6 feet from the vertex and parallel to the base.

3. Find the convex surface of a right triangular pyramid, each side of the base being 4 feet and the slant height 12 feet.

4. A right pyramid whose altitude is 8 feet and whose base is a square each side of which is 4 feet, is cut by a plane parallel to the base and 2 feet from the vertex; required the convex surface of the frustum included between the base and the cutting plane.

5. The three concurrent edges of a rectangular parallelopipedon are 4, 6, and 8 feet; find the length of the diagonal.

6. Of two rectangular parallelopipedons having equal bases, the altitude of the first is 12 feet and its volume is 275 cubic feet; the altitude of the second is 8 feet—find its volume.

7. Two rectangular parallelopipedons having equal altitudes are respectively 80 and 45 cubic feet in volume, and the area of the base of the first is 12 square feet; find the base of the second and the altitude of both.

8. Find the volume of a triangular prism whose base is an equilateral triangle of which the altitude is 3 feet, the altitude of the prism being 8 feet.

9. The volumes of two pyramids having equal altitudes are respectively 60 and 115 cubic yards and the base of the smaller is 8 square yards; find the base of the larger.

10. Given a pyramid whose volume is 512 cubic feet and altitude 8 feet; find the volume of a similar pyramid whose altitude is 12 feet, and find also the area of the base of each.

11. Find the volume of the frustum of a right triangular pyramid with each side of the lower base 6 feet and each side of the upper base 4 feet, the altitude being 5 feet.

12. Find the volume of the pyramid of which the frustum given in the last example is a frustum.

[Find the radii of the inscribed circles of the upper and lower bases (B. IV., P. VI., C. 2); then the altitude of the pyramid, slant height, and the two radii form two similar triangles from which the altitude may be found.]

13. Given two similar prisms; the base of the first contains 30 square yards and its altitude is 8 yards; the altitude of the second prism is 6 yards—find its volume and the area of its base.

14. A pyramid, whose base is a regular pentagon of which the apothem is 3.5 feet, contains 129 cubic feet; find the volume of a similar pyramid, the apothem of whose base is 4 feet.

15. Show that the four diagonals of a parallelopipedon bisect each other in a common point.

16. Show that the two lines joining the points of the opposite faces of a parallelopipedon, in which the diagonals of those faces intersect, bisect each other at the point in which the diagonals of the parallelopipedon intersect.

17. Show that two regular polyedrons of the same kind are similar.

18. Show that the surfaces of any two similar polyedrons are to each other as the squares of any two homologous edges

BOOK VIII.

THE CYLINDER, THE CONE, AND THE SPHERE.

DEFINITIONS.

1. A CYLINDER is a volume which may be generated
by a rectangle revolving about one of its sides as an *axis*.

Thus, if the rectangle ABCD be turned about the side
AB, as an axis, it will generate the cylinder FGCQ–P.

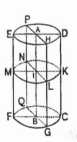

The fixed line AB is called *the axis of
the cylinder;* the curved surface generated
by the side CD, opposite the axis, is called
the convex surface of the cylinder; the equal
circles FGCQ, and EHDP, generated by the
remaining sides BC and AD, are called *bases
of the cylinder;* and the perpendicular distance between the planes of the bases is
called *the altitude of the cylinder.*

The line DC, which generates the convex surface, is, in
any position, called an *element of the surface;* the elements are all perpendicular to the planes of the bases,
and any one of them is equal to the altitude of the cylinder.

Any line of the generating rectangle ABCD, as IK,
which is perpendicular to the axis, will generate a circle
whose plane is perpendicular to the axis, and which is
equal to either base: hence, any section of a cylinder by
a plane perpendicular to the axis, is a circle equal to
either base. Any section, FCDE, made by a plane through
the axis, is a rectangle double the generating rectangle.

2. SIMILAR CYLINDERS are those which may be generated by similar rectangles revolving about homologous sides.

The axes of similar cylinders are proportional to the
radii of their bases (B. IV., D. 1) ; they are also proportional to any other homologous lines of the cylinders.

3. A prism is said to be *inscribed in a
cylinder*, when its bases are inscribed in
the bases of the cylinder. In this case,
the cylinder is said to be circumscribed
about the prism.

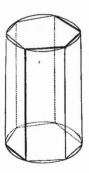

The lateral edges of the inscribed prism
are elements of the surface of the circumscribing cylinder.

4. A prism is said to be *circumscribed
about a cylinder*, when its bases are circumscribed about
the bases of the cylinder. In this case, the cylinder is said
to be *inscribed in the prism.*

The straight lines which join the corresponding points of contact in the upper
and lower bases, are common to the surface of the cylinder and to the lateral
faces of the prism, and they are the only
lines which are common. The lateral faces
of the prism are *tangent* to the cylinder
along these lines, which are then called
elements of contact.

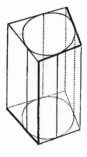

5. A CONE is a volume which may be generated by a
right-angled triangle revolving about one of the sides adjacent to the right angle, as an axis.

Thus, if the triangle SAB, right-angled at A, be turned about the side SA, as an axis, it will generate the cone S–CD,BE.

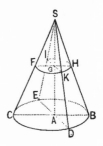

The fixed line SA, is called *the axis of the cone;* the curved surface generated by the hypothenuse SB, is called *the convex surface of the cone;* the circle generated by the side AB, is called *the base of the cone;* and the point S, is called *the vertex of the cone;* the distance from the vertex to any point in the circumference of the base, is called *the slant height of the cone;* and the perpendicular distance from the vertex to the plane of the base, is called *the altitude of the cone.*

The line SB, which generates the convex surface, is, in any position, called an *element of the surface;* the elements are all equal, and any one is equal to the slant height; the axis is equal to the altitude.

Any line of the generating triangle SAB, as GH, which is perpendicular to the axis, generates a circle whose plane is perpendicular to the axis: hence, any section of a cone by a plane perpendicular to the axis, is a circle. Any section SBC, made by a plane through the axis, is an isosceles triangle, double the generating triangle.

6. A TRUNCATED CONE is that portion of a cone included between the base and any plane which cuts the one.

When the cutting plane is parallel to the plane of the base, the truncated cone is called a FRUSTUM OF A CONE, and the intersection of the cutting plane with the cone is called the *upper base* of the frustum; the base of the cone is called the *lower base* of the frustum.

If the trapezoid HGAB, right-angled at A and G, be revolved about AG, as an axis, it will generate a frustum of a cone, whose bases are ECDB and FKH, whose altitude is AG, and whose slant height is BH.

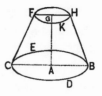

7. SIMILAR CONES are those which may be generated by similar right-angled triangles revolving about homologous sides.

The axes of similar cones are proportional to the radii of their bases (B. IV., D. 1); they are also proportional to any other homologous lines of the cones.

8. A pyramid is said to be *inscribed in a cone*, when its base is inscribed in the base of the cone, and when its vertex coincides with that of the cone.

The lateral edges of the inscribed pyramid are elements of the surface of the circumscribing cone.

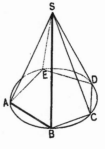

9. A pyramid is said to be *circumscribed about a cone*, when its base is circumscribed about the base of the cone, and when its vertex coincides with that of the cone.

In this case, the cone is *inscribed in the pyramid.*

The lateral faces of the circumscribing pyramid are tangent to the surface of the inscribed cone, along lines which are called *elements of contact.*

10. A frustum of a pyramid is *inscribed in a frustum of a cone,* when its bases are inscribed in the bases of the frustum of the cone.

The lateral edges of the inscribed frustum of a pyramid are elements of the surface of the circumscribing frustum of a cone.

11. A frustum of a pyramid is circumscribed about a frustum of a cone, when its bases are circumscribed about those of the frustum of the cone.

Its lateral faces are tangent to the surface of the frustum of the cone, along lines which are called *elements of contact.*

12. A SPHERE is a volume bounded by a surface, every point of which is equally distant from a point within called the *centre*. A sphere may be generated by a semicircle revolving about its diameter as an axis.

13. A RADIUS of a sphere is a straight line drawn from the centre to any point of the surface. A DIAMETER is a straight line through the centre, limited by the surface.

All the radii of a sphere are equal: the diameters are also equal, and each is double the radius.

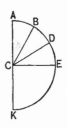

14. A SPHERICAL SECTOR is a volume generated by a sector of the semicircle that generates the sphere. The surface generated by the arc of the circular sector is *the base of the sector.* The other bounding surfaces are either surfaces of cones or planes. The spherical sector generated by ACB is bounded by the surface generated by the arc AB and the conic surface generated by BC; the sector generated by BCD is bounded by the surface generated by BD and the conic surfaces generated by BC and DC, and so on.

15. A plane is TANGENT TO A SPHERE when it touches it in a single point.

16. A ZONE is a portion of the surface of a sphere included between two parallel planes. The bounding lines

of the sections are called *bases* of the zone, and the distance between the planes is called the *altitude* of the zone.

If one of the planes is tangent to the sphere, the zone has but one base.

17. A SPHERICAL SEGMENT is a portion of a sphere included between two parallel planes. The sections made by the planes are called *bases* of the segment, and the distance between them is called the *altitude of the segment*.

If one of the planes is tangent to the sphere, the segment has but one base.

The CYLINDER, the CONE, and the SPHERE, are sometimes called THE THREE ROUND BODIES.

PROPOSITION I.   THEOREM.

The convex surface of a cylinder is equal to the circumference of its base multiplied by its altitude.

Let ABD be the base of a cylinder whose altitude is H: then is its convex surface equal to the circumference of its base multiplied by the altitude.

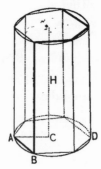

For, inscribe in the cylinder a prism whose base is a regular polygon. The convex surface of this prism is equal to the perimeter of its base multiplied by its altitude (B. VII., P. I.), whatever may be the number of sides of its base. But, when the number of sides is infinite (B. V., P. X., Sch.), the convex surface of the prism coincides with that of the cylinder, the perimeter

of the base of the prism coincides with the circumference
of the base of the cylinder, and the altitude of the prism
is the same as that of the cylinder: hence, the convex
surface of the cylinder is equal to the circumference of
its base multiplied by its altitude; *which was to be proved.*

Cor. The convex surfaces of cylinders having equal
altitudes are to each other as the circumferences of their
bases.

PROPOSITION II. THEOREM.

*The volume of a cylinder is equal to the product of its
base and altitude.*

Let ABD be the base of a cylinder whose altitude is
H; then is its volume equal to the product of its base
and altitude.

For, inscribe in it a prism whose base
is a regular polygon. The volume of
this prism is equal to the product of
its base and altitude (B. VII., P. XIV.),
whatever may be the number of sides
of its base. But, when the number of
sides is infinite, the prism coincides with
the cylinder, the base of the prism with
the base of the cylinder, and the altitude of the prism is the same as that

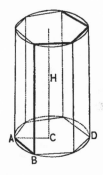

of the cylinder: hence, the volume of the cylinder is
equal to the product of its base and altitude; *which was
to be proved.*

Cor. 1. Cylinders are to each other as the products of
their bases and altitudes; cylinders having equal bases are
to each other as their altitudes; cylinders having equal
altitudes are to each other as their bases.

Cor. 2. Similar cylinders are to each other as the cubes of their altitudes, or as the cubes of the radii of their bases.

For, the bases are as the squares of their radii (B. V., P. XIII.), and the cylinders being similar, these radii are to each other as their altitudes (D. 2): hence, the bases are as the squares of the altitudes; therefore, the bases multiplied by the altitudes, or the cylinders themselves, are as the cubes of the altitudes.

PROPOSITION III.   THEOREM.

The convex surface of a cone is equal to the circumference of its base multiplied by half its slant height.

Let S–ACD be a cone whose base is ACD, and whose slant height is SA: then is its convex surface equal to the circumference of its base multiplied by half its slant height.

For, inscribe in it a right pyramid. The convex surface of this pyramid is equal to the perimeter of its base multiplied by half its slant height (B. VII., P. IV.), whatever may be the number of sides of its base. But when the number of sides of the base is infinite, the convex surface coincides with that of the cone, the perimeter of the

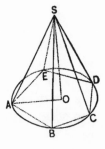

base of the pyramid coincides with the circumference of the base of the cone, and the slant height of the pyramid is equal to the slant height of the cone: hence, the convex surface of the cone is equal to the circumference of its base multiplied by half its slant height; *which was to be proved.*

PROPOSITION IV. THEOREM.

The convex surface of a frustum of a cone is equal to half the sum of the circumferences of its two bases multiplied by its slant height.

Let BIA–D be a frustum of a cone, BIA and EGD its two bases, and EB its slant height: then is its convex surface equal to half the sum of the circumferences of its two bases multiplied by its slant height.

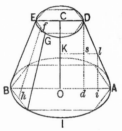

For, inscribe in it the frustum of a right pyramid. The convex surface of this frustum is equal to half the sum of the perimeters of its bases, multiplied by the slant height (B. VII., P. IV., C.), whatever may be the number of its lateral faces. But when the number of these faces is infinite, the convex surface of the frustum of the pyramid coincides with that of the cone, the perimeters of its bases coincide with the circumferences of the bases of the frustum of the cone, and its slant height is equal to that of the cone: hence, the convex surface of the frustum of a cone is equal to half the sum of the circumferences of its bases multiplied by its slant height; *which was to be proved.*

Scholium. From the extremities A and D, and from the middle point *l*, of a line AD, let the lines AO, DC, and *l*K be drawn perpendicular to the axis OC: then will *l*K be equal to half the sum of AO and DC. For, draw D*d* and *li*, perpendicular to AO: then, because A*l* is equal to *l*D, we shall have A*i* equal to *id* (B. IV., P. XV.), and consequently to *ls*; that is, AO exceeds *l*K as much as *l*K exceeds DC : hence, *l*K is equal to the half sum of AO and DC.

Now, if the line AD be revolved about OC, as an axis, it will generate the surface of a frustum of a cone whose slant height is AD; the point *l* will generate a circumference which is equal to half the sum of the circumferences generated by A and D: hence, *if a straight line is revolved about another straight line, it generates a surface whose measure is equal to the product of the generating line and the circumference generated by its middle point.*

This proposition holds true when the line AD meets OC, and also when AD is parallel to OC.

PROPOSITION V.   THEOREM.

The volume of a cone is equal to its base multiplied by one third of its altitude.

Let ABDE be the base of a cone whose vertex is S, and whose altitude is S*o* ; then is its volume equal to the base multiplied by one third of the altitude.

For, inscribe in the cone a right pyra-mid.   The volume of this pyramid is equal to its base multiplied by one third of its altitude (B. VII., P. XVII.), what-ever may be the number of its lateral faces.   But, when the number of lateral faces is infinite, the pyramid coincides with the cone, the base of the pyramid coincides with that of the cone, and their 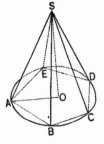 altitudes are equal : hence, the volume of a cone is equal to its base multiplied by one third of its altitude ; *which was to be proved.*

Cor. 1. A cone is equal to one third of a cylinder having an equal base and an equal altitude.

Cor. 2. Cones are to each other as the products of their bases and altitudes. Cones having equal bases are to each other as their altitudes. Cones having equal altitudes are to each other as their bases.

PROPOSITION VI. THEOREM.

The volume of a frustum of a cone is equal to the sum of the volumes of three cones, having for a common altitude the altitude of the frustum, and for bases the lower base of the frustum, the upper base of the frustum, and a mean proportional between the bases.

Let BIA be the lower base of a frustum of a cone, EGD its upper base, and OC its altitude: then is its volume equal to the sum of three cones whose common altitude is OC, and whose bases are the lower base, the upper base, and a mean proportional between them.

For, inscribe a frustum of a right pyramid in the given frustum. The volume of this frustum is equal to the sum of the volumes of three pyramids whose common altitude is that of the frustum, and whose bases are the lower base, the upper base, and a mean proportional between the two (B. VII., P. XVIII.), whatever may

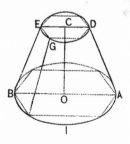

be the number of lateral faces. But when the number of faces is infinite, the frustum of the pyramid coincides with the frustum of the cone, its bases with the bases of the cone, the three pyramids become cones, and their altitudes are equal to that of the frustum: hence, the volume of the frustum of a cone is equal to the sum of the volumes of three cones whose common altitude is that of the frustum, and whose bases are the lower base of the frustum, the upper base of the frustum, and a mean proportional between them; *which was to be proved.*

PROPOSITION VII. THEOREM.

Any section of a sphere made by a plane is a circle.

Let C be the centre of a sphere, CA one of its radii, and AMB any section made by a plane: then is this section a circle.

For, draw a radius CO perpendicular to the cutting plane, and let it pierce the plane of the section at O. Draw radii of the sphere to any two points M, M', of the curve which bounds the section, and join these points with O: then, because the radii CM, CM' are equal, the points M, M', will be equally 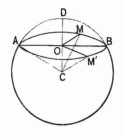 distant from O (B. VI., P. V., C.); hence, the section is a circle; *which was to be proved.*

Cor. 1. When the cutting plane passes through the centre of the sphere, the radius of the section is equal to that of the sphere; when the cutting plane does not pass through the centre of the sphere, the radius of the section will be less than that of the sphere.

A section whose plane passes through the centre of the sphere, is called a *great circle* of the sphere. A section whose plane does not pass through the centre of the sphere, is called a *small circle* of the sphere. All great circles of the same, or of equal spheres, are equal.

Cor. 2. Any great circle divides the sphere, and also the surface of the sphere, into equal parts. For, the parts may be so placed as to coincide, otherwise there would be some points of the surface unequally distant from the centre, which is impossible.

Cor. 3. The centre of a sphere, and the centre of any small circle of that sphere, are in a straight line perpendicular to the plane of the circle.

Cor. 4. The square of the radius of any small circle is equal to the square of the radius of the sphere diminished by the square of the distance from the centre of the sphere to the plane of the circle (B. IV., P. XI., C. 1): hence, circles which are equally distant from the centre, are equal; and of two circles which are unequally distant from the centre, that one is the less whose plane is at the greater distance from the centre.

Cor. 5. The circumference of a great circle may always be made to pass through any two points on the surface of a sphere. For, a plane can always be passed through these points and the centre of the sphere (B. VI., P. II.), and its section will be a great circle. If the two points are the extremities of a diameter, an infinite number of planes can be passed through them and the centre of the sphere (B. VI., P. I., S.); in this case, an infinite number of great circles can be made to pass through the two points.

Cor. 6. The bases of a zone are the circumferences of circles (D. 16), and the bases of a segment of a sphere are circles.

PROPOSITION VIII.   THEOREM.

Any plane perpendicular to a radius of a sphere at its outer extremity, is tangent to the sphere at that point.

Let C be the centre of a sphere, CA any radius, and FAG a plane perpendicular to CA at A: then is the plane FAG tangent to the sphere at A.

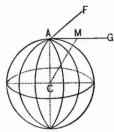

For, from any other point of the plane, as M, draw the line MC: then because CA is a perpendicular to the plane, and CM an oblique line, CM is greater than CA (B. VI., P. V.): hence, the point M lies without the sphere. The plane FAG, therefore, touches the sphere at A, and consequently is tangent to it at that point; *which was to be proved.*

Scholium. It may be shown, by a course of reasoning analogous to that employed in Book III., Propositions XI., XII., XIII., and XIV., that two spheres may have any one of six positions with respect to each other, viz. :

1°. When the distance between their centres is greater than the sum of their radii, *they are external one to the other :*

2°. When the distance is equal to the sum of their radii, *they are tangent externally :*

3°. When this distance is less than the sum, and greater than the difference of their radii, *they intersect each other :*

4°. When this distance is equal to the difference of their radii, *they are tangent internally :*

5°. When this distance is less than the difference of their radii, *one is wholly within the other :*

6°. When this distance is equal to zero, *they have a common centre*, or are *concentric*.

DEFINITIONS.

1°. If a semi-circumference is divided into equal arcs, the chords of these arcs form half of the perimeter of a regular·inscribed polygon; this half perimeter is called *a regular semi-perimeter*. The figure bounded by the regular semi-perimeter and the diameter of the semi-circum_ference is called *a regular semi-polygon*. The diameter itself is called the *axis* of the semi-polygon.

2°. If lines are drawn from the extremities of any side perpendicular to the axis, the intercepted portion of the axis is called the *projection* of that side.

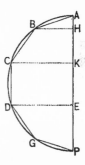

The broken line ABCDGP is a regular semi-perimeter; the figure bounded by it and the diameter AP, is a regular semi-polygon, AP is its axis, HK is the projection of the side BC, and the axis, AP, is the projection of the entire semi-perimeter.

PROPOSITION IX. LEMMA.

If a regular semi-polygon is revolved about its axis, the surface generated by the semi-perimeter is equal to the axis multiplied by the circumference of the inscribed circle.

Let ABCDEF be a regular semi-polygon, AF its axis, and ON its apothem: then is the surface generated by the regular semi-perimeter equal to AF × *circ.* ON.

From the extremities of any side, as DE, draw DI and‛ EH perpendicular to AF; draw also NM perpendicular to AF, and EK perpendicular to DI. Now, the surface generated by DE is equal to DE × *circ.* NM (P. IV., S.). But, because the triangles EDK and ONM are similar (B. IV., P. XXI.), we have,

 DE : EK or IH :: ON : NM :: *circ.* ON : *circ.* NM ;

whence,

$$DE \times circ. \ NM = IH \times circ. \ ON ;$$

that is, the surface generated by any side is equal to the projection of that side multiplied by the circumference of the inscribed circle: hence, the surface generated by the entire semi-perimeter is equal to the sum of the projections of its sides, or the axis, multiplied by the circumference of the inscribed circle; *which was to be proved.*

Cor. The surface generated by any portion of the perimeter, as CDE, is equal to its projection PH, multiplied by the circumference of the inscribed circle.

PROPOSITION X.   THEOREM.

The surface of a sphere is equal to its diameter multiplied by the circumference of a great circle.

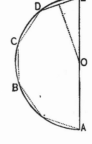

Let ABCDE be a semi-circumference, O its centre, and AE its diameter: then is the surface of the sphere generated by revolving the semi-circumference about AE, equal to AE $\times$ *circ.* OE.

For, the semi-circumference may be regarded as a regular semi-perimeter with an infinite number of sides, whose axis is AE, and the radius of whose inscribed circle is OE: hence (P. IX.), the surface generated by it is equal to AE $\times$ *circ.* OE; *which was to be proved.*

Cor. 1. The circumference of a great circle is equal to $2\pi OE$ (B. V., P. XVI.): hence, the area of the surface of the sphere is equal to $2OE \times 2\pi OE$, or to $4\pi \overline{OE}^2$, that is, *the area of the surface of a sphere is equal to four great circles.*

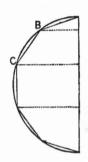

Cor. 2. The surface generated by any arc of the semicircle, as BC, is a zone, whose altitude is equal to the projection of that arc on the diameter. But, the arc BC is a portion of a semi-perimeter having an infinite number of sides, and the radius of whose inscribed circle is equal to that of the sphere: hence (P. IX., C.), the surface of a zone is equal to its altitude multiplied by the circumference of a great circle of the sphere.

Cor. 3. Zones, on the same sphere, or on equal spheres, are to each other as their altitudes.

PROPOSITION XI. LEMMA.

If a triangle and a rectangle having the same base and equal altitudes, are revolved about the common base, the volume generated by the triangle is one third of that generated by the rectangle.

Let ABC be a triangle, and EFBC a rectangle, having the same base BC, and an equal altitude AD, and let them both be revolved about BC: then is the volume generated by ABC one third of that generated by EFBC.

For, the cone generated by the right-angled triangle ADB, is equal to one third of the cylinder generated by the rectangle ADBF (P. V., C. 1), and the cone generated by the triangle ADC, is equal to one third of the cylinder generated by the rectangle ADCE.

When AD falls within the triangle, the sum of the cones generated by ADB and ADC, is equal to the volume generated by the triangle ABC ; and the sum of the cylinders generated by ADBF and ADCE, is equal to the volume generated by the rectangle EFBC.

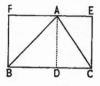

When AD falls without the triangle, the difference of the cones generated by ADB and ADC, is equal to the volume generated by ABC ; and the dif-ference of the cylinders generated by ADBF and ADCE, is equal to the volume generated by EFBC : hence, in either case, the volume generated by the tri-angle ABC, is equal to one third of the volume generated by the rectangle EFBC ; *which was to be proved.*

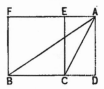

Cor. The volume of the cylinder generated by EFBC, is equal to the product of its base and altitude, or to $\pi \overline{AD}^2 \times BC$: hence, the volume generated by the triangle ABC, is equal to $\frac{1}{3}\pi \overline{AD}^2 \times BC$.

PROPOSITION XII. LEMMA.

If an isosceles triangle is revolved about a straight line passing through its vertex, the volume generated is equal to the surface generated by the base multiplied by one third of the altitude.

Let CAB be an isosceles triangle, C its vertex, AB its base, CI its altitude, and let it be revolved about the line CD, as an axis: then is the volume generated equal to *surf.* $AB \times \frac{1}{3}CI$. There may be three cases:

Suppose the base, when produced, to meet the axis at D; draw AM, IK, and BN, perpendicular to CD, and BO parallel to DC. Now, the volume generated by CAB is equal to the difference of the volumes generated by CAD and CBD; hence (P. XI., C.),

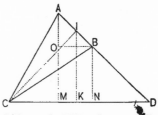

$$vol. \text{ CAB} = \tfrac{1}{3}\pi\overline{\text{AM}}^2 \times \text{CD} - \tfrac{1}{3}\pi\overline{\text{BN}}^2 \times \text{CD} = \tfrac{1}{3}\pi\left(\overline{\text{AM}}^2 - \overline{\text{BN}}^2\right) \times \text{CD}.$$

But, $\overline{\text{AM}}^2 - \overline{\text{BN}}^2$ is equal to (AM + BN) (AM − BN) (B. IV., P. X.); and because AM + BN is equal to 2IK (P. IV., S.), and AM − BN to AO, we have,

$$vol. \text{ CAB} = \tfrac{2}{3}\pi \text{ IK} \times \text{AO} \times \text{CD}.$$

But, the right-angled triangles AOB and CDI are similar (B. IV., P. XVIII.); hence,

$$\text{AO} : \text{AB} :: \text{CI} : \text{CD}; \quad \text{or,} \quad \text{AO} \times \text{CD} = \text{AB} \times \text{CI}.$$

Substituting, and changing the order of the factors, we have,

$$vol. \text{ CAB} = \text{AB} \times 2\pi \text{ IK} \times \tfrac{1}{3}\text{CI}.$$

But, $\text{AB} \times 2\pi \text{ IK} =$ the surface generated by AB; hence,

$$vol. \text{ CAB} = surf. \text{ AB} \times \tfrac{1}{3}\text{CI}.$$

2°. Suppose the axis to coincide with one of the equal sides.

Draw CI perpendicular to AB, and AM and IK, perpendicular to CB. Then,

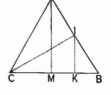

$$vol. \text{ CAB} = \tfrac{1}{3}\pi\overline{\text{AM}}^2 \times \text{CB} = \tfrac{1}{3}\pi \text{AM} \times \text{AM} \times \text{CB}.$$

But, since AMB and CIB are similar,

$$\text{AM} : \text{AB} :: \text{CI} : \text{CB}; \quad \text{whence,} \quad \text{AM} \times \text{CB} = \text{AB} \times \text{CI}.$$

Also, AM = 2IK; hence, by substitution, we have,

$$vol. \text{ CAB} = \text{AB} \times 2\pi \text{ IK} \times \tfrac{1}{3}\text{CI} = surf. \text{ AB} \times \tfrac{1}{3}\text{CI}.$$

3°. Suppose the base to be parallel to the **axis**.

Draw AM and BN perpendicular
to the axis.   The volume generated
by CAB, is equal to the cylinder
generated by the rectangle ABNM,
diminished by the sum of the
cones generated by the triangles
CAM and CBN; hence,

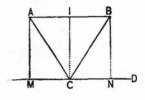

$$vol. \; \text{CAB} \; = \; \pi \overline{CI}^2 \times AB - \tfrac{1}{3}\pi \overline{CI}^2 \times AI - \tfrac{1}{3}\pi \overline{CI}^2 \times IB.$$

But the sum of AI and IB is equal to AB: hence, we
have, by reducing, and changing the order of the factors,

$$vol. \; \text{CAB} \; = \; AB \times 2\pi CI \times \tfrac{1}{3}CI.$$

But $AB \times 2\pi CI$ is equal to the surface generated by AB;
consequently,

$$vol. \; \text{CAB} \; = \; surf. \; AB \times \tfrac{1}{3}CI \,;$$

hence, in all cases, the volume generated by CAB is equal
to *surf.* $AB \times \tfrac{1}{3}CI$; *which was to be proved.*

PROPOSITION XIII.   LEMMA.

*If a regular semi-polygon is revolved about its axis, the
volume generated is equal to the surface generated by
the semi-perimeter multiplied by one third of the apothem.*

Let FBDG be a regular semi-polygon,
FG its axis, OI its apothem, and let the
semi-polygon be revolved about FG: then
is the volume generated equal to *surf.*
FBDG $\times \tfrac{1}{3}$OI.

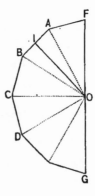

For, draw lines from the vertices to
the centre O. These lines will divide the
semi-polygon into isosceles triangles whose
bases are sides of the semi-polygon, and
whose altitudes are each equal to OI.

Now, the sum of the volumes generated by these triangles is equal to the volume generated by the semi-polygon. But, the volume generated by any triangle, as OAB, is equal to *surf.* AB × ⅓OI (P. XII.); hence, the volume generated by the semi-polygon is equal to *surf.* FBDG × ⅓OI; *which was to be proved.*

Cor. The volume generated by a portion of the semi-polygon, OABC, limited by OC, OA, drawn to vertices is equal to *surf.* ABC × ⅓OI.

PROPOSITION XIV.   THEOREM.

The volume of a sphere is equal to its surface multiplied by one third of its radius.

Let ACE be a semicircle, AE its diameter, O its centre, and let the semicircle be revolved about AE: then is the volume generated equal to the surface generated by the semi-circumference multiplied by one third of the radius OA.

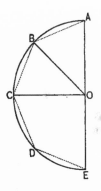

For, the semicircle may be regarded as a regular semi-polygon having an infinite number of sides, whose semi-perimeter coincides with the semi-circumference, and whose apothem is equal to the radius: hence (P. XIII.), the volume generated by the semicircle is equal to the surface generated by the semicircumference multiplied by one third of the radius; *which was to be proved.*

Cor. 1. Any portion of the semicircle, as OBC, bounded by two radii, will generate a volume equal to the surface generated by the arc BC multiplied by one third of the radius (P. XIII., C.). But this portion of the semicircle is a circular sector, the volume which it generates is a spherical sector, and the surface generated by the arc is a zone: hence, the *volume of a spherical sector is equal to the zone which forms its base multiplied by one third of the radius.*

Cor. 2. If we denote the volume of a sphere by V, and its radius by R, the area of the surface will be equal to $4\pi R^2$ (P. X., C. 1), and the volume of the sphere will be equal to $4\pi R^2 \times \frac{1}{3}R$; consequently, we have,

$$V = \tfrac{4}{3}\pi R^3.$$

Again, if we denote the diameter of the sphere by D, we shall have R equal to $\frac{1}{2}D$, and R^3 equal to $\frac{1}{8}D^3$, and consequently,

$$V = \tfrac{1}{6}\pi D^3;$$

hence, *the volumes of spheres are to each other as the cubes of their radii, or as the cubes of their diameters.*

Scholium. If the figure EBDF, formed by drawing lines from the extremities of the arc BD perpendicular to CA, be revolved about CA, as an axis, it will generate a segment of a sphere whose volume may be found by adding to the spherical sector generated by CDB, the cone generated by CBE, and subtracting from their sum the cone generated by

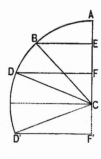

CDF. If the arc BD is so taken that the points E and F fall on opposite sides of the centre C, the latter cone must be added, instead of subtracted. The area of the zone BD is equal to $2\pi CD \times EF$ (P. X., C. 2); hence,

segment EBDF $= \tfrac{1}{3}\pi\,(2\overline{CD}^2 \times EF + \overline{BE}^2 \times CE \mp \overline{DF}^2 \times CF).$

PROPOSITION XV. THEOREM.

The surface of a sphere is to the entire surface of the circumscribed cylinder, including its bases, as 2 is to 3; and the volumes are to each other in the same ratio.

Let PMQ be a semicircle, and PADQ a rectangle, whose sides PA and QD are tangent to the semicircle at P and Q, and whose side AD, is tangent to the semicircle at M. If the semicircle and the rectangle be revolved about PQ, as an axis, the former will generate a sphere, and the latter a circumscribed cylinder.

1°. The surface of the sphere is to the entire surface of the cylinder, as 2 is to 3.

For, the surface of the sphere is equal to four great circles (P. X., C. 1), the convex surface of the cylinder is equal to the circumference of its base multiplied by its altitude (P. I.); that is, it is equal to the circumference of a great circle multiplied by its diameter, or to four great circles (B. V., P. XV.); adding to this the two bases, 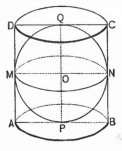 each of which is equal to a great circle, we have the entire surface of the cylinder equal to six great circles: hence, the surface of the sphere is to the entire surface of the circumscribed cylinder, as 4 is to 6, or as 2 is to 3; *which was to be proved.*

2°. The volume of the sphere is to the volume of the cylinder as 2 is to 3.

For, the volume of the sphere is equal to $\frac{4}{3}\pi R^3$ (P. XIV., C. 2); the volume of the cylinder is equal to its base multiplied by its altitude (P. II.); that is, it is equal to

$\pi R^2 \times 2R$, or to $\frac{4}{3}\pi R^3$: hence, the volume of the sphere is to that of the cylinder as 4 is to 6, or as 2 is to 3; *which was to be proved.*

Cor. The surface of a sphere is to the entire surface of a circumscribed cylinder, as the volume of the sphere is to the volume of the cylinder.

Scholium. Any polyedron which is circumscribed about a sphere, that is, whose faces are all tangent to the sphere, may be regarded as made up of pyramids, whose bases are the faces of the polyedron, whose common vertex is at the centre of the sphere, and each of whose altitudes is equal to the radius of the sphere. But, the volume of any one of these pyramids is equal to its base multiplied by one third of its altitude: hence, the volume of a circumscribed polyedron is equal to its surface multiplied by one third of the radius of the inscribed sphere.

Now, because the volume of the sphere is also equal to its surface multiplied by one third of its radius, it follows that the volume of a sphere is to the volume of any circumscribed polyedron, as the surface of the sphere is to the surface of the polyedron.

Polyedrons circumscribed about the same, or about equal spheres, are proportional to their surfaces.

GENERAL FORMULAS.

If we denote the convex surface of a cylinder by S, its volume by V, the radius of its base by R, and its altitude by H, we have (P. I., II.),

$$S = 2\pi R \times H \quad \cdots \cdots \cdots \cdots \quad (1.)$$
$$V = \pi R^2 \times H \quad \cdots \cdots \cdots \cdots \quad (2.)$$

If we denote the convex surface of a cone by S, its volume by V, the radius of its base by R, its altitude by H, and its slant height by H', we have (P. III., V.),

$$S = \pi R \times H' \quad \cdots \cdots \cdots \cdots \quad (3.)$$

$$V = \pi R^2 \times \tfrac{1}{3}H \quad \cdots \cdots \cdots \cdots \quad (4.)$$

If we denote the convex surface of a frustum of a cone by S, its volume by V, the radius of its lower base by R, the radius of its upper base by R', its altitude by H, and its slant height by H', we have (P. IV., VI.),

$$S = \pi (R + R') \times H' \quad \cdots \cdots \cdots \cdots \quad (5.)$$

$$V = \tfrac{1}{3}\pi (R^2 + R'^2 + R \times R') \times H \quad \cdots \cdots \quad (6.)$$

If we denote the surface of a sphere by S, its volume by V, its radius by R, and its diameter by D, we have (P. X., C. 1, XIV., C. 2, XIV., C. 1),

$$S = 4\pi R^2 \quad \cdots \cdots \cdots \cdots \cdots \quad (7.)$$

$$V = \tfrac{4}{3}\pi R^3 = \tfrac{1}{6}\pi D^3 \quad \cdots \cdots \cdots \cdots \quad (8.)$$

If we denote the radius of a sphere by R, the area of any zone of the sphere by S, its altitude by H, and the volume of the corresponding spherical sector by V, we shall have (P. X., C. 2, XIV., C. 1),

$$S = 2\pi R \times H \quad \cdots \cdots \cdots \cdots \cdots \quad (9.)$$

$$V = \tfrac{2}{3}\pi R^2 \times H \quad \cdots \cdots \cdots \cdots \quad (10.)$$

If we denote the volume of the corresponding spherical segment by V, its altitude by H, the radius of its upper base by R', the radius of its lower base by R'', the distance of its upper base from the centre by H', and of its lower base from the centre by H'', we shall have (P. XIV., S.):

$$V = \tfrac{1}{2}\pi \left(2R^2 \times H + R'^2 H' \mp R''^2 \times H''\right) \quad \cdots \quad (11.)$$

EXERCISES.

1. The radius of the base of a cylinder is 2 feet, and its altitude 6 feet; find its entire surface, including the bases.

2. The volume of a cylinder, of which the radius of the base is 10 feet, is 6283.2 cubic feet; find the volume of a similar cylinder of which the diameter of the base is 16 feet, and find also the altitude of each cylinder.

3. Two similar cones have the radii of the bases equal, respectively, to $4\frac{1}{4}$ and 6 feet, and the convex surface of the first is 667.59 square feet; find the convex surface of the second and the volume of both.

4. A line 12 feet long is revolved about another line as an axis; the distance of one extremity of the line from the axis is 4 feet and of the other extremity 6 feet; find the area of the surface generated.

5. Find the convex surface and the volume of the frustum of a cone the altitude of which is 6 feet, the radius of the lower base being 4 feet and that of the upper base 2 feet.

6. Find the surface and the volume of the cone of which the frustum in the preceding example is a frustum.

7. A small circle, the radius of which is 4 feet, is 3 feet from the centre of a sphere; find the circumference of a great circle of the same sphere.

8. The radius of a sphere is 10 feet; find the area of a small circle distant from the centre 6 feet.

9. Find the area of the surface generated by the semi-perimeter of a regular semihexagon revolving about its axis, the radius of the inscribed circle being 5.2 feet and the axis 12 feet.

10. The area of the surface generated by the semi-

perimeter of a regular semioctagon revolved about an axis is 178.2426 square feet, and the radius of the inscribed circle is 3.62 feet; find the axis.

11. An isosceles triangle, whose base is 8 feet and altitude 9 feet, is revolved about a line passing through its vertex and parallel to its base; how many cubic feet in the volume generated?

12. The altitude of a zone is 3 feet and the radius of the sphere is 5 feet; find the area of the zone and the volume of the corresponding spherical sector.

13. Find the surface and the volume of a sphere whose radius is 4 feet.

14. The radius of a sphere is 5 feet; how many cubic feet in a spherical segment whose altitude is 7 feet and the distance of whose lower base from the centre of the sphere is 3 feet?

15. A cone such that the diameter of its base is equal to its slant height is circumscribed about a sphere; show that the surface of the sphere is to the entire surface of the cone, including its base, as 4 is to 9, and that the volumes are in the same ratio.

16. The radius of a sphere is 6 feet; find the entire surface and the volume of the circumscribing cylinder.

17. A cone, with the diameter of the base and the slant height equal, is circumscribed about a sphere whose radius is 5 feet; find the entire surface and the volume of the cone.

18. A cone, with the diameter of the base and the slant height equal, and a cylinder, are circumscribed about a sphere; what relation exists between the entire surfaces and the volumes of the cylinder, the cone and the sphere?

19. The edge of a regular octaedron is 10 feet, and the radius of the inscribed sphere is 4.08 feet; find the volume of the octaedron.

BOOK IX.

DEFINITIONS.

1. A SPHERICAL ANGLE is the amount of divergence of the arcs of two great circles of a sphere meeting at a point. The arcs are called *sides* of the angle, and their point of intersection is called the *vertex* of the angle.

The measure of a spherical angle is the same as that of the diedral angle included between the planes of its sides. Spherical angles may be *acute, right,* or *obtuse.*

2. A SPHERICAL POLYGON is a portion of the surface of a sphere bounded by arcs of three or more great circles. The bounding arcs are called *sides* of the polygon, and the points in which the sides meet are called *vertices* of the polygon. Each side is taken less than a semi-circumference.

Spherical polygons are classified in the same manner as plane polygons.

3. A SPHERICAL TRIANGLE is a spherical polygon of three sides.

Spherical triangles are classified in the same manner as plane triangles.

4. A LUNE is a portion of the surface of a' sphere bounded by semi-circumferences of two great circles.

5. A SPHERICAL WEDGE is a portion of a sphere bounded by a lune and two semicircles which intersect in a diameter of the sphere.

6. A SPHERICAL PYRAMID is a portion of a sphere bounded by a spherical polygon and sectors of circles whose common centre is the centre of the sphere.

The spherical polygon is called the *base* of the pyramid, and the centre of the sphere is called the *vertex* of the pyramid.

7. A POLE OF A CIRCLE is a point, on the surface of the sphere, equally distant from all the points of the circumference of the circle.

8. A DIAGONAL of a spherical polygon is an arc of a great circle joining the vertices of any two angles which are not consecutive.

PROPOSITION I. THEOREM.

Any side of a spherical triangle is less than the sum of the two others.

Let ABC be a spherical triangle situated on a sphere whose centre is O: then is any side, as AB, less than the sum of the sides AC and BC.

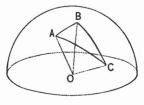

For, draw the radii OA, OB, and OC: these radii form the edges of a triedral angle whose vertex is O, and the plane angles included between them are measured by the arcs AB, AC, and BC (B. III., P. XVII., Sch.). But any plane angle, as AOB, is less than the sum of the plane angles AOC and BOC (B. VI., P. XIX.): hence, the arc AB is less than the sum of the arcs AC and BC; *which was to be proved.*

Cor. 1. Any side AB, of a spherical polygon ABCDE, is less than the sum of all the other sides.

For, draw the diagonals AC and AD, dividing the polygon into triangles. The arc AB is less than the sum of AC and BC, the arc AC is less than the sum of AD and DC, and the arc AD is less than the sum of DE and EA; hence, AB is less than the sum of BC, CD, DE, and EA.

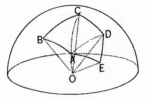

Cor. 2. The arc of a small circle, on the surface of a sphere, is greater than the arc of a great circle joining its two extremities.

For, divide the arc of the small circle into equal parts, and through the two extremities of each part suppose the arc of a great circle to be drawn. The sum of these arcs, whatever may be their number, will be greater than the arc of the great circle joining the given points (C. 1). But when this number is infinite, each arc of the great circle will coincide with the corresponding arc of the small circle, and their sum is equal to the entire arc of the small circle, which is, consequently, greater than the arc of the great circle.

Cor. 3. The shortest distance from one point to another on the surface of a sphere, is measured on the arc of a great circle joining them.

PROPOSITION II.   THEOREM.

The sum of the sides of a spherical polygon is less than the circumference of a great circle.

Let ABCDE be a spherical polygon situated on a sphere whose centre is O : then is the sum of its sides less than the circumference of a great circle.

For, draw the radii OA, OB, OC, **OD**, and **OE**: these radii form the edges of a polyedral angle whose vertex is at O, and the angles included be-

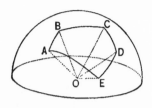

tween them are measured by the arcs AB,. BC, CD, DE, and EA. But the sum of these angles is less than four right angles (B. VI., P. XX.): hence, the sum of the arcs which measure them is less than the circumference of a great circle; *which was to be proved.*

PROPOSITION III. THEOREM.

If a diameter of a sphere is drawn perpendicular to the plane of any circle of the sphere, its extremities are poles of that circle.

Let C be the centre of a sphere, FNG any circle of the sphere, and DE a diameter of the sphere perpendicular to the plane of FNG: then are its extremities, D and E, poles of the circle FNG.

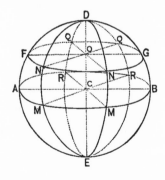

The diameter DE, being per-pendicular to the plane of FNG, must pass through the centre O (B. VIII., P. VII., C. 3). If arcs of great circles DN, DF DG, &c., are drawn from D to different points of the circum-ference FNG, and chords of these arcs are drawn, these chords are equal (B. VI., P. V.), consequently, the arcs them-selves are equal. But these arcs are the shortest lines that can be drawn from the point D to the different points of the circumference (P. I., C. 3): hence, the point D is equally distant from all the points of the circumference, and consequently is a pole of the circle (D. 7). In like manner, it may be shown that the point E is also a pole of the circle: hence, both D and E are poles of the circle FNG; *which was to be proved.*

Cor. 1. Let AMB be a great circle perpendicular to DE: then are the angles DCM, ECM, &c., right angles; and consequently, the arcs DM, EM, &c., are each equal to a quadrant (B. III., P. XVII., S.): hence, the two poles of a great circle are at equal distances from the circumference.

Cor. 2. The two poles of a small circle are at unequal distances from the circumference, the sum of the distances being equal to a semi-circumference.

Cor. 3. If any point, as M, in the circumference of a great circle, is joined with either pole by the arc of a great circle, such arc is perpendicular to the circumference AMB, since its plane passes through CD, which is perpendicular to AMB. Conversely: if MN is perpendicular to the arc AMB, it passes through the poles D and E: for, the plane of MN being perpendicular to AMB and passing through C, contains CD, which is perpendicular to the plane AMB (B. VI., P. XVII., C.).

Cor. 4. If the distance of a point D from each of the points A and M, in the circumference of a great circle, is equal to a quadrant, the point D is the pole of the arc AM (the arc AM is supposed to be either less or greater than a semi-circumference).

For, let C be the centre of the sphere, and draw the radii CD, CA, CM. Since the angles ACD, MCD, are right angles, the line CD is perpendicular to the two straight lines CA, CM: it is, therefore, perpendicular to their plane

(B. VI., P. IV.): hence, the point D is the pole of the arc AM.

Scholium. The properties of these poles enable us to describe arcs of a circle on the surface of a sphere, with the same facility as on a plane surface. For, by turning the arc DF about the point D, the extremity F will describe the small circle FNG; and by turning the quadrant DFA round the point D, its extremity A will describe an arc of a great circle.

PROPOSITION IV. THEOREM.

The angle formed by arcs of two great circles, is equal to that formed by the tangents to these arcs at their point of intersection, and is measured by the arc of a great circle described from the vertex as a pole, and limited by the sides, produced if necessary.

Let the angle BAC be formed by the two arcs AB, AC: then is it equal to the angle FAG formed by the tangents AF, AG, and is measured by the arc DE of a great circle, described about A as a pole.

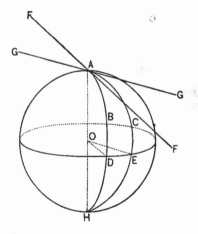

For, the tangent AF, drawn in the plane of the arc AB, is perpendicular to the radius AO; and the tangent AG, drawn in the plane of the arc AC, is perpendicular to the same radius AO: hence, the angle FAG is equal to the angle contained by the planes ABDH, ACEH (B. VI., D. 4); which is that of the arcs AB, AC. Now, if the arcs AD and AE are both quadrants, the lines OD, OE, are perpendicular to OA, and the angle DOE is equal to the angle of the planes ABDH, ACEH: hence, the arc DE is the measure of the angle contained by these planes, or of the angle CAB; *which was to be proved.*

Cor. 1. The angles of spherical triangles may be compared by means of the arcs of great circles described from their vertices as poles, and included between their sides.

A spherical angle can always be constructed equal to a given spherical angle.

Cor. 2. Vertical angles, such as ACP and BCN, are equal; for either of them is the angle formed by the two planes ACB, PCN. When two arcs ACB, PCN, intersect, the sum of two adjacent angles, as ACP, PCB, is equal to two right angles.

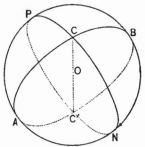

PROPOSITION V. THEOREM.

If from the vertices of the angles of a spherical triangle, as poles, arcs be described forming a second spherical triangle, the vertices of the angles of this second triangle are respectively poles of the sides of the first.

From the vertices A, B, C, as poles, let the arcs EF, FD, DE, be described, forming the triangle DFE: then are the vertices D, E, and F, respectively poles of the sides BC, AC, AB.

For, the point A being the

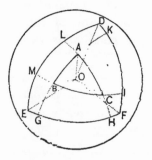

pole of the arc EF, the distance AE is a quadrant; the point C being the pole of the arc DE, the distance CE is likewise a quadrant: hence, the point E is at a quadrant's distance from the points A and C: hence, it is the pole of the arc AC (P. III., C. 4). It may be shown, in like manner, that D is the pole of the arc BC, and F that of the arc AB; *which was to be proved.*

Cor. The triangle ABC, may be described by means of DEF, as DEF is described by means of ABC. Triangles so related that any vertex of either is the pole of the side opposite it in the other, are called *polar triangles.*

PROPOSITION VI. THEOREM.

Any angle, in one of two polar triangles, is measured by a semi-circumference, minus *the side lying opposite to it in the other triangle.*

Let ABC, and EFD, be any two polar triangles on a sphere whose centre is O: then is any angle in either triangle measured by a semi-circumference, minus the side lying opposite to it in the other triangle.

For, produce the sides AB, AC, if necessary, till they meet EF in G and H. The point A being the pole of the arc GH, the angle A is measured by that arc (P. IV.). But, since E is the pole of AH, the arc EH is a quadrant; and since F is the pole of AG, FG is a quad-

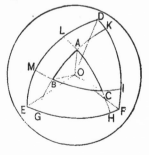

rant: hence, the sum of the arcs EH and GF is equal to a semi-circumference. But, the sum of the arcs EH and

GF is equal to the sum of the arcs EF and GH: hence, the arc GH, which measures the angle A, is equal to a semi-circumference minus the arc EF. In like manner, it may be shown, that any other angle, in either triangle, is measured by a semi-circumference minus the side lying opposite to it in the other triangle; *which was to be proved*

Cor. 1. Beside the triangle DEF, three other triangles, polar to ABC, may be formed by the intersection of the arcs DE, EF, DF, prolonged. But the proposition is applicable only to the central triangle, ABC, which is distinguished from the three others by the circumstance, that the vertices A 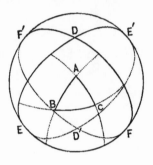 and D lie on the same side of BC; B and E, on the same side of AC; C and F, on the same side of AB. The polar triangles ABC and DEF are called *supplemental* triangles, any part of either being the supplement of the part opposite it in the other.

Cor. 2. Arcs of great circles, drawn from corresponding vertices of two supplemental polar triangles perpendicular to the respective sides opposite, are supplements of each other. For, from A draw the arc of a great circle, AN, perpendicular to BC; it must, when prolonged, pass through D, the pole of BC, and 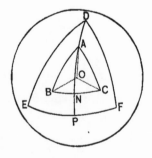 must also, when prolonged to P, be perpendicular to EF (P. III., C. 3): DN and AP being quadrants (P. III. C. 1), DP and AN are supplements of each other.

PROPOSITION VII. THEOREM.

If from the vertices of any two angles of a spherical tri-
angle, as poles, arcs of circles are described passing through
the vertex of the third angle; and if from the second
point in which these arcs intersect, arcs of great circles
are drawn to the vertices, used as poles, the parts of the
triangle thus formed are equal to those of the given tri-
angle, each to each.

Let ABC be a spherical triangle situated on a sphere
whose centre is O, CED and CFD arcs of circles described
about B and A as poles, and let DA and DB be arcs of
great circles: then are the parts of the triangle ABD equal
to those of the given triangle ABC, each to each.

For, by construction, the side
AD is equal to AC, the side BD is
equal to BC, and the side AB is
common: hence, the sides are
equal, each to each. Draw the
radii OA, OB, OC, and OD. The
radii OA, OB, and OC, form the

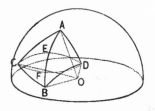

edges of a triedral angle whose vertex is O; and the radii
OA, OB, and OD, form the edges of a second triedral angle
whose vertex is also at O; and the plane angles formed
by these edges are equal, each to each: hence, the planes
of the equal angles are equally inclined to each other
(B. VI., P. XXI.). But, the angles made by these planes
are equal to the corresponding spherical angles; consequently, the angle BAD is equal to BAC, the angle ABD to
ABC, and the angle ADB to ACB: hence, the parts of the
triangle ABD are equal to the parts of the triangle ACB,
each to each; *which was to be proved.*

Scholium 1. The triangles ABC and ABD, are not, in general, capable of superposition, but their parts are *symmetrically* disposed with respect to AB. *Triangles which have all the parts of the one equal to all the parts of the other, each to each, but are not capable of superposition, are called symmetrical triangles.*

Scholium 2. If symmetrical triangles are isosceles, they can be so placed as to coincide throughout: hence, they are *equal in area.*

PROPOSITION VIII.   THEOREM.

If two spherical triangles, on the same, or on equal spheres, have two sides and the included angle of the one equal to two sides and the included angle of the other, each to each, the remaining parts are equal, each to each.

Let the spherical triangles ABC and EFG, on the sphere whose centre is O, have the side EF equal to AB, the side EG equal to AC, and the angle FEG equal to BAC: then is the side FG equal to BC, the angle EFG to ABC, and the angle EGF to ACB.

For, draw the radii OE, OF, OG, OA, OB, and OC, forming the trie- dral angles O–EFG and O–ABC. Since the sides EF and EG are equal, respectively, to the sides AB and AC, the plane angles EOF and

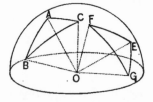

EOG are equal, respectively, to the plane angles AOB and AOC; and as the spherical angles FEG and BAC are equal, the inclination of the faces EOF and EOG of the triedral angle O–EFG, is equal to the inclination of the faces AOB and AOC of the triedral angle O–ABC; therefore (B. VI., P. XXI., C.), the angle FOG is equal to BOC, and the

side FG equals the side BC : again, since the angle EOF is
equal to AOB, FOG to BOC, and GOE to COA, the planes
of the equal angles are equally inclined to each other
(B. VI., P. XXI.), and, consequently (D. 1), the angle EFG
is equal· to ABC, and EGF to ACB—hence, the remaining
parts of the triangles are equal, each to each; *which was
to be proved.*

PROPOSITION IX.   THEOREM.

*If two spherical triangles on the same, or on equal spheres,
have two angles and the included side of the one equal
to two angles and the included side of the other, each
to each, the remaining parts are equal, each to each.*

Let the spherical triangles ABC and EFG, on the sphere
whose centre is O, have the angle FEG equal to BAC, the
angle EFG equal to ABC, and the
side EF equal to AB: then is the
side EG equal to AC, the side FG
to BC, and the angle FGE to BCA.

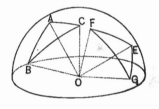

For, draw radii, as before, forming the triedral angles O–EFG and
O–ABC. Since the side EF is equal
to AB, the plane angle EOF is equal to AOB; as the angle
FEG is equal to BAC, and EFG to ABC, the inclination of
the face EOF, of the triedral angle O–EFG, to each of the
faces EOG and FOG, is equal, respectively, to the inclination of the face AOB, of the triedral angle O–ABC, to each
of the faces AOC and BOC, and hence (B. VI., P. XXI.,
S. 2), the · plane angles EOG and GOF are equal, respectively, to AOC and COB; therefore, the sides EG and GF
are equal to the sides AC and CB, and the angle FGE to
BCA; *which was to be proved.*

PROPOSITION X.   THEOREM.

If two spherical triangles on the same, or on equal spheres, have their sides equal, each to each, their angles are equal, each to each, the equal angles lying opposite the equal sides.

Let the spherical triangles EFG and ABC, on the sphere whose centre is O, have the side EF equal to AB, EG equal to AC, and FG equal to BC : then the angle FEG is equal to BAC, EFG to ABC, and EGF to ACB, and the equal angles lie opposite the equal sides.

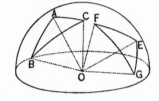

For, draw the radii, as before, forming the triedral angles O–EFG and O–ABC.   Because the sides of the triangles are respectively equal, the plane angle EOF is equal to AOB, FOG to BOC, and GOE to COA.   Hence (B. VI., P. XXI.), the planes of the equal angles are equally inclined to each other, and, consequently, the spherical angle EFG is equal to spherical angle ABC, FEG to BAC, and EGF to ACB, the equal angles lying opposite the equal sides; *which was to be proved.*

Note.—The triangle EFG is equal in all respects to either ABC or its symmetrical triangle.

PROPOSITION XI.   THEOREM.

In any isosceles spherical triangle, the angles opposite the equal sides are equal; and conversely, if two angles of a spherical triangle are equal, the triangle is isosceles.

1°.   Let ABC be a spherical triangle, on a sphere whose centre is O, having the side AB equal to AC : then is the angle C equal to the angle B.

For, draw the arc of a great circle from the vertex A, to the middle point D, of the base BC: then in the two triangles ADB and ADC, we shall have the side AB equal to AC, by hypothesis, the side BD equal to DC, by construction, and the side AD common; consequently, the triangles have their angles equal, each to each (P. X.): hence, the angle C is equal to the angle B; *which was to be proved.*

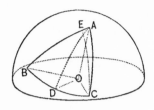

2°. Let ABC be a spherical triangle having the angle C equal to the angle B: then is the side AB equal to the side AC, and consequently the triangle is isosceles.

For, suppose that AB and AC are not equal, but that one of them, as AB, is the greater. On AB lay off the arc BE equal to AC, and draw the arc of a great circle from E to C: then in the triangles ACB and EBC, we shall have the side AC equal to EB, by construction, the side BC common, and the included angle ACB equal to the included angle EBC, by hypothesis; hence, the remaining parts of the triangles are equal, each to each, and consequently, the angle ECB is equal to the angle ABC. But, the angle ACB is equal to ABC, by hypothesis, and therefore, the angle ECB is equal to ACB, or a part is equal to the whole, which is impossible: hence, the supposition that AB and AC are unequal, is absurd; they are therefore equal. and consequently, the triangle ABC is isosceles; *which was to be proved.*

Cor. The triangles ADB and ADC, having all of their parts equal, each to each, the angle ADB is equal to ADC, and the angle DAB is equal to DAC; that is, *if an arc of a great circle is drawn from the vertex of an isosceles*

*spherical triangle to the middle of its base, it is perpen-
dicular to the base, and bisects the vertical angle of the triangle.*

PROPOSITION XII. THEOREM.

*In any spherical triangle, the greater side is opposite the
greater angle; and conversely, the greater angle is opposite the greater side.*

1°. Let ABC be a spherical triangle, on a sphere whose
centre is O, in which the angle A is greater than the
angle B: then is the side BC
greater than the side AC.

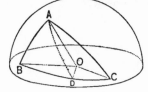

For, draw the arc AD, making
the angle BAD equal to ABD; then
is AD equal to BD (P. XI.). But,
the sum of AD and DC is greater
than AC (P. I.); or, putting for AD
its equal BD, we have the sum of BD and DC, or BC,
greater than AC; *which was to be proved.*

2°. In the triangle ABC, let the side BC be greater
than AC: then is the angle A greater than the angle B.

For, if the angles A and B were equal, the sides BC
and AC would be equal; or if the angle A were less than
the angle B, the side BC would be less than AC, either of
which conclusions contradicts the hypothesis, and is impossible: hence, the angle A is greater than the angle B;
which was to be proved.

PROPOSITION XIII. THEOREM.

If two triangles on the same, or on equal spheres, are mutually equiangular, they are also mutually equilateral.

Let the spherical triangles A and B be mutually equiangular: then are they also mutually equilateral.

For, let P be the supplemental polar triangle of A, and Q, the supplemental polar triangle of B: then, because the triangles A and B are mutually equiangular, their supplemental triangles P

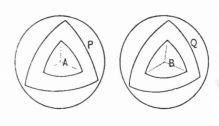

and Q must be mutually equilateral (P. VI.), and consequently mutually equiangular (P. X.). But, the triangles P and Q being mutually equiangular, their supplemental triangles A and B are mutually equilateral (P. VI.); *which was to be proved.*

Scholium. Two plane triangles that are mutually equiangular are not necessarily mutually equilateral; that is, they may be similar without being equal. Two spherical triangles on the same or on equal spheres can not be similar without being equal in all respects.

PROPOSITION XIV. THEOREM.

The sum of the angles of a spherical triangle is less than six right angles, and greater than two right angles.

Let ABC be a spherical triangle, on a sphere whose centre is O, and DEF its supplemental triangle: then is the sum of the angles A, B, and C, less than six right angles and greater than two.

For, any angle, as A, being measured by a semi-circumference, minus the side EF (P. VI.), is less than two right angles: hence, the sum of the three angles is less than six right angles. Again, because the measure of each angle is equal to a semi-circumference minus the side lying opposite to it, in the supplemental triangle, the measure of the sum of the three angles is equal to three semi-circumferences, minus the sum of the sides of the supplemental triangle DEF. But the latter sum is less than a circumference; consequently, the measure of the sum of the angles A, B, and C, is greater than a semi-circumference, and therefore the sum of the angles is greater than two right angles: hence, the sum of the angles A, B, and C, is less than six right angles and greater than two; *which was to be proved.*

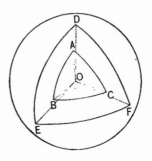

Cor. 1. The sum of the three angles of a spherical triangle is not constant, like that of the angles of a rectilineal triangle, but varies between two right angles and six, without ever reaching either of these limits. Two angles, therefore, do not serve to determine the third.

Cor. 2. A spherical triangle may have two, or even three of its angles right angles; also two, or even three of its angles obtuse.

Cor. 3. If a triangle, ABC, is *bi-rectangular*, that is, has two right angles B and C, the vertex A is the pole of the other side BC, and AB, AC, will be quadrants.

For, since the arcs AB and AC are perpendicular to BC, each must pass through its pole (P. III., Cor. 3): hence, their intersection A is that pole, and consequently, AB and AC are quadrants.

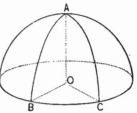

If the angle A is also a right angle, the triangle ABC is *tri-rectangular;* each of its angles is a right angle, and its sides are quadrants. Four tri-rectangular triangles make up the surface of a hemisphere, and eight the entire surface of a sphere.

Scholium. The right angle is taken as the unit of measure of spherical angles, and is denoted by 1.

The excess of the sum of the angles of a spherical triangle over two right angles, is called the *spherical excess.* If we denote the spherical excess by E, and the three angles expressed in terms of the right angle, as a unit, by A, B, and C, we have,

$$E = A + B + C - 2.$$

The *spherical excess* of any spherical polygon is equal to the excess of the sum of its angles over two right angles taken as many times, less two, as the polygon has sides. If we denote the spherical excess by E, the sum of the angles by S, and the number of sides by n, we have,

$$E = S - 2 (n - 2) = S - 2n + 4.$$

PROPOSITION XV. THEOREM.

Any lune is to the surface of the sphere, as the arc which measures its angle is to the circumference of a great circle; or, as the angle of the lune is to four right angles.

Let AMBN be a lune, and MON the angle of the lune; then is the area of the lune to the surface of the sphere, as the arc MN is to the circumference of a great circle MNPQ; or, as the angle MON is to four right angles (B. III., P. XVII., C. 2).

In the first place, suppose the arc MN and the circumference MNPQ to be commensurable. For example, let them be to each other as 5 is to 48. Divide the circumference MNPQ into 48 equal parts, beginning at M; MN will contain five of these parts. Join each point of division with the points A and B, by a quadrant; there will be formed 96 equal isosceles

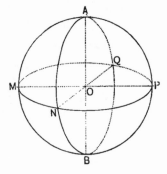

spherical triangles (P. VII., S. 2) on the surface of the sphere, of which the lune will contain 10; hence, in this case, the area of the lune is to the surface of the sphere, as 10 is to 96, or as 5 is to 48; that is, as the arc MN is to the circumference MNPQ, or as the angle of the lune is to four right angles.

In like manner, the same relation may be shown to exist when the arc MN, and the circumference MNPQ are to each other as any other whole numbers.

If the arc MN, and the circumference MNPQ, are not commensurable, the same relation may be shown to exist by a course of reasoning entirely analogous to that employed in Book IV., Proposition III. Hence, in all cases, the area of a lune is to the surface of the sphere, as the arc measuring the angle is to the circumference of a great circle; or, as the angle of the lune is to four right angles; *which was to be proved.*

Cor. 1. Lunes, on the same or on equal spheres, are to each other as their angles.

Cor. 2. If we denote the area of a tri-rectangular triangle by T, the area of a lune by L, and the angle of the lune by A, the right angle being denoted by 1, we have,

$$L \; : \; 8T \; :: \; A \; : \; 4 ;$$

whence,

$$L = T \times 2A ;$$

hence, the area of a lune is equal to the area of a tri-rectangular triangle multiplied by twice the angle of the lune.

Scholium. The spherical wedge, whose angle is MON, is to the entire sphere, as the angle of the wedge is to four right angles, as may be shown by a course of reasoning entirely analogous to that just employed: hence, we infer that the volume of a spherical wedge is equal to the lune which forms its base, multiplied by one third of the radius.

PROPOSITION XVI. THEOREM.

Symmetrical triangles are equal in area.

Let ABC and DEF be symmetrical triangles, on a sphere whose centre is O, the side DE being equal to AB, the side DF to AC, and the side EF to BC: then are the triangles equal in area.

For, conceive a small circle to be drawn through A, B, and C, and let P be its pole; draw arcs of great circles from P to A, B, and C: these arcs will be equal (D. 7). Draw the arc of a great circle FQ, making the angle DFQ equal to ACP, and lay off on it FQ equal to CP; draw arcs of great circles QD and QE.

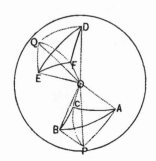

In the triangles PAC and FDQ, we have the side FD equal to AC, by hypothesis; the side FQ equal to PC, by construction, and the angle DFQ equal to ACP, by construction: hence (P. VIII.), the side DQ is equal to AP, the angle FDQ to PAC, and the angle FQD to APC. Now, because the triangles QFD and PAC are isosceles and equal in all their parts, they may be placed so as to co-incide throughout, the base FD falling on AC, DQ on CP, and FQ on AP: hence, they are equal in area.

If we take from the angle DFE the angle DFQ, and from the angle ACB the angle ACP, the remaining angles QFE and PCB, will be equal. In the triangles FQE and PCB, we have the side QF equal to PC, by construction, the side FE equal to BC, by hypothesis, and the angle QFE equal to PCB, from what has just been shown: hence, the triangles are equal in all their parts, and being isosceles, they may be placed so as to coincide through-out, the side QE falling on PC, and the side QF on PB; these triangles are, therefore, equal in area.

In the triangles QDE and PAB, we have the sides QD, QE, PA, and PB, all equal, and the angle DQE equal to APB, because they are the sums of equal angles: hence, the triangles are equal in all their parts, and because they are isosceles, they may be so placed as to coincide throughout, the side QD falling on PB, and the side QE on PA; these triangles are, therefore, equal in area.

Hence, the sum of the triangles QFD and QFE, is equal to the sum of the triangles PAC and PBC. If from the former sum we take away the triangle QDE, there will remain the triangle DFE; and if from the latter sum we take away the triangle PAB, there will remain the triangle ABC: hence, the triangles ABC and DEF are equal in area.

If the point P falls within the triangle ABC, the point Q will fall within the triangle DEF, and we shall have the triangle DEF equal to the sum of the triangles QFD, QFE, and QDE, and the triangle ABC equal to the sum of the equal triangles PAC, PBC, and PAB. Hence, in either case, the triangles ABC and DEF are equal in area; *which was to be proved.*

PROPOSITION XVII. THEOREM.

If the circumferences of two great circles intersect on the surface of a hemisphere, the sum of the opposite triangles thus formed is equal to a lune, whose angle is equal to that formed by the circles.

Let the circumferences ACB, PCN, intersect on the surface of a hemisphere whose centre is O: then is the sum of the opposite triangles ACP, NCB, equal to the lune whose angle is NCB.

For, produce the arcs CB, CN, on the other hemisphere till they meet at D. Now, since ACB and

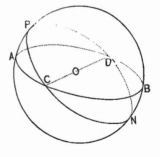

CBD are semi-circumferences, if we take away the common

part CB, we shall have BD equal to AC. For a like reason, we have DN equal to CP, and BN equal to AP:
hence, the two triangles ACP, BND,
have their sides respectively equal:
they are therefore symmetrical;
consequently, they are equal in
area (P. XVI.). But the sum of
the triangles BDN, BCN, is equal
to the lune CBDNC, whose angle
is NCB: hence, the sum of ACP
and NCB is equal to the lune
whose angle is NCB; *which was to be proved.*

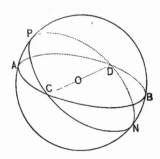

Scholium. It is evident that the two spherical pyramids,
which have the triangles ACP, NCB, for bases, are together
equal to the spherical wedge whose angle is NCB.

PROPOSITION XVIII. THEOREM.

*The area of a spherical triangle is equal to its spherical
excess multiplied by a tri-rectangular triangle.*

Let ABC be a spherical triangle on a sphere whose centre is Q: then is its surface equal to

$$(A + B + C - 2) \times T.$$

For, produce its sides till they
meet the great circle DEFG, drawn
at pleasure, without the triangle.
By the last theorem, the two triangles ADE, AGH, are together equal
to the lune whose angle is A; but
the area of this lune is equal to
$2A \times T$ (P. XV., C. 2): hence, the
sum of the triangles ADE and AGH,

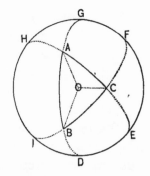

is equal to $2A \times T$. In like manner, it may be shown that the sum of the triangles BFG and BID is equal to $2B \times T$, and that the sum of the triangles CIH and CFE is equal to $2C \times T$.

But the sum of these six triangles exceeds the hemisphere, or four times T, by twice the triangle ABC. We therefore have,

$$2 \times area \text{ ABC} = 2A \times T + 2B \times T + 2C \times T - 4T;$$

or, by reducing and factoring,

$$area \text{ ABC} = (A + B + C - 2) \times T;$$

which was to be proved.

Scholium 1. The same relation which exists between the spherical triangle ABC, and the tri-rectangular triangle, exists also between the spherical pyramid which has ABC for its base, and the tri-rectangular pyramid. The triedral angle of the pyramid is to the triedral angle of the tri-rectangular pyramid, as the triangle ABC to the tri-rectangular triangle. From these relations, the following consequences are deduced:

1°. Triangular spherical pyramids are to each other as their bases; and since a polygonal pyramid may always be divided into triangular pyramids, it follows that any two spherical pyramids are to each other as their bases.

2°. Polyedral angles at the centre of the same, or of equal spheres, are to each other as the spherical polygons intercepted by their faces.

Scholium 2. A triedral angle whose faces are perpendicular to each other, is called a *right triedral angle;* and if the vertex is at the centre of a sphere, its faces intercept a tri-rectangular triangle. The right triedral angle is taken as the unit of polyedral angles, and the tri-rectangular spherical triangle is taken as its measure. If the vertex of a polyedral angle is taken as the centre of a sphere, the portion of the surface intercepted by its faces is the measure of the polyedral angle, a tri-rectangular triangle of the same sphere being the unit.

PROPOSITION XIX. THEOREM.

The area of a spherical polygon is equal to its spherical excess multiplied by the tri-rectangular triangle.

Let ABCDE be a spherical polygon on a sphere whose centre is O, the sum of whose angles is S, and the number of whose sides is n: then is its area equal to

$$(S - 2n + 4) \times T.$$

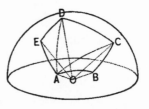

For, draw the diagonals AC, AD, dividing the polygon into spherical triangles: there are $n - 2$ such triangles. Now, the area of each triangle is equal to its spherical excess into the tri-rectangular triangle: hence, the sum of the areas of all the triangles, or the area of the polygon, is equal to the sum of all the angles of the triangles, or the sum of the angles of the polygon diminished by $2\,(n-2)$, into the tri-rectangular triangle; or,

$$area \ ABCDE = [S - 2\,(n - 2)] \times T;$$

whence, by reduction,

$$area \ ABCDE = (S - 2n + 4) \times T;$$

which was to be proved.

GENERAL SCHOLIUM 1.

From any point P on a hemisphere, two arcs of a great circle, PC and PD, can always be drawn, which shall be perpendicular to the circumference of the base of the hemi-
sphere, and they will in general
be unequal. Now, it may be
proved, by a course of reasoning
analogous to that employed in
Book I., Proposition XV.:

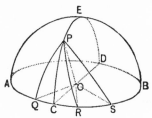

1°. That the shorter of the
two arcs, PC, is the shortest arc that can be drawn from
the given point to the circumference; and, therefore, that
the longer of the two, PED, is the longest arc that can
be drawn from the given point to the circumference:

2°. That two oblique arcs, PQ and PR, drawn from
the same point, to points of the circumference at equal
distances from the foot of the perpendicular, are equal:

3°. That of two oblique arcs, PR and PS, drawn from
the same point, that is the longer which meets the circumference at the greater distance from the foot of the
perpendicular.

GENERAL SCHOLIUM 2.

The arc of a great circle drawn perpendicular to an
arc of a second great circle of a sphere, passes through
the poles of the second arc (P. III., C. 3). The measure
of a spherical angle is the arc of a great circle included
between the sides of the angle and at the distance of a
quadrant from its vertex (P. IV.). It is evident, therefore,

that the pole of either side of an *acute* spherical angle lies *without* the sides of the angle; and that the pole of either side of an *obtuse* spherical angle lies *within* the sides of the angle.

Now, let A be an acute spherical angle, ST its measure, MN any
arc of a great circle, other than
ST, drawn perpendicular to the
side AQ, and included between the
two sides AQ and AR, and P the
pole of the side AQ: and

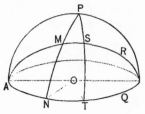

Let B be an obtuse spherical angle, CD its measure, EF any arc of a great circle, other than CD, drawn perpendicular to the side BH, and in-
cluded between the two sides BH
and BG, and P′ the pole of the
side BH: then

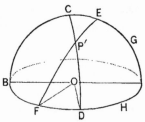

It may readily be shown (P.
III., C. 1, and Gen. S. I., 1°),

1°. That ST is longer than MN,
and, hence, is the *longest* arc of a great circle that can
be drawn perpendicular to the side AQ and included between the two sides AQ and AR: and

2°. That CD is shorter than EF, and, hence, is the
shortest arc of a great circle that can be drawn perpendicular to the side BH and included between the two sides
BH and BG.

EXERCISES.

1. The sides of a spherical triangle are 80°, 100°, and 110°; find the angles of its supplemental triangle, and the angles of each of its polar triangles.

2. Find the area of a tri-rectangular triangle, on a sphere whose diameter is 8 feet.

3. Find the area of a tri-rectangular triangle, on a sphere whose surface and volume may be expressed by the same number.

4. The angle of a lune, on a sphere whose radius is 5 feet, is 50°; find the area of the lune and the volume of the corresponding wedge.

5. The area of a lune is 33.5104 square feet and the angle of the lune is 60°; find the surface and the volume of the sphere.

6. Show that if two spherical triangles on unequal spheres are mutually equiangular, they are similar.

7. Show how to circumscribe a circle about a given spherical triangle.

8. Show how to inscribe a circle in a given spherical triangle.

9. Show that the intersection of the surfaces of two spheres is a circle, and that the line which joins the centres of two intersecting spheres is perpendicular to the circle in which their surfaces intersect.

10. Show that two spherical pyramids of the same or equal spheres, which have symmetrical triangles for bases, are equal in volume. [Proof analogous to that in P. XVI.]

11. The circumferences of two great circles intersect on the surface of a hemisphere whose diameter is 10 feet, and the acute angle formed by them is 40°; find the sum of the opposite triangles thus formed and the sum of the corresponding spherical pyramids.

12. Show that the volume of a triangular spherical pyramid is equal to its base multiplied by one third the radius of the sphere.

13. Show that the volume of any spherical pyramid is equal to its base multiplied by one third the radius of the sphere.

14. Find the volume of a spherical pyramid whose base is a tri-rectangular triangle, the diameter of the sphere being 8 feet.

15. The angles of a triangle, on a sphere whose radius is 9 feet, are 100°, 115°, and 120°; find the area of the triangle and the volume of the corresponding spherical pyramid.

16. A spherical pyramid, of a sphere whose diameter is 10 feet, has for its base a triangle of which the angles are 60°, 80°, and 85°; what is its ratio to a pyramid whose base is a tri-rectangular triangle of the same sphere?

17. The sum of the angles of a regular spherical octagon is 1140°, and the radius of the sphere is 12 feet; find the area of the octagon.

18. The volume of a spherical pyramid, whose base is an equiangular triangle, is 84.8232 cubic feet, and the radius of the sphere is 6 feet; find one of the angles of the base.

19. Given a spherical angle of 40°; what is the number of degrees in the longest arc of a great circle that can be drawn perpendicular to either side of the angle and included between the two sides?

20. Given a spherical angle of 115°; what is the number of degrees in the shortest arc of a great circle that can be drawn perpendicular to either side of the angle and included between the two sides?

APPENDIX.

ADDITIONAL DEFINITIONS.

1. The DISTANCE of a point from a line is measured on a perpendicular to that line.

2. The BISECTRIX of an angle is a line that divides the angle into two equal parts.

3. A MEDIAN is a line drawn from any vertex of a triangle to the middle of the opposite side.

4. The PROJECTION of a point, on a line, is the foot of a perpendicular drawn from the point to the line.

5. The PROJECTION of one straight line on another, is that part of the second line which is contained between the projections of the two extreme points of the first line, upon the second.

PROPOSITIONS.

I. THEOREM.—Show that the bisectrices of two adjacent angles are perpendicular to each other.

II. THEOREM.—Show that the perimeter of any triangle is greater than the sum of the distances from any point within the triangle to its three vertices, and less than twice that sum.

III. Theorem.—Show that the angle between the bisectrices of two consecutive angles of any quadrilateral, is equal to one half the sum of the other two angles.

IV. Theorem.—Show that any point in the bisectrix of an angle is equally distant from the sides of the angle.

V. Theorem.—If two sides of a triangle are prolonged beyond the third side, show that the bisectrices of this included angle and of the exterior angles all meet in the same point.

VI. Theorem.—Show that the projection of a line on a parallel line, is equal to the line itself; and that the projection of a line on a line to which it is oblique, is less than the line itself.

VII. Theorem.—If a line is drawn through the point of intersection of the diagonals of a parallelogram and limited by the sides of the parallelogram, show that the line is bisected at the point.

VIII. Theorem.—The bisectrices of the four angles of any parallelogram form, by their intersection, a rectangle whose diagonals are parallel to the sides of the given parallelogram.

IX. Theorem.—Show that the sum of the distances from any point in the base of an isosceles triangle to the two other sides, is equal to the distance from the vertex of either angle at the base to the opposite side.

X. Theorem.—Show that the middle point of the hypothenuse of any right-angled triangle is equally distant from the three vertices of the triangle.

XI. PROBLEM.—Draw two lines that shall divide a given right angle into three equal parts.

XII. THEOREM.—Draw a line AP through the vertex A of a triangle ABF and perpendicular to the bisectrix of the angle A; construct a triangle PBF, having its vertex P on AP, and its base coinciding with that of the given triangle: then show that the perimeter of PBF is greater than that of ABF.

XIII. THEOREM.—Let an altitude of the triangle ABC be drawn from the vertex A, and also the bisectrix of the angle A; then show that their included angle is equal to half the difference of the angles B and C.

XIV. PROBLEM.—Given two lines that would meet, if sufficiently prolonged: then draw the bisectrix of their included angle, without finding its vertex.

XV. PROBLEM.—From two points on the same side of a given line, to draw two lines that shall meet each other at some point of the given line, and make equal angles with that line.

XVI. THEOREM.—Show that the sum of the lines drawn to a point of a given line, from two given points, is the least possible when these lines are equally inclined to the given line.

XVII. PROBLEM.—From two given points, on the same side of a given line, draw two lines meeting on the given line and equal to each other.

XVIII. PROBLEM.—Through a given point A, draw a line that shall be equally distant from two given points, B and C.

XIX. PROBLEM.—Through a given point, draw a line cutting the sides of a given angle and making the interior angles equal to each other.

XX. PROBLEM.—Draw a line PQ parallel to the base BC of a triangle ABC, so that PQ shall be equal to the sum of BP and CQ.

XXI. PROBLEM.—In a given isosceles triangle, draw a line that shall cut off a trapezoid whose base is the base of the given triangle and whose three other sides shall be equal to each other.

XXII. THEOREM.—If two opposite sides of a parallelogram are bisected, and lines are drawn from the points of bisection to the vertices of the opposite angles, show that these lines divide the diagonal, which they intersect, into three equal parts.

XXIII. PROBLEM.—Construct a triangle, having given the two angles at the base and the sum of the three sides.

XXIV. PROBLEM.—Construct a triangle, having given one angle, one of its including sides, and the sum of the two other sides.

XXV. PROBLEM.—Construct an equilateral triangle, having given one of its altitudes.

XXVI. THEOREM.—Show that the three altitudes of a triangle all intersect in a common point.

XXVII. THEOREM.—If one of the acute angles of a right-angled triangle is double the other, show that the hypothenuse is double the smaller side about the right angle.

XXVIII. THEOREM.—Let a median be drawn from the vertex of any angle A of a triangle ABC: then show that the angle A is a right angle when the median is equal to half the side BC, an acute angle when the median is greater than half of BC, and an obtuse angle when the median is less than half of BC.

XXIX. THEOREM.—Let any quadrilateral be circumscribed about a circle: then show that the sum of two opposite sides is equal to the sum of the other two opposite sides.

XXX. PROBLEM.—Draw a straight line tangent to two given circles.

XXXI. PROBLEM.—Through a given point P, draw a circle that shall be tangent to a given line CB, at a given point B.

XXXII. THEOREM.—Let two circles intersect each other, and through either point of intersection let diameters of the circles be drawn: then show that the other extremities of these diameters and the other point of intersection lie in the same straight line.

XXXIII. PROBLEM.—Through two given points A and B, draw a circle that shall be tangent to a given line CP.

XXXIV. PROBLEM.—Draw a circle that shall be tangent to a given circle C, and also to a given line DP, at a given point P.

XXXV. PROBLEM.—Draw a circle that shall be tangent to a given line TP, and also to a given circle C, at a given point Q.

XXXVI. PROBLEM.—Draw a circle that shall pass through a given point Q, and be tangent to a given circle C, at a given point P.

XXXVII. PROBLEM.—Draw a circle, with a given radius, that shall be tangent to a given line DP, and to a given circle C.

XXXVIII. PROBLEM.—Find a point in the prolongation of any diameter of a given circle, such that a tangent from it to the circumference shall be equal to the diameter of the circle.

XXXIX. THEOREM.—Show that when two circles intersect each other, the longest common secant that can be drawn through either point of intersection, is parallel to the line joining the centres of the circles.

XL. PROBLEM.—Construct the greatest possible equilateral triangle whose sides shall pass through three given points A, B, and C, not in the same straight line.

XLI. THEOREM.—Show that the bisectrices of the four angles of any quadrilateral intersect in four points, all of which lie on the circumference of the same circle.

XLII. THEOREM.—If two circles touch each other externally, and if two common secants are drawn through the point of contact and terminating in the concave arcs, show that the lines joining the extremities of these secants, in the two circles, are parallel.

XLIII. THEOREM.—Let an equilateral triangle be inscribed in a circle, and let two of the subtended arcs be bisected by a chord of the circle: then show that the sides of the triangle divide the chord into three equal parts.

XLIV. PROBLEM.—Find a point, within a triangle, such that the angles formed by drawing lines from it to the three vertices of the triangle shall be equal to each other.

XLV. PROBLEM.—Inscribe a circle in a quadrant of a given circle.

XLVI. PROBLEM.—Through a given point P, within a given angle ABC, draw a circle that shall be tangent to both sides of that angle.

XLVII. THEOREM.—Show that the middle points of the sides of any quadrilateral are the vertices of an inscribed parallelogram.

XLVIII. PROBLEM.—Inscribe in a given triangle, a triangle whose sides shall be parallel to the sides of a second given triangle.

XLIX. PROBLEM.—Through a point P, within a given angle, draw a line such that it and the parts of the sides that are intercepted shall contain a given area.

L. PROBLEM.—Construct a parallelogram whose area and perimeter are respectively equal to the area and perimeter of a given triangle.

LI. PROBLEM.—Inscribe a square in a semicircle; that is, a square two of whose vertices are in the diameter, and the other two in the semi-circumference.

LII. PROBLEM.—Through a given point P draw a line cutting a triangle, so that the sum of the perpendiculars to it, from the two vertices on one side of the line, shall be equal to the perpendicular to it from the vertex, on the other side of the line.

LIII. THEOREM.—Show that the line which joins the middle points of two opposite sides of any quadrilateral, bisects the line joining the middle points of the two diagonals.

LIV. THEOREM.—If from the extremities of one of the oblique sides of a trapezoid, lines are drawn to the middle point of the opposite side, show that the triangle thus formed is equal to one half the given trapezoid.

LV. PROBLEM.—Find a point in the base of a triangle, such that the lines drawn from it, parallel to and limited by the other sides of the triangle, shall be equal to each other.

LVI. THEOREM.—Show that the line drawn from the middle of the base of any triangle to the middle of any line of the triangle parallel to the base, will pass through the opposite vertex, if sufficiently produced.

LVII. THEOREM.—Show that the three medians of any triangle meet in a common point.

LVIII. THEOREM.—On the sides AB and AC of any triangle ABC, construct any two parallelograms ABDE and ACFG; prolong the sides DE and FG till they meet in H; draw HA, and on the third side BC of the triangle, construct a parallelogram two of whose sides are parallel and equal to HA: then show that the parallelogram on BC is equal to the sum of the parallelograms on AB and AC.

LIX. THEOREM.—Assuming the principle demonstrated in the last proposition, deduce from it the truth that the square on the hypothenuse of a right-angled triangle is equal to the sum of the squares on the two other sides.

LX. THEOREM.—If from the middle of the base of a right-angled triangle, a line is drawn perpendicular to the hypothenuse dividing it into two segments, show that the difference of the squares of these segments is equal to the square of the other side about the right angle.

LXI. THEOREM.—If lines are drawn from any point P to the four vertices of a rectangle, show that the sum of the squares of the two lines drawn to the extremities of one diagonal, is equal to the sum of the squares of the two lines drawn to the extremities of the other diagonal.

LXII. THEOREM.—Let a line be drawn from the centre of a circle to any point of any chord; then show that the square of this line, plus the rectangle of the segments of the chord, is equal to the square of the radius.

LXIII. PROBLEM.—Draw a line from the vertex of any scalene triangle to a point in the base, such that this line shall be a mean proportional between the segments into which it divides the base.

LXIV. THEOREM.—Show that the sum of the squares of the diagonals of any quadrilateral is equal to the sum of the squares of the four sides of the quadrilateral, diminished by four times the square of the distance between the middle points of the diagonals.

LXV. PROBLEM.—Construct an equilateral triangle equal in area to any given isosceles triangle.

LXVI. THEOREM.—In a triangle ABC, let two lines be drawn from the extremities of the base BC, intersecting at any point P on the median through A, and meeting the opposite sides in the points E and D: show that DE is parallel to BC.

APPLICATION OF ALGEBRA TO GEOMETRY.

To solve a geometrical problem by means of algebra, draw a figure which shall contain all the given and required parts and also such other lines as may be necessary to establish the relations between them; then denote the given parts by leading letters, and the required parts by final letters of the alphabet: next consider the relations between the given and required parts and express these relations by equations, taking care to have as many independent equations as there are parts to be determined (Bourdon, Art. 92). The solution of these equations will give the values of the required parts.

To indicate the method of proceeding, the solution of the first problem is given.

LXVII. PROBLEM.—In a right-angled triangle ABC, given the base BA and the sum of the hypothenuse and the perpendicular, to find the hypothenuse and the perpendicular.

Solution. Denote BA by c, BC by x, AC by y, and the sum of BC and AC by s.

Then, $x + y = s.$ · · · · · · · · (1.)

From B. IV., P. XI., $x^2 = y^2 + c^2.$ · · · · · · · (2.)

From (1), we have, $x = s - y.$

Squaring, $x^2 = s^2 - 2sy + y^2.$ · · · · (3.)

Subtracting (2) from (3), $\quad 0 = s^2 - 2sy - c^2$.

Transposing and dividing, $\quad y = \dfrac{s^2 - c^2}{2s}$,

whence, $\qquad\qquad\qquad x = s - \dfrac{s^2 - c^2}{2s} = \dfrac{s^2 + c^2}{2s}$.

If $c = 3$ and $s = 9$, we have $x = 5$ and $y = 4$.

LXVIII. PROBLEM.—In a right-angled triangle, given the hypothenuse and the sum of the sides about the right angle, to find these sides.

LXIX. PROBLEM.—In a rectangle, given the diagonal and the perpendicular, to find the sides.

LXX. PROBLEM.—Given the base and perpendicular of a triangle, to find the side of an inscribed square.

LXXI. PROBLEM.—In an equilateral triangle, given the distances from a point within the triangle to each of the three sides, to find one of the equal sides.

LXXII. PROBLEM.—In a right-angled triangle, given the base and the difference between the hypothenuse and the perpendicular, to find the sides.

LXXIII. PROBLEM.—In a right-angled triangle, given the hypothenuse and the difference between the base and the perpendicular, to determine the triangle.

LXXIV. PROBLEM.—Having given the area of a rectangle inscribed in a given triangle, to determine the sides of the rectangle.

LXXV. PROBLEM.—In a triangle, having given the ratio of the two sides together with both segments of the base made by a perpendicular from the vertex, to determine the triangle.

LXXVI. PROBLEM.—In a triangle, having given the base, the sum of the two other sides, and the length of a line drawn from the vertex to the middle of the base; to find the sides of the triangle.

LXXVII. PROBLEM.—In a triangle, having given the two sides about the vertical angle, together with the line bisecting that angle and terminating in the base; to find the base.

LXXVIII. PROBLEM.—To determine a right-angled triangle, having given the lengths of two lines drawn from the vertices of the acute angles to the middle points of the opposite sides.

LXXIX. PROBLEM.—To determine a right-angled triangle, having given the perimeter and the radius of the inscribed circle.

LXXX. PROBLEM.—To determine a triangle, having given the base, the perpendicular, and the ratio of the two sides.

LXXXI. PROBLEM.—To determine a right-angled triangle, having given the hypothenuse and the side of the inscribed square.

LXXXII. PROBLEM.—To determine the radii of three equal circles, described within and tangent to a given circle, and also tangent to each other.

LXXXIII. PROBLEM.—In a right-angled triangle, having given the perimeter and the perpendicular let fall from the right angle on the hypothenuse, to determine the triangle.

LXXXIV. PROBLEM.—To determine a right-angled triangle, having given the hypothenuse and the difference of two lines drawn from the two acute angles to the centre of the inscribed circle.

LXXXV. PROBLEM.—To determine a triangle, having given the base, the perpendicular, and the difference of the two other sides.

LXXXVI. PROBLEM.—To determine a triangle, having given the base, the perpendicular, and the rectangle of the two sides.

LXXXVII. PROBLEM.—To determine a triangle, having given the lengths of three lines drawn from the three angles to the middle of the opposite sides.

LXXXVIII. PROBLEM.—In a triangle, having given the three sides, to find the radius of the inscribed circle.

LXXXIX. PROBLEM.—To determine a right-angled triangle, having given the side of the inscribed square and the radius of the inscribed circle.

XC. PROBLEM.—To determine a right-angled triangle, having given the hypothenuse and the radius of the inscribed circle.

TRIGONOMETRY

AND

MENSURATION.

INTRODUCTION TO TRIGONOMETRY.

LOGARITHMS.

1. The LOGARITHM of a given number is the *exponent* of the power to which it is necessary to raise a *fixed numbe.* to produce the given number.

The *fixed number* is called THE BASE OF THE SYSTEM. Any positive number, except 1, may be taken as the base of a system. In the common system, to which alone reference is here made, the base is 10. Every number is, therefore, regarded as some power of 10, and the *exponent* of that power is the *logarithm* of the number.

2. If we denote any positive number by n, and the corresponding exponent of 10 by x, we shall have the exponential equation,

$$10^x = n. \quad \cdots \quad \cdots \quad (1.)$$

In this equation, x is, by definition, the logarithm of n, which may be expressed thus,

$$x = \log n. \quad \cdots \quad \cdots \quad (2.)$$

3. If a number is an exact power of 10, its logarithm is a *whole number.* Thus, 100, being equal to 10^2, has for its logarithm 2. If a number is not an exact power of 10, its logarithm is composed of two parts, a *whole number* called the CHARACTERISTIC, and a *decimal* part called the MANTISSA. Thus, 225 being greater than 10^2 and less than 10^3, its logarithm is found to be 2.352183, of which 2 is the *characteristic* and .352183 is the *man tissa.*

4. If, in the equation,

$$\log (10)^p = p, \quad \cdots \quad \cdots \quad (3.)$$

we make p successively equal to 0, 1, 2, 3, &c., and also equal to -0, -1, -2, -3, &c., we may form the following

TABLE.

log 1 = 0	
log 10 = 1	log .1 = -1
log 100 = 2	log .01 = -2
log 1000 = 3	log .001 = -3
&c., &c.	&c., &c.

If a number lies between 1 and 10, its logarithm lies between 0 and 1, that is, it is equal to 0 *plus* a decimal; if a number lies between 10 and 100, its logarithm is equal to 1 *plus* a decimal; if between 100 and 1000, its logarithm is equal to 2 *plus* a decimal; and so on; hence, we have the following

RULE.—*The characteristic of the logarithm of an entire number is positive, and numerically 1 less than the number of places of figures in the given number.*

If a decimal fraction lies between .1 and 1, its logarithm lies between -1 and 0, that is, it is equal to -1 *plus* a decimal; if a number lies between .01 and .1, its logarithm is equal to -2 *plus* a decimal; if between .001 and .01, its logarithm is equal to -3 *plus* a decimal; and so on: hence, the following

RULE.—*The characteristic of the logarithm of a decimal fraction is negative, and numerically 1 greater than the number of 0's that immediately follow the decimal point.*

The characteristic alone is negative, *the mantissa being always positive.* This fact is indicated by writing the negative sign over the characteristic: thus, $\overline{2}.371465$, is equivalent to $-2 + .371465$.

Note.—It is to be observed, that the characteristic of the logarithm of a mixed number is the same as that of its entire part. Thus, the characteristic of the logarithm of 725.4275 is the same as the characteristic of the logarithm of 725.

GENERAL PRINCIPLES.

5. Let m and n denote any two numbers, and x and y their logarithms. We shall have, from the definition of a logarithm, the following equations,

$$10^x = m. \quad \cdots \cdots \cdots \quad (4.)$$

$$10^y = n. \quad \cdots \cdots \cdots \quad (5.)$$

Multiplying (4) and (5), member by member, we have

$$10^{x+y} = mn;$$

whence, by the definition,

$$x + y = \log (mn). \quad \cdots \cdots \quad (6.)$$

That is, *the logarithm of the product of two numbers is equal to the sum of the logarithms of the numbers.*

6. Dividing (4) by (5), member by member, we have

$$10^{x-y} = \frac{m}{n};$$

whence, by the definition,

$$x - y = \log \left(\frac{m}{n}\right). \quad \cdots \cdots \quad (7.)$$

That is, *the logarithm of a quotient is equal to the logarithm of the dividend diminished by that of the divisor.*

7. Raising both members of (4) to the power denoted by p, we have,

$$10^{xp} = m^p;$$

whence, by the definition,

$$xp = \log m^p. \quad \cdot \quad \cdot \quad \cdot \quad \cdot \quad \cdot \quad (8.)$$

That is, *the logarithm of any power of a number is equal to the logarithm of the number multiplied by the exponent of the power.*

8. Extracting the root, indicated by r, of both members of (4), we have

$$10^{\frac{x}{r}} = \sqrt[r]{m};$$

whence, by the definition,

$$\frac{x}{r} = \log \sqrt[r]{m}. \quad \cdot \quad \cdot \quad \cdot \quad \cdot \quad \cdot \quad (9.)$$

That is, *the logarithm of any root of a number is equal to the logarithm of the number divided by the index of the root.*

The preceding principles enable us to abbreviate the operations of multiplication and division, by converting them into the simpler ones of addition and subtraction.

TABLE OF LOGARITHMS.

9. A TABLE OF LOGARITHMS is a table containing a set of numbers and their logarithms, so arranged that, having given any one of the numbers, we can find its logarithm; or, having the logarithm, we can find the corresponding number.

In the table appended, the complete logarithm is given for all numbers from 1 up to 100. For other numbers, the mantissas alone are given; the characteristic may be found by one of the rules of Art. 4.

Before explaining the use of the table, it is to be shown that the *mantissa* of the logarithm of any number is not changed by multiplying or dividing the number by any *exact* power of 10.

Let n represent any number whatever, and 10^p any power of 10, p being any whole number, either positive or negative. Then, in accordance with the principles of Arts. 5 and 3, we shall have

$$\log (n \times 10^p) = \log n + \log 10^p = p + \log n;$$

but p is, by hypothesis, a whole number: hence, the *decimal* part of the $\log (n \times 10^p)$ is the same as that of $\log n$; *which was to be proved.*

Hence, in finding the mantissa of the logarithm of a number, the position of the decimal point may be changed at pleasure. Thus, the mantissa of the logarithm of 456357, is the same as that of the number 4563.57; and the mantissa of the logarithm of 759 is the same as that of 7590.

MANNER OF USING THE TABLE.

1°. *To find the logarithm of a number less than* 100.

10. Look on the first page, in the column headed "N," for the given number; the number opposite is the logarithm required. Thus,

$$\log 67 = 1.826075.$$

2°. *To find the logarithm of a number between* 100 *and* 10,000.

11. Find the characteristic by the first rule of Art. 4.

To determine the mantissa, find in the column headed "N" the left-hand three figures of the given number; then pass along the horizontal line in which these figures are found, to the column headed by the fourth figure of the given number, and take out the four figures found there; pass back again to the column headed "0," and there will be found in this column, either upon the horizontal line of the first three figures or a few lines above it, a number consisting of six figures, the left-hand two figures of which must be prefixed to the four already taken out. Thus,

$$\log 8979 = 3.953228.$$

If, however, any *dots* are found at the place of the four figures first taken out, or if in returning to the "0" column any dots are passed, the two figures to be prefixed are the left-hand two of the six figures of the "0" column *immediately below*. Dots in the number taken out must be replaced by zeros. Thus,

$$\log 3098 = 3.491081,$$
$$\log 2188 = 3.340047.$$

NOTE.—The above method of finding the mantissa assumes that the given number has *four* places of figures. If, therefore, the number lies between 100 and 1000, and has but *three* places of figures, find the characteristic by the first rule of Art. 4, and *then*, to find the mantissa, fill out the given number to *four* places of figures (or conceive it to be so filled out) by annexing 0 (see Art. 9), and find the mantissa corresponding to the resulting number, as above.

3°. *To find the logarithm of a number greater than* 10,000.

12. Find the characteristic by the first rule of Art. 4.

To find the mantissa: set aside all of the given number except the left-hand four figures, and find the mantissa corresponding to these four, as in Art. 11; multiply the corresponding *tabular difference*, found in column "D," by the part of the number set aside, and discard as many of the right-hand figures of the product as there are figures in the multiplier, and add the result thus obtained to the mantissa already found. If the left-hand figure of those discarded is 5 or more, increase the number added by 1.

Note.—It is to be observed that the *tabular difference*, found in column "D," is *millionths*, and not a whole number; and that, therefore, the result to be added "to the mantissa already found" is *millionths*.

Example.—To find the logarithm of 672887: the characteristic is 5; set aside 87, and the mantissa corresponding to 6728 is .827886; the corresponding tabular difference is 65, which multiplied by 87, the part of the number set aside, gives 5655; as there are two figures in the multiplier, discard the right-hand two figures of this product, leaving 56; but as the left-hand figure of those discarded is 5, call the result 57 (which is *millionths*); adding this 57 to the mantissa already found, will give .827943 for the required mantissa; hence,

$$\log 672887 = 5.827943.$$

The explanation of the method just given is briefly this: for the purpose of finding the mantissa, the given number is conceived to be a *mixed* one, thus, 6728.87, the mantissa not being affected by the position of the decimal point (see Art. 9). The numbers in the column

"D" are the differences between the logarithms of two consecutive whole numbers. In the example just given, the mantissa of the logarithm of 6728 is .827886, and that of 6729 is .827951, and their difference is 65 millionths; 87 hundredths of this difference is 57 millionths; hence, the mantissa of the logarithm of 6728.87 is found by adding 57 millionths to .827886. The principle employed is, that the differences of numbers are proportional to the differences of their logarithms, when these differences are small.

4°. *To find the logarithm of a decimal.*

13. Find the characteristic by the second rule of Art. 4.

To find the mantissa, drop the decimal point, and consider the decimal a whole number. Find the mantissa of the logarithm of this number as in preceding articles, and it will be the mantissa required. Thus,

$$\log .0327 = \overline{2}.514548,$$

$$\log .378024 = \overline{1}.577520.$$

NOTE.—To find the logarithm of a *mixed number*, find the characteristic by the Note, Art. 4; then drop the decimal point and proceed as above.

5°. *To find the number corresponding to a given logarithm.*

14. The rule is the reverse of those just given. Look in the table for the mantissa of the given logarithm. If it can not be found, take out the next less mantissa, and also the corresponding number, which set aside. Find the difference between the mantissa taken out and that of the given logarithm; annex any number of 0's, and divide this result by the corresponding number in the column "D." Annex the quotient to the number set aside, and then, if the characteristic is *positive*, point off, from the left hand, a number of places of figures equal to the characteristic plus 1; the result will be the number required.

If the characteristic is *negative*, prefix to the figures obtained a number of 0's one less than the number of units in the negative characteristic and to the whole prefix a decimal point; the result, a pure decimal, will be the number required.

Examples.

1. Let it be required to find the number corresponding to the logarithm 5.233568.

The next less mantissa in the table is 233504; the corresponding number is 1712, and the tabular difference is 253.

Operation.

Given mantissa, · · · · · 233568
Next less mantissa, · · · · 233504 · · 1712
253) 6400000 (25296

∴ The required number is 171225.296.

The number corresponding to the logarithm $\overline{2}$.233568 is .0171225.

2. What is the number corresponding to the logarithm $\overline{2}$.785407?   *Ans.* .06101084.

3. What is the number corresponding to the logarithm $\overline{1}$.846741?   *Ans.* .702653.

MULTIPLICATION BY MEANS OF LOGARITHMS.

15. From the principle proved in Art. 5, we deduce the following

RULE.—*Find the logarithms of the factors, and take their*

sum; then find the number corresponding to the resulting logarithm, and it will be the product required.

Examples.

1. Multiply 23.14 by 5.062.

Operation.

log 23.14 · · · 1.364363
log 5.062 · · · 0.704322
2.068685   ∴ 117.1347, product.

2. Find the continued product of 3.902, 597.16, and 0.0314728.

Operation.

log   3.902 · · · 0.591287
log   597.16 · · · 2.776091
log 0.0314728 · · · $\overline{2}$.497936
1.865314   ∴ 73.3354, product.

Here, the $\overline{2}$ cancels the $+2$, and the 1 carried from the decimal part is set down.

3. **Find the continued product of 3.586, 2.1046, 0.8372, and 0.0294.**   *Ans.* 0.1857615.

DIVISION BY MEANS OF LOGARITHMS.

16. From the principle proved in Art. 6, we have the following

RULE.—*Find the logarithms of the dividend and divisor, and subtract the latter from the former; then find the number corresponding to the resulting logarithm, and it will be the quotient required.*

Examples.

1. Divide 24163 by 4567.

Operation.

log 24163 · · · 4.383151
log 4567 · · · 3.659631
0.723520   ∴ 5.29078, quotient.

2. Divide 0.7438 by 12.9476.

Operation.

log 0.7438 · · · $\bar{1}$.871456
log 12.9476 · · · 1.112189
$\bar{2}$.759267   ∴ 0.057447, quotient.

Here, 1 taken from $\bar{1}$, gives $\bar{2}$ for a result. The subtraction, as in this case, is always to be performed in the algebraic sense.

3. Divide 37.149 by 523.76.   *Ans.* 0.0709274.

The operation of division, particularly when combined with that of multiplication, can often be simplified by using the principle of

THE ARITHMETICAL COMPLEMENT.

17. The ARITHMETICAL COMPLEMENT of a logarithm is the result obtained by subtracting it from 10. Thus, 8.130456 is the arithmetical complement of 1.869544. The arithmetical complement of a logarithm may be written out *by commencing at the left hand and subtracting each figure from 9, until the last significant figure is reached, which must be taken from* 10. The arithmetical complement is denoted by the symbol (a. c.)

Let a and b represent any two logarithms whatever, and $a - b$ their difference. Since we may add 10 to, and subtract it from, $a - b$, without altering its value, we have,

$$a - b = a + (10 - b) - 10. \quad \cdot \quad \cdot \quad \cdot \quad (10.)$$

But $10 - b$ is, by definition, the arithmetical complement of b: hence, Equation (10) shows that the difference between two logarithms is equal to *the first,* plus *the arithmetical complement of the second,* minus 10.

Hence, to divide one number by another by means of the arithmetical complement, we have the following

RULE.—*Find the logarithm of the dividend, and the arithmetical complement of the logarithm of the divisor, add them together, and diminish the sum by* 10; *the number corresponding to the resulting logarithm will be the quotient required.*

Examples.

1. Divide 327.5 by 22.07

Operation.

log 327.5 · · ·	2.515211
(a. c.) log 22.07 · · ·	8.656198
	1.171409

∴ 14.839, quotient.

The operation of subtracting 10 is performed mentally.

2. Divide 37.149 by 523.76.   *Ans.* 0.0709273.

3. Divide the product of 358884 and 5672, by the product of 89721 and 42.056.

log 358884 · · ·	5.554954
log 5672 · · · ·	3.753736
(a. c.) log 89721 · · · ·	5.047106
(a. c.) log 42.056 · · ·	8.376182
	2.731978

∴ 539.48, result.

20 is here subtracted, as (a. c.) has been twice used.

4. Solve the proportion,

$$3976 \; : \; 7952 \; :: \; 5903 \; : \; x.$$

Applying logarithms, the logarithm of the 4th term is equal to the sum of the logarithms of the 2d and 3d terms, minus the logarithm of the 1st: Or, *the arithmetical complement of the logarithm of the 1st term, plus the logarithm of the 2d term, plus the logarithm of the 3d term, minus* 10, *is equal to the logarithm of the 4th term.*

Operation.

(a. c.) log 3976 · · ·	6.400554	
log 7952 · · ·	3.900476	
log 5903 · · ·	3.771073	
log x · · ·	4.072103	∴ x = 11806.

RAISING TO POWERS BY MEANS OF LOGARITHMS.

18. From Article 7, we have the following

RULE.—*Find the logarithm of the number, and multiply it by the exponent of the power; then find the number corresponding to the resulting logarithm, and it will be the power required.*

Examples.

1. Find the 5th power of 9.

Operation.

log 9 · · · 0.954243
            5
         ———————
         4.771215   ∴ 59049, power

2. Find the 7th power of 8.   *Ans.* 2097154, nearly.

EXTRACTING ROOTS BY MEANS OF LOGARITHMS.

19. From the principle proved in Art. 8, we have the following

RULE.—*Find the logarithm of the number, and divide it by the index of the root; then find the number corresponding to the resulting logarithm, and it will be the root required.*

Examples.

1. Find the cube root of 4096.

The logarithm of 4096 is 3.612360, and one third of this is 1.204120. The corresponding number is 16, which is the root sought.

If the characteristic of the logarithm of the given number is *negative* and not *exactly* divisible by the index of the root, add to it such *negative* quantity as shall make it exactly divisible, and add also to the mantissa a numerically equal *positive* quantity.

2. Find the 4th root of .00000081.

The logarithm of .00000081 is $\overline{7}.908485$, which is equal to $\overline{8} + 1.908485$, and one fourth of this is $\overline{2}.477121$.

The number corresponding to this logarithm is .03 ; hence, .03 is the root required.

PLANE TRIGONOMETRY.

20. PLANE TRIGONOMETRY is that branch of Mathematics which treats of the *solution* of plane triangles.

In every plane triangle there are six parts: *three sides* and *three angles*. When three of these parts are given, one being a side, the remaining parts may be found by computation. The operation of finding the unknown parts is called the *solution* of the triangle.

21. A plane angle is measured by the arc of a circle included between its sides, the centre of the circle being at the vertex, and its radius being equal to 1.

Thus, if the vertex A is taken as a centre, and the radius AB is equal to 1, the intercepted arc BC measures the angle A (B. III., P. XVII., S.).

Let ABCD represent a circle whose radius is equal to 1, and AC, BD, two diameters perpendicular to each other. These diameters
divide the circumference into four equal
parts, called *quadrants;* and because
each of the angles at the centre is a
right angle, it follows that a *right
angle* is measured by *a quadrant.* An

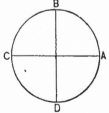

acute angle is measured by an arc *less* than a quadrant, and an *obtuse angle*, by an arc *greater* than a quadrant.

22. In Geometry, the unit of angular measure is a *right angle;* so in Trigonometry, the primary unit is a *quadrant,* which is the measure of a right angle.

For convenience, the quadrant is divided into 90 equal parts, each of which is called a degree; each degree into 60 equal parts, called minutes; and each minute into 60 equal parts, called seconds. Degrees, minutes, and seconds, are denoted by the symbols °, ', ". Thus, the expression 7° 22' 33", is read, 7 degrees, 22 minutes, and 33 seconds. Fractional parts of a second are expressed decimally.

A quadrant contains 324,000 seconds, and an arc of 7° 22' 33" contains 26553 seconds; hence, the angle measured by the latter arc is the $\frac{26553}{324000}$ part of a right angle. In like manner, any angle may be expressed in terms of a right angle.

23. The *complement of an arc* is the difference between that arc and 90°. The *complement of an angle* is the difference between that angle and a right angle.

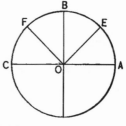

Thus, EB is the complement of AE, and FB is the complement of CF. In like manner, the angle EOB is the complement of the angle AOE, and FOB is the complement of COF

In a right-angled triangle, the acute angles are complements of each other.

24. The *supplement of an arc* is the difference between that arc and 180°. The *supplement of an angle* is the difference between that angle and two right angles.

Thus, EC is the supplement of AE, and FC the supplement of AF. In like manner, the angle EOC is the supplement of the angle AOE, and FOC the supplement of AOF.

In any plane triangle, any angle is the supplement of the sum of the two others.

25. Instead of the arcs themselves, certain *functions* of the arcs, as explained below, are used. A *function* of a quantity is something which depends upon that quantity for its value.

The following functions are the only ones needed for solving triangles:

26. The *sine* of an arc is the distance of one extremity of the arc from the diameter through the other extremity.

Thus, PM is the sine of AM, and P'M' is the sine of AM'.

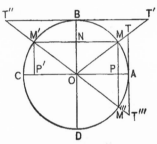

If AM is equal to M'C, AM and AM' are supplements of each other; and because MM' is parallel to AC, PM is equal to P'M' (B. I., P. XXIII.): hence, *the sine of an arc is equal to the sine of its supplement.*

27. The *cosine* of an arc is the sine of the complement of the arc, "complement sine" being contracted into cosine.

Thus, NM is the cosine of AM, and NM' is the cosine of AM'. These lines are respectively equal to OP and OP'.

It is evident, from the equal triangles ONM and ONM', that NM is equal to NM'; hence, *the cosine of an arc is equal to the cosine of its supplement.*

28. The *tangent* of an arc is the perpendicular to the radius at one extremity of the arc, limited by the prolongation of the diameter drawn to the other extremity.

Thus, AT is the tangent of
the arc AM, and AT''' is the tangent of the arc AM'.

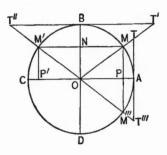

If AM is equal to M'C, AM
and AM' are supplements of each
other. But AM''' and AM' are
also supplements of each other:
hence, the arc AM is equal to
the arc AM''', and the corresponding angles, AOM and AOM''', are
also equal. The right-angled triangles AOT and AOT''' have
a common base AO, and the angles at the base equal;
consequently, the remaining parts are respectively equal:
hence, AT is equal to AT'''. But AT is the tangent of AM,
and AT''' is the tangent of AM': hence, *the tangent of an
arc is equal to the tangent of its supplement.*

29. The *cotangent* of an arc is the tangent of its
complement, "complement tangent" being contracted into
cotangent.

Thus, BT' is the cotangent of the arc AM, and BT'' is
the cotangent of the arc AM'.

It is evident, from the equal triangles OBT' and OBT'',
that BT' is equal to BT''; hence, *the cotangent of an arc
is equal to the cotangent of its supplement.*

When it is stated that the cotangent, tangent, &c., of
an arc are equal respectively to the cotangent, tangent,
&c., of its supplement, the *numerical values* only of the
functions are referred to; no account being taken of the
algebraic signs ascribed to the several functions in the
different quadrants, as will be explained hereafter.

The sine, cosine, tangent, and cotangent of an arc, a,
are, for convenience, written sin a, cos a, tan a, and cot a,

These functions of an arc have been defined on the supposition that the radius of the arc is equal to 1; in this case, they may also be considered as functions of the *angle* which the arc measures.

Thus, PM, NM, AT, and BT', are respectively the sine, cosine, tangent, and cotangent of the *angle* AOM, as well as of the arc AM.

30. It is often convenient to use some other radius than 1; in such case, the functions of the arc to the radius 1, may be reduced to corresponding functions, to the radius R, R denoting *any* radius.

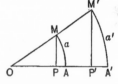

Let AOM represent any angle, AM an arc described from O as a centre with the radius 1, PM its sine; A'M' an arc described from O as a centre, with any radius R, and P'M' its sine.

Then, because OPM and OP'M' are similar triangles, we shall have,

$$\text{OM} \ : \ \text{PM} \ :: \ \text{OM}' \ : \ \text{P}'\text{M}',$$

or,

$$1 \ : \ \text{PM} \ :: \ \text{R} \ : \ \text{P}'\text{M}';$$

whence,

$$\text{PM} \ = \ \frac{\text{P}'\text{M}'}{\text{R}},$$

and

$$\text{P}'\text{M}' \ = \ \text{PM} \times \text{R};$$

and similarly for each of the other functions: hence,

Any function of an arc whose radius is 1, is equal to the corresponding function of an arc whose radius is R divided by that radius. Also, *any function of an arc whose radius is R, is equal to the corresponding function of an arc whose radius is 1 multiplied by the radius R.*

By means of this principle, formulas may be rendered homogeneous in terms of any radius,

TABLE OF NATURAL SINES.

31. A NATURAL SINE, COSINE, TANGENT, or COTANGENT, is the sine, cosine, tangent, or cotangent of an arc whose radius is 1.

A TABLE OF NATURAL SINES, COSINES, &c., is a table by means of which the natural sine, cosine, tangent, or cotangent of any arc, or angle, may be found.

Such a table might be used for all the purposes of trigonometrical computation, but it is usually found more convenient to employ a table of logarithmic sines, as explained in the next article.

TABLE OF LOGARITHMIC SINES.

32. A LOGARITHMIC SINE, COSINE, TANGENT, or COTANGENT is the logarithm of the sine, cosine, tangent, or cotangent of an arc whose radius is 10,000,000,000. This value of the radius is taken simply for convenience in making the table, its logarithm being 10.

A TABLE OF LOGARITHMIC SINES is a table from which the logarithmic sine, cosine, tangent, or cotangent of any arc, or angle, may be found.

Any *logarithmic* function of an arc, or angle, may be found by multiplying the corresponding *natural* function by 10,000,000,000 (Art. 30), and then taking the logarithm of the result; or more simply, by taking the logarithm of the corresponding *natural* function, and then adding 10 to the result (Art. 5).

33. In the table appended, the logarithmic functions are given for every *minute* from 0° up to 90°. In addition, their rates of change for each *second* are given in the column headed "D."

The method of computing the numbers in the column headed "D," will be understood from a single example. The logarithmic sines of 27° 34', and of 27° 35', are, respectively, 9.665375 and 9.665617. The difference between their mantissas is 242 millionths; this, divided by 60, the number of seconds in one minute, gives 4.03 millionths, which is the change in the mantissa for 1", between the limits 27° 34' and 27° 35'.

For the sine and cosine, there are separate columns of differences, which are written to the right of the respective columns; but for the tangent and cotangent there is but a single column of differences, which is written between them. The logarithm of the tangent increases just as fast as that of the cotangent decreases, and the reverse, their sum being always equal to 20. The reason of this is, that the product of the tangent and cotangent is always equal to the square of the radius; hence, the sum of their logarithms must always be equal to twice the logarithm of the radius, or 20.

The arc, or angle, obtained by taking the degrees from the *top* of the page and the minutes from the *left*-hand column, is the complement of that obtained by taking the degrees from the *bottom* of the page, and the minutes from the *right*-hand column on the same horizontal line. But, by definition, the cosine and the cotangent of an arc, or angle, are, respectively, the sine and the tangent of the complement of that arc, or angle (Arts. 26 and 28): hence, the columns designated *sine* and *tang* at the *top* of the page, are designated *cosine* and *cotang* at the *bottom*.

USE OF THE TABLE.

To find the logarithmic functions of an arc, or angle, which is expressed in degrees and minutes.

34. If the arc, or angle, is less than 45°, look for the degrees at the *top* of the page, and for the minutes in the *left*-hand column; then follow the corresponding horizontal line till you come to the column designated at the *top* by *sine, cosine, tang,* or *cotang,* as the case may be; the number there found is the logarithm required. Thus,

$$\log \sin 19° 55' \quad \cdot \quad \cdot \quad \cdot \quad 9.532312$$
$$\log \tan 19° 55' \quad \cdot \quad \cdot \quad \cdot \quad 9.559097$$

If the arc, or angle, is 45° or more, look for the degrees at the *bottom* of the page, and for the minutes in the *right*-hand column; then follow the corresponding horizontal line backward till you come to the column designated at the *bottom* by *sine, cosine, tang,* or *cotang,* as the case may be; the number there found is the logarithm required. Thus,

$$\log \cos 52° 18' \quad \cdot \quad \cdot \quad \cdot \quad 9.786416$$
$$\log \tan 52° 18' \quad \cdot \quad \cdot \quad \cdot \quad 10.111884$$

To find the logarithmic functions of an arc or angle which is expressed in degrees, minutes, and seconds.

35. Find the logarithm corresponding to the degrees and minutes as before; then multiply the corresponding number taken from the column headed "D," which is *millionths,* by the number of seconds, and add the product to the preceding result for the sine or tangent, and subtract it therefrom for the cosine or cotangent.

Examples.

1. Find the logarithmic sine of 40° 26′ 28″.

Operation.

log sin 40° 26′ · · · · · · · · · · · · · 9.811952
Tabular difference   2.47
No. of seconds   28
Product   · · ·   69.16   to be added   · ·   69
log sin 40° 26′ 28″ · · · · · · · · · · 9.812021

The same rule is followed for decimal parts, as in Art. 12.

2. Find the logarithmic cosine of 53° 40′ 40″.

Operation.

log cos 53° 40′ · · · · · · · · · · · · · 9.772675
Tabular difference   2.86
No. of seconds   40
Product   · · ·   114.40   to be subtracted   114
log cos 53° 40′ 40″ · · · · · · · · · · 9.772561

If the arc or angle is greater than 90°, find the required function of its supplement (Arts. 26 and 28).

3. Find the logarithmic tangent of 118° 18′ 25″.

Operation.

                                    180°
Given arc · · · · · · ·   118° 18′ 25″
Supplement · · · · · ·   61° 41′ 35″
log tan 61° 41′ · · · · · · · · · · · · 10.268556
Tabular difference   5.04
No. of seconds   35
Product   · · ·   176.40   to be added   176
log tan 118° 18′ 25″ · · · · · · · · · 10.268732

4. Find the logarithmic sine of 32° 18′ 35″.

Ans. 9.727945.

5. Find the logarithmic cosine of 95° 18′ 24″.

Ans. 8.966080.

6. Find the logarithmic cotangent of 125° 23′ 50″.

Ans. 9.851619.

To find the arc or angle corresponding to any logarithmic function.

36. This is done by reversing the preceding rule :

Look in the proper column of the table for the given logarithm ; if it is found there, the degrees are to be taken from the top or bottom, and the minutes from the left or right hand column, as the case may be. If the given logarithm is not found in the table, then find the next less logarithm, and take from the table the corresponding degrees and minutes, and set them aside. Subtract the logarithm found in the table from the given logarithm, and divide the remainder by the corresponding tabular difference. The quotient will be seconds, which must be *added* to the degrees and minutes set aside in the case of a sine or tangent, and *subtracted* in the case of a cosine or a cotangent.

Examples.

1. Find the arc or angle corresponding to the logarithmic sine 9.422248.

Operation.

Given logarithm · · · 9.422248
Next less in table · · · 9.421857 · · · 15° 19′
Tabular difference   7.68)   391.00 (51″, to be added.

Hence, the required arc is 15° 19′ 51″.

2. Find the arc or angle corresponding to the logarithmic cosine 9.427485.

Operation.

Given logarithm · · · 9.427485
Next less in table· · · 9.427354 · · · 74° 29'
Tabular difference   7.58)   131.00 (17'', to be subt.

Hence, the required arc is 74° 28' 43''.

3. Find the arc or angle corresponding to the logarithmic sine 9.880054.   *Ans.* 49° 20' 50''.

4. Find the arc or angle corresponding to the logarithmic cotangent 10.008688.   *Ans.* 44° 25' 37''.

5. Find the arc or angle corresponding to the logarithmic cosine 9.944599.   *Ans.* 28° 19' 45''.

SOLUTION OF RIGHT-ANGLED TRIANGLES.

37. In what follows, the three angles of every triangle are designated by the capital letters A, B, and C, A denoting the right angle; and the sides lying opposite the angles by the corresponding small letters *a*, *b*, and *c*. Since the order in which these letters are placed may be changed, without affecting the demonstration, it follows that whatever is proved with the letters placed in any given order, will be equally true when the letters are correspondingly placed in any other order.

Let CAB represent any triangle, right-angled at A. With C as a centre, and a radius CD, equal to 1, describe the arc DG, and draw GF and DE perpendicular to CA: then will FG be the sine of the angle C, CF will be its cosine, and DE its tangent.

Since the three triangles CFG, CDE, and CAB are similar (B. IV., P. XVIII.), we may write the proportions,

CB : AB :: CG : FG, or, $a : c :: 1 : \sin C$,

CB : CA :: CG : CF, or, $a : b :: 1 : \cos C$,

CA : AB :: CD : DE, or, $b : c :: 1 : \tan C$;

hence, we have (B. II., P. I.),

$$c = a \sin C \cdots (1.)$$
$$b = a \cos C \cdots (2.)$$
$$c = b \tan C \cdots (3.)$$

$$\therefore$$

$$\sin C = \frac{c}{a}, \cdots (4.)$$
$$\cos C = \frac{b}{a}, \cdots (5.)$$
$$\tan C = \frac{c}{b}, \cdots (6.)$$

Translating these formulas into ordinary language, we have the following

PRINCIPLES.

1. *The perpendicular of any right-angled triangle is equal to the hypothenuse multiplied by the sine of the angle at the base.*

2. *The base is equal to the hypothenuse multiplied by the cosine of the angle at the base.*

3. *The perpendicular is equal to the base multiplied by the tangent of the angle at the base.*

4. *The sine of the angle at the base is equal to the perpendicular divided by the hypothenuse.*

5. *The cosine of the angle at the base is equal to the base divided by the hypothenuse.*

6. *The tangent of the angle at the base is equal to the perpendicular divided by the base.*

Either side about the right angle may be regarded as the base; the other is then to be taken as the perpendicular. B may be substituted for C in the formulas, provided that, at the same time, b is substituted for c, and c for b: from (4), (5), (6), we may thus obtain,

$$\sin B = \frac{b}{a}, \quad \cdots \quad \cdots \quad (4'.)$$

$$\cos B = \frac{c}{a}, \quad \cdots \quad \cdots \quad (5'.)$$

$$\tan B = \frac{b}{c}. \quad \cdots \quad \cdots \quad (6'.)$$

From the relations shown in (4), (5), (6), (4'), (5'), (6'), the natural functions of the acute angles of a right-angled triangle are sometimes defined as *ratios:* thus, of either of such angles,

the *sine* is the ratio of the *hypothenuse*
to the *side opposite;*

the *cosine* is the ratio of the *hypothenuse*
to the *side adjacent;*

the *tangent* is the ratio of the *side adjacent*
to the *side opposite.*

Formulas (1) to (6) are sufficient for the solution of every case of right-angled triangles. They are in proper form for use with a table of *natural* functions: when a table of *logarithmic* functions is used, as is done in this book, they must be made homogeneous in terms of R, R being equal to 10,000,000,000, as stated in Art. 32. The formulas may be made homogeneous by the principle of Art. 30; thus, for example, the second member of (4), being the value of sin C when the radius is 1, must be multiplied by R for the value of sin C when the radius is R, giving

$$\sin C = \frac{Rc}{a};$$

whence, by solving with reference to c,

$$c = \frac{a \sin C}{R}.$$

In like manner, the remaining formulas may be made homogeneous, giving

$$c = \frac{a \sin C}{R} \quad \cdot \ \cdot \ \cdot \quad (7.) \qquad \sin C = \frac{Rc}{a} \quad \cdot \ \cdot \ \cdot \quad (10.)$$

$$b = \frac{a \cos C}{R} \quad \cdot \ \cdot \ \cdot \quad (8.) \qquad \cos C = \frac{Rb}{a} \quad \cdot \ \cdot \ \cdot \quad (11.)$$

$$c = \frac{b \tan C}{R} \quad \cdot \ \cdot \ \cdot \quad (9.) \qquad \tan C = \frac{Rc}{b} \quad \cdot \ \cdot \ \cdot \quad (12.)$$

In applying logarithms to these formulas, care must be taken to observe the principles of logarithms (Arts. 5 and 6), giving, for example (as logarithm of R is 10),

$$\log c = \log a + \log \sin C - 10,$$

$$\log \sin C = \log c + 10 - \log a$$
$$= \log c + (\text{a. c.}) \log a \ (\text{see Art. 11}); \ \&c.$$

In solving right-angled triangles, four cases arise:

CASE I.

Given the hypothenuse and one of the acute angles, to find the remaining parts.

38. The other acute angle may be found by subtracting the given one from 90° (Art. 23).

The sides about the right angle may be found by formulas (7) and (8).

Examples.

1. Given $a = 749$, and $C = 47° 03' 10''$; required B, c, and b.

Operation.

$$B = 90° - 47° 03' 10'' = 42° 56' 50''.$$

Applying logarithms to formula (7), we have,

$$\log c = \log a + \log \sin C - 10 ;$$

log a   (749) · · · · 2.874482
log sin C (47° 03' 10'') · 9.864501
   log c · · · · · · 2.738983   ∴   c = 548.255.

[The 10 is subtracted mentally.]

Applying logarithms to formula (8), we have,

$$\log b = \log a + \log \cos C - 10 ;$$

log a   (749) · · · · 2.874482
log cos C (47° 03' 10'') · 9.833354
   log b · · · · · · 2.707836   ∴   b = 510.31.

Ans. B $= 42° 56' 50''$, $b = 510.31$, and $c = 548.255$.

2. Given $a = 439$, and $B = 27° 38' 50''$, to find C, c, and b.
Ans. C $= 62° 21' 10''$, $b = 203.708$, and $c = 388.875$.

3. Given $a = 125.7$ yds., and $B = 75° 12'$, to find the other parts.
Ans. C $= 14° 48'$, $b = 121.53$ yds., and $c = 32.11$ yds.

4. Given $a = 7.521$ ft., and $C = 57° 34' 48''$, to find the other parts.
Ans. B $= 32° 25' 12''$, $c = 6.348$ ft., $b = 4.032$ ft.

CASE II.

Given one of the sides about the right angle and one of the acute angles, to find the remaining parts.

39. The other acute angle may be found by subtracting the given one from 90°.

The hypothenuse may be found by formula (7), and the unknown side about the right angle by formula (8).

Examples.

1. Given $c = 56.293$, and $C = 54° 27' 39''$, to find B, a, and b.

Operation.

$$B = 90° - 54° 27' 39'' = 35° 32' 21''.$$

Applying logarithms to formula (7), we have

$$\log a = \log c + 10 - \log \sin C\,;$$

but, $10 - \log \sin C = $ (a. c.) of $\log \sin C$; whence,

$$\begin{aligned}
&\log c \quad\;\; (56.293) \;\cdot\;\cdot\;\cdot\;\; 1.750454 \\
&\text{(a. c.) } \log \sin C \;\; (54° 27' 39'') \;\cdot\;\; 0.089527 \\
&\qquad \log a \;\cdot\;\cdot\;\cdot\;\cdot\;\cdot\;\cdot\;\; \underline{1.839981} \quad \therefore \;\; a = 69.18
\end{aligned}$$

Applying logarithms to formula (8), we have

$$\log b = \log a + \log \cos C - 10\,;$$

$$\begin{aligned}
&\log a \quad\;\; (69.18) \;\cdot\;\cdot\;\cdot\;\; 1.839981 \\
&\log \cos C \;\; (54° 27' 39'') \;\cdot\;\; 9.764370 \\
&\qquad \log b \;\cdot\;\cdot\;\cdot\;\cdot\;\cdot\;\cdot\;\; \underline{1.604351} \;\; \therefore \;\; b = \cdot 40.2114.
\end{aligned}$$

Ans. $B = 35° 32' 21''$, $a = 69.18$, and $b = 40.2114$.

2. Given $c = 358$, and $B = 28° 47'$, to find C, a, and b.
   Ans. $C = 61° 13'$, $a = 408.466$, and $b = 196.676$.

3. Given $b = 152.67$ yds., and $C = 50° 18' 32''$, to find the other parts.

Ans. $B = 39° 41' 28''$, $c = 183.95$, and $a = 239.05$.

4. Given $c = 379.628$, and $C = 39° 26' 16''$, to find B, a, and b.

Ans. $B = 50° 33' 44''$, $a = 597.613$, and $b = 461.55$.

CASE III.

Given the two sides about the right angle, to find the remaining parts.

40. The angle at the base may be found by formula (12), and the solution may be completed as in Case II.

Examples.

1. Given $b = 26$, and $c = 15$, to find C, B, and a.

Operation.

Applying logarithms to formula (12), we have

$$\log \tan C = \log c + 10 - \log b;$$

$\log c$ (15) · · · ·	1.176091	
(a. c.) $\log b$ (26) · · · ·	8.585027	
$\log \tan C$ · ·	9.761118	∴ $C = 29° 58' 54''$.

[From Art. 28, it is evident that $\log \tan C$ here found corresponds to *two* angles, viz., $29° 58' 54''$, and $180° - 29° 58' 54''$, or $150° 1' 6''$. As, however, the triangle is *right-angled*, the angle C is *acute*, and the *smaller* value must be taken.]

$$B = 90° - C = 60° 01' 06''.$$

As in Case II.,

$$\log a = \log c + 10 - \log \sin C;$$

$$\begin{array}{lll}
\log c \cdot \cdot \cdot (15) \cdot \cdot & 1.176091 \\
\text{(a. c.) } \log \sin C \ (29°\ 58'\ 54'') & 0.301271 \\
\log a \cdot \cdot \cdot \cdot \cdot \cdot & \underline{1.477362} & \therefore\ a = 30.017.
\end{array}$$

Ans. C $= 29°\ 58'\ 54''$, B $= 60°\ 01'\ 06''$, and $a = 30.017$.

2. Given $b = 1052$ yds., and $c = 347.21$ yds., to find B, C, and a.

B $= 71°\ 44'\ 05''$, C $= 18°\ 15'\ 55''$, and $a = 1107.82$ yds.

3. Given $b = 122.416$, and $c = 118.297$, to find B, C, and a.

B $= 45°\ 58'\ 50''$, C $= 44°\ 1'\ 10''$, and $a = 170.235$.

4. Given $b = 103$, and $c = 101$, to find B, C, and a.

B $= 45°\ 33'\ 42''$, C $= 44°\ 26'\ 18''$, and $a = 144.256$.

CASE IV.

Given the hypothenuse and either side about the right angle, to find the remaining parts.

41. The angle at the base may be found by one of formulas (10) and 11), and the remaining side may then be found by one of formulas (7) and (8).

Examples.

1. Given $a = 2391.76$, and $b = 385.7$, to find C, B, and c.

Operation.

Applying logarithms to formula (11), we have

$$\log \cos C = \log b + 10 - \log a;$$

$$\log b \ (385.7) \ \cdot \ \cdot \ \cdot \ \ 2.586250$$
$$(\text{a. c.}) \log a \ (2391.76) \ \cdot \ \cdot \ \ 6.621282$$
$$\log \cos C \ \cdot \ \cdot \ \cdot \ \ 9.207532 \ \ \therefore \ \ C = 80° \ 43' \ 11'';$$

$$B = 90° - 80° \ 43' \ 11'' = 9° \ 16' \ 49''.$$

From formula (7), we have

$$\log c = \log a + \log \sin C - 10;$$

$$\log a \qquad (2391.76) \ \cdot \ \ 3.378718$$
$$\log \sin C \ \ (80° \ 43' \ 11'') \ \ 9.994278$$
$$\log c \ \cdot \ \cdot \ \cdot \ \cdot \ \cdot \ \ 3.372996 \qquad \therefore \ \ c = 2360.45.$$

Ans. B = 9° 16′ 49″, C = 80° 43′ 11″, and c = 2360.45.

2. Given $a = 127.174$ yds., and $c = 125.7$ yds., to find C, B, and *b*.

Operation.

From formula (10), we have

$$\log \sin C = \log c + 10 - \log a;$$

$$\log c \ (125.7) \ \cdot \ \cdot \ \cdot \ \ 2.099335$$
$$(\text{a. c.}) \log a \ (127.174) \ \cdot \ \cdot \ \ 7.895602$$
$$\log \sin C \ \cdot \ \cdot \ \cdot \ \ 9.994937 \ \ \therefore \ \ C = 81° \ 16' \ 6'';$$

$$B = 90° - 81° \ 16' \ 6'' = 8° \ 43' \ 54''.$$

From formula (8), we have

$$\log b = \log a + \log \cos C - 10;$$

$$\log a \qquad (127.174) \ \cdot \ \cdot \ \ 2.104398$$
$$\log \cos C \ \ (81° \ 16' \ 6'') \ \cdot \ \ 9.181292$$
$$\log b \ \cdot \ \cdot \ \cdot \ \cdot \ \cdot \ \cdot \ \ 1.285690 \qquad \therefore \ \ b = 19.3.$$

Ans. B = 8° 43′ 54″, C = 81° 16′ 6″, and b = 19.3 yds.

3. Given $a = 100$, and $b = 60$, to find B, C, and c.

   Ans. B $= 36° 52' 11''$, C $= 53° 7' 49''$, and $c = 80$.

4. Given $a = 19.209$, and $c = 15$, to find B, C, and b.

   Ans. B $= 38° 39' 30''$, C $= 51° 20' 30''$, $b = 12$.

SOLUTION OF OBLIQUE-ANGLED TRIANGLES.

42. In the solution of oblique-angled triangles, *four* cases may arise. We shall discuss these cases in order.

CASE I.

Given one side and two angles, to determine the remaining parts.

43. Let ABC represent any oblique-angled triangle. From the vertex C, draw CD perpendicular to the base, forming two right-angled triangles ACD and BCD. Assume the notation of the figure.

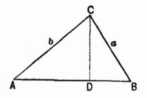

From formula (1), we have

$$CD = b \sin A,$$

$$CD = a \sin B.$$

Equating these two values, we have,

$$b \sin A = a \sin B;$$

whence (B. II., P. II.),

$$a \; : \; b \; :: \; \sin A \; : \; \sin B. \quad \cdot \; \cdot \; \cdot \quad (13.)$$

Since a and b are any two sides, and A and B the angles lying opposite to them, we have the following principle :

The sides of a plane triangle are proportional to the sines of their opposite angles.

It is to be observed that formula (13) is true for any value of the radius. Hence, to solve a triangle, when a side and two angles are given:

First find the third angle, by subtracting the sum of the given angles from 180°; then find each of the required sides by means of the principle just demonstrated.

Examples.

1. Given B = 58° 07′, C = 22° 37′, and a = 408, to find A, b, and c.

Operation.

$$\begin{array}{l} \text{B} \quad \cdots \quad \cdots \quad 58° \ 07′ \\ \text{C} \quad \cdots \quad \cdots \quad \underline{22° \ 37′} \\ \text{A} \quad \cdots \quad 180° - 80° \ 44′ = 99° \ 16′. \end{array}$$

To find *b*, write the proportion,

$$\sin A \ : \ \sin B \ :: \ a \ : \ b;$$

that is, *the sine of the angle opposite the given side, is to the sine of the angle opposite the required side, as the given side is to the required side.*

Applying logarithms, we have (Ex. 4, P. 15)

$$\log b = (\text{a. c.}) \log \sin A + \log \sin B + \log a - 10 ;$$

$$\begin{array}{l} (\text{a. c.}) \log \sin A \ (99° \ 16′) \quad \cdots \quad 0.005705 \\ \log \sin B \ (58° \ 07′) \quad \cdots \quad 9.928972 \\ \log a \quad (408) \quad \cdots \quad \underline{2.610660} \\ \log b \quad \cdots \quad \cdots \quad \underline{2.545337} \quad \therefore b = 351.024. \end{array}$$

In like manner,

$$\sin A \ : \ \sin C \ :: \ a \ : \ c;$$

and $\log c = $ (a. c.) $\log \sin$ A $+ \log \sin$ C $+ \log a - 10$;

(a. c.) $\log \sin$ A (99° 16′) · · · 0.005705

$\quad\quad$ $\log \sin$ C (22° 37′) · · · 9.584968

$\quad\quad$ $\log a$ $\quad$ (408) · · · · 2.610660

$\quad\quad\quad$ $\log c$ · · · · · · · 2.201333 ∴ $c = 158.976$.

$\quad\quad$ *Ans.* A $= 99°$ 16′, $b = 351.024$, and $c = 158.976$.

2. Given A $= 38°$ 25′, B $= 57°$ 42′, and $c = 400$, to find C, a, and b.

$\quad\quad\quad$ *Ans.* C $= 83°$ 53′, $a = 249.974$, $b = 340.04$.

3. Given A $= 15°$ 19′ 51″, C $= 72°$ 44′ 05″, and $c =$ 250.4 yds., to find B, a, and b.

Ans. B $= 91°$ 56′ 04″, $a = 69.328$ yds., $b = 262.066$ yds.

4. Given B $= 51°$ 15′ 35″, C $= 37°$ 21′ 25″, and $a =$ 305.296 ft., to find A, b, and c.

$\quad\quad$ *Ans.* A $= 91°$ 23′, $b = 238.1978$ ft., $c = 185.3$ ft.

CASE II.

Given two sides and an angle opposite one of them, to find the remaining parts.

44. The solution, in this case, is commenced by finding a second angle by means of formula (13), after which we may proceed as in CASE I.; or, the solution may be completed by a continued application of formula (13).

Examples.

1. Given A $= 22°$ 37′, $b = 216$, and $a = 117$, to find B, C, and c.

From formula (13), we have

$$a : b :: \sin A : \sin B;$$

that is, *the side opposite the given angle, is to the side opposite the required angle, as the sine of the given angle is to the sine of the required angle.*

Whence, by the application of logarithms,

$$\log \sin B = (a.\ c.) \log a + \log b + \log \sin A - 10;$$

(a. c.) $\log a$	(117)	· ·	7.931814
$\log b$	(216)	· ·	2.334454
$\log \sin A$	(22° 37')	·	9.584968
$\log \sin B$	· · ·		9.851236

$\therefore$ B $= 45°\ 13'\ 55''$, and B$' = 134°\ 46'\ 05''$.

Hence, we find two values of B, which are supplements of each other, because the sine of any angle is equal to the sine of its supplement. This would seem to indicate that the problem admits of two solutions. It now remains to determine under what conditions there will be *two solutions, one solution,* or *no solution.*

There may be two cases: the given angle may be *acute,* or it may be *obtuse.*

Represent the given parts of the triangle by A, *a, b.* The particular letters employed are of no consequence in the discussion, and, therefore, in the results, C or B may be substituted for A, provided that, at the same time, like changes are made in the corresponding small letters.

1st Case: A < 90°.

Let ABC represent the triangle, in which the angle A, and the sides a and b are given. From C let fall a perpendicular upon AB, prolonged if necessary, and denote its length by p. We shall have, from formula (1), Art. 37,

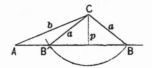

$$p = \frac{b \sin A}{R};$$

from which the value of p may be computed.

If a is greater than p and less than b, there will be *two solutions*. For, if with C as a centre, and a as a radius, an arc be described, it will cut the line AB in two points, B and B′, each of which being joined with C, will give a triangle, and we shall thus have two triangles, ABC and AB′C, which will conform to the conditions of the problem.

In this case, the angles B′ and B, of the two triangles AB′C and ABC, will be supplements of each other.

If $a = p$, there will be but *one solution*. For, in this case, the arc will be tangent to AB, the two points B and B′ will unite, and there will be but one triangle formed.

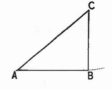

In this case, the angle ABC will be equal to 90°.

If a is greater than both p and b, there will also be but one solution. For, although the arc cuts AB in two points, and consequently gives two triangles, only one of them, ABC, conforms to the conditions of the problem.

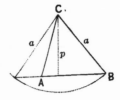

In this case, the angle ABC will be less than A and consequently acute.

If $a < p$, there will be *no solution.* For, the arc can neither cut AB nor be tangent to it.

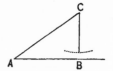

2d Case: A > 90°.

When the given angle A is obtuse, the angle ABC will be acute; the side a will be greater than b, and there will be but *one solution.*

(See B. III., Prob. XI., S.)

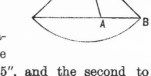

In the example under consideration, there are two solutions, the first corresponding to B = 45° 13′ 55″, and the second to B′ = 134° 46′ 05″.

In the first case, we have

A · · · · · · · 22° 37′
B · · · · · · · 45° 13′ 55″
C · · · 180° − 67° 50′ 55″ = 112° 09′ 05″.

To find c, we have

$$\sin B \ : \ \sin C \ :: \ b \ : \ c;$$

and $\log c = $ (a. c.) $\log \sin B + \log \sin C + \log b - 10$;

(a. c.) log sin B (45° 13′ 55″) · 0.148764
log sin C (112° 09′ 05″) · 9.966700
log b · (216) · · · · · 2.334454
log c · · · · · · · 2.449918 ∴ c = 281.785.

Ans. B = 45° 13′ 55″, C = 112° 09′ 05″, and c = 281.785.

In the second case, we have,

A · · · · · ·   22° 37′
B′ · · · · · ·   134° 46′ 05″
C′ · · · 180° − $\overline{157°\ 23′\ 05″}$ = 22° 36′ 55″;

and as before,

(a. c.) log sin B′ (134° 46′ 05″) ·   0.148764
       log sin C′   (22° 36′ 55″) ·   9.584943
       log b · · · (216) · · $\underline{2.334454}$
          log $c′$ · · · · · · · ·   2.068161 ∴ $c′$ = 116.993.

Ans. B′ = 134° 46′ 05″, C′ = 22° 36′ 55″, and $c′$ = 116.993.

2. Given A = 32°, a = 40, and b = 50, to find B, C, and c.

$Ans.$ $\begin{cases} \text{B} = 41°\ 28′\ 59″, & \text{C} = 106°\ 31′\ 01″, & c = 72.368. \\ \text{B}′ = 138°\ 31′\ 01″, & \text{C}′ = 9°\ 28′\ 59″, & c′ = 12.436. \end{cases}$

3. Given B = 18° 52′ 13″, b = 27.465 yds., and a = 13.189 yds., to find A, C, and c.

Ans. A = 8° 56′ 05″, C = 152° 11′ 42″, c = 39.611 yds.

4. Given C = 32° 15′ 26″, b = 176.21 ft., and c = 94.047 ft., to find B, A, and a.

Ans. B = 90°, A = 57° 44′ 34″, a = 149.014 ft.

CASE III.

Given two sides and their included angle, to find the remaining parts.

45. The solution, in this case, is begun by finding the half sum and the half difference of the two required angles. The half sum of these angles may be found by subtracting the given angle from 180°, and dividing the remainder by 2; the half difference may be found by means of the following principle, now to be demonstrated, viz. :

In any plane triangle, the sum of the sides including any angle, is to their difference, as the tangent of half the sum of the two other angles, is to the tangent of half their difference.

Let ABC represent any plane triangle, c and b any two sides, and A their included angle. Then we are to show that

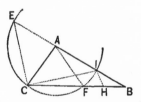

$$c + b \; : \; c - b \; :: \; \tan \tfrac{1}{2}(C + B) \; : \; \tan \tfrac{1}{2}(C - B).$$

With A as a centre, and b, the shorter of the two sides, as a radius, describe a semicircle meeting AB in I, and the prolongation of AB in E. Draw EC and CI, and through I draw IH parallel to EC. Since the angle ECI is inscribed in a semicircle, it is a right angle (B. III., P. XVIII., C. 2); hence, EC is perpendicular to CI, at the point C; and since IH is parallel to EC, it is also perpendicular to CI.

The inscribed angle CIE is half the angle at the centre, CAE, intercepting the same arc CE. Since the angle CAE is exterior to the triangle ABC, we have (B. I., P. XXV., C. 6),

$$CAE = C + B;$$

hence, $$CIE = \tfrac{1}{2}(C + B).$$

AC and AF, being radii of the same circle, are equal to each other, and therefore (B. I., P. XI.), the angle AFC is equal to the angle C; but the angle AFC is exterior to the triangle FBA, and hence we have

$$AFC \quad or \quad C = FAB + B;$$

hence, $$FAB = C - B.$$

But the inscribed angle, ICH, is half the angle at the centre, FAB, intercepting the same arc FI; hence,

$$ICH = \tfrac{1}{2}(C - B).$$

From the two right-angled triangles ICE and ICH, we have (formula 3, Art. 37),

$$EC = IC \tan CIE$$
$$= IC \tan \tfrac{1}{2}(C + B)$$

and
$$IH = IC \tan ICH$$
$$= IC \tan \tfrac{1}{2}(C - B);$$

hence, we have, after omitting the equal factor IC (B. II., P. VII.),

$$EC : IH :: \tan \tfrac{1}{2}(C + B) : \tan \tfrac{1}{2}(C - B).$$

The triangles ECB and IHB being similar (B. IV., P. XXI.),

$$EC \ : \ IH \ :: \ EB \ : \ IB,$$

or, since $$EB = c + b,$$

and $$IB = c - b,$$

$$EC \ : \ IH \ :: \ c + b \ : \ c - b.$$

Combining the preceding proportions, we have

$$c + b \ : \ c - b \ :: \ \tan \tfrac{1}{2}(C + B) \ : \ \tan \tfrac{1}{2}(C - B); \ \cdot \ (14.)$$

which was to be proved.

By means of (14), the half difference of the two required angles may be found. Knowing the half sum and the half difference, the greater angle is found by adding the half difference to the half sum, and the less angle is found by subtracting the half difference from the half sum. Then the solution is completed as in Case L

Examples.

1. Given $c = 540$, $b = 450$, and $A = 80°$, to find B C, and a.

Operation.

$$c + b = 990;$$

$$c - b = 90;$$

$$\tfrac{1}{2}(C + B) = \tfrac{1}{2}(180° - 80°)$$

$$= 50°.$$

Applying logarithms to formula (14), we have

$$\log \tan \tfrac{1}{2} (C - B) = (\text{a. c.}) \log (c + b) + \log (c - b)$$
$$+ \log \tan \tfrac{1}{2} (C + B) - 10 \, ;$$

$$\begin{aligned}
(\text{a. c.}) \log (c + b) \; \cdot \; \cdot \; (990) \quad & 7.004365 \\
\log (c - b) \; \cdot \; \cdot \quad (90) \quad & 1.954243 \\
\log \tan \tfrac{1}{2} (C + B) \; (50°) \; & \underline{10.076187} \\
\log \tan \tfrac{1}{2} (C - B) \qquad & \underline{9.034795} \; \therefore \; \tfrac{1}{2}(C - B) = 6° \, 11'.
\end{aligned}$$

$$C = 50° + 6° \, 11' = 56° \, 11' ;$$

$$B = 50° - 6° \, 11' = 43° \, 49'.$$

From formula (13), we have

$$\sin C \; : \; \sin A \; :: \; c \; : \; a \, ;$$

whence,

$$\begin{aligned}
(\text{a. c.}) \log \sin C \; (56° \, 11') \; \cdot \; & 0.080492 \\
\log \sin A \; (80°) \quad \cdot \; \cdot \; & 9.993351 \\
\log c \; \cdot \; (540) \quad \cdot \; \cdot \; & 2.732394 \\
\log a \; \cdot \; \cdot \; \cdot \; \cdot \; \cdot \; & \underline{2.806237} \; \therefore \; a = 640.082.
\end{aligned}$$

Ans. B = 43° 49', C = 56° 11', a = 640.082.

2. Given $c = 1686$ yds., $b = 960$ yds., and $A = 128° \, 04'$, to find B, C, and a.

Ans. B = 18° 21' 21", C = 33° 34' 39", a = 2400 yds.

3. Given $a = 18.739$ yds., $c = 7.642$ yds., and $B = 45° \, 18' \, 28"$, to find A, b, and C.

Ans. A = 112° 34' 13", C = 22° 07' 19", b = 14.426 yds.

4. Given $a = 464.7$ yds., $b = 289.3$ yds., and $C = 87° \, 03' \, 48''$, to find A, B, and c.

Ans. A $= 60° \, 13' \, 39''$, B $= 32° \, 42' \, 33''$, $c = 534.66$ yds.

5. Given $a = 16.9584$ ft., $b = 11.9613$ ft., and $C = 60° \, 43' \, 36''$, to find A, B, and c.

Ans. A $= 76° \, 04' \, 12''$, B $= 43° \, 12' \, 12''$, $c = 15.22$ ft.

6. Given $a = 3754$, $b = 3277.628$, and $C = 57° \, 53' \, 17''$, to find A, B, and c.

Ans. A $= 68° \, 02' \, 25''$, B $= 54° \, 04' \, 18''$, $c = 3428.512$.

CASE IV.

*Given the three sides of a triangle, to find the remaining parts.**

46. Let ABC represent any plane triangle, of which BC is the longest side. Draw AD perpendicular to the base, dividing it into two segments CD and BD.

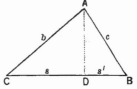

[The longest side is taken as the base, to make it certain that the perpendicular from the vertex shall fall on the base, and *not* on the base *produced*.]

From the right-angled triangles CAD and BAD, we have
$$\overline{AD}^2 = \overline{AC}^2 - \overline{DC}^2,$$

and
$$\overline{AD}^2 = \overline{AB}^2 - \overline{BD}^2.$$

---

* The angles may be found by formula (A) or (B), Lemma, Art. 97, Mensuration.

Equating these values of $\overline{AD}^2$, we have,

$$\overline{AC}^2 - \overline{DC}^2 = \overline{AB}^2 - \overline{BD}^2 ;$$

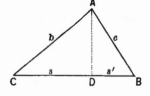

whence, by transposition,

$$\overline{AC}^2 - \overline{AB}^2 = \overline{DC}^2 - \overline{BD}^2.$$

Hence (B. IV., P. X), we have

$$(AC + AB)(AC - AB) = (DC + BD)(DC - BD).$$

Converting this equation into a proportion (B. II., P. II.), we have

$$DC + BD \; : \; AC + AB \; :: \; AC - AB \; : \; DC - BD ;$$

or, denoting the greater segment by s and the less segment by s', and the sides of the triangle by a, b, and c,

$$s + s' \; : \; b + c \; :: \; b - c \; : \; s - s' ; \quad \cdot \quad (15.)$$

that is, if in any plane triangle, a line be drawn from the vertex perpendicular to the base, dividing it into two segments; then,

The sum of the two segments, or the whole base, is to the sum of the two other sides, as the difference of these sides is to the difference of the segments.

The half difference of the segments added to the half sum gives the greater segment, and the half difference subtracted from the half sum gives the less segment. [The greater segment is, of course, adjacent to the greater side.] We shall then have two right-angled triangles, in each of which we know the hypothenuse and the base;

hence, the angles of these triangles may be found, and consequently, those of the given triangle.

Examples.

1. Given $a = 40$, $b = 34$, and $c = 25$, to find A, B, and C.

Operation.

Applying logarithms to formula (15), we have

$$\log (s - s') = (\text{a. c.}) \log (s + s') + \log (b+c) + \log (b-c) - 10 ;$$

(a. c.) $\log (s + s')$ · ·	(40)	· ·	8.397940
$\log (b + c)$ · ·	(59)	· ·	1.770852
$\log (b - c)$ · ·	(9)	· ·	0.954243
$\log (s - s')$		· · · ·	1.123035 ∴ $s - s' = 13.275$.

$$s = \tfrac{1}{2}(s + s') + \tfrac{1}{2}(s - s') = 26.6375.$$

$$s' = \tfrac{1}{2}(s + s') - \tfrac{1}{2}(s - s') = 13.3625.$$

From formula (11), we find

$\log \cos C = \log s + (\text{a. c.}) \log b \quad \therefore \quad C = 38° 25' 20''$, and
$\log \cos B = \log s' + (\text{a. c.}) \log c \quad \therefore \quad B = \underline{57° 41' 25''}$
$\phantom{\log \cos B = \log s' + (\text{a. c.}) \log c \quad \therefore \quad B = } \underline{96° 06' 45''}$

$$A = 180° - 96° 06' 45'' = 83° 53' 15''.$$

2. Given $a = 6$, $b = 5$, and $c = 4$, to find A, B and C.

Ans. A = 82° 49' 09'', B = 55° 46' 16'', C = 41° 24' 35''.

3. Given $a = 71.2$ yds., $b = 64.8$ yds., and $c = 37$ yds., to find A, B, and C.

Ans. A = 84° 01' 53'', B = 64° 50' 51'', C = 31° 07' 16''.

PROBLEMS.

1. Knowing the distance AB, equal to 600 yards, and the angles BAC = 57° 35′, ABC = 64° 51′, find the two distances AC and BC.

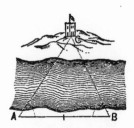

$$Ans. \begin{cases} AC = 643.49 \text{ yds.,} \\ BC = 600.11 \text{ yds.} \end{cases}$$

2. At what horizontal distance from a column, 200 feet high, will it subtend an angle of 31° 17′ 12″?

Ans. 329.114 ft.

3. Required the height of a hill D above a horizontal plane AB, the distance between A and B being equal to 975 yards,

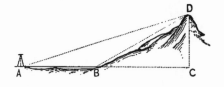

and the angles of elevation at A and B being respectively 15° 36′ and 27° 29′.

Ans. DC = 587.61 yds.

4. The distances AC and BC are found by measurement to be respectively, 588 feet and 672 feet, and their included angle 55° 40′. Required the distance AB.

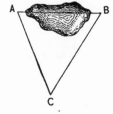

Ans. 592.967 ft.

5. Being on a horizontal plane, and wanting to ascertain the height of a tower, standing on the top of an inaccessible hill, there were measured, the angle of elevation of the top of the hill 40°, and of the top of the tower 51°; then measuring in a direct line 180 feet

farther from the hill, the angle of elevation of the top of the tower was 33° 45'; required the height of the tower.

Ans. 83.998 ft.

6. Wanting to know the horizontal distance between two inaccessible objects E and W, the following measurements were made:

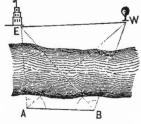

viz. :
$$
\begin{cases}
AB & = & 536 \text{ yards} \\
BAW & = & 40° \ 16' \\
WAE & = & 57° \ 40' \\
ABE & = & 42° \ 22' \\
EBW & = & 71° \ 07'
\end{cases}
$$

Required the distance EW.   *Ans.* 939.617 yds.

7. Wanting to know the horizontal distance between two inaccessible objects A and B, and not finding any station from which both of them could be seen, two points C and D were chosen at a distance from each other equal to 200 yards; from the former of these points, A could be seen, and from the

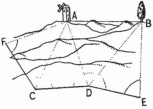

latter, B; and at each of the points C and D a staff was set up. From C a distance CF was measured, not in the direction DC, equal to 200 yards, and from D, a distance DE equal to 200 yards, and the following angles taken:

AFC = 83° 00',   BDE = 54° 30',   ACD = 53° 30',

BDC = 156° 25',   ACF = 54° 31',   BED = 88° 30'.

Required the distance AB.   *Ans.* 345.459 yds.

8. The distances AB, AC, and BC, between the points A, B, and C, are known; viz.: AB = 800 yds., AC = 600 yds., and BC = 400 yds. From a fourth point P, the angles APC and BPC are measured; viz.:

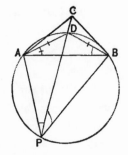

$$APC = 33° 45',$$
and $$BPC = 22° 30'.$$

Required the distances AP, BP, and CP.

$$Ans. \begin{cases} AP = \quad 710.198 \text{ yds.} \\ BP = \quad 934.289 \text{ yds.} \\ CP = 1042.524 \text{ yds.} \end{cases}$$

This problem is used in locating the position of buoys in maritime surveying, as follows. Three points, A, B, and C, on shore are known in position. The surveyor stationed at a buoy P, measures the angles APC and BPC. The distances AP, BP, and CP, are then found as follows:

Suppose the circumference of a circle to be described through the points A, B, and P. Draw CP, cutting the circumference in D, and draw the lines DB and DA.

The angles CPB and DAB, being inscribed in the same segment, are equal (B. III., P. XVIII., C. 1); for a like reason, the angles CPA and DBA are equal: hence, in the triangle ADB, we know two angles and one side; we may, therefore, find the side DB. In the triangle ACB, we know the three sides, and we may compute the angle B. Subtracting from this the angle DBA, we have the angle DBC. Now, in the triangle DBC, we have two sides and their included angle, and we can find the angle DCB. Finally, in the triangle CPB, we have two angles and one side, from which data we can find CP and BP. In like manner, we can find AP.

ANALYTICAL TRIGONOMETRY.

47. ANALYTICAL TRIGONOMETRY is that branch of Mathematics which treats of the general properties and relations of trigonometrical functions.

DEFINITIONS AND GENERAL PRINCIPLES.

48. Let ABCD represent a circle whose radius is 1, and suppose its circumference to be divided into four equal parts, by the diameters AC and BD drawn perpendicular to each other. The horizontal diameter AC is called the *initial diameter ;* the vertical diameter BD is called the *secondary diameter ;* the point A, from which arcs are usually reckoned, is called the *origin of arcs,* and the point B, 90° distant, is called the *secondary origin.* Arcs estimated from A, around toward B, that is, in a direction contrary to that of the motion of the hands of a watch, are considered *positive ;* consequently, those reckoned in a contrary direction must be regarded as *negative.*

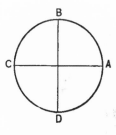

The arc AB, is called the *first quadrant ;* the arc BC, the *second quadrant ;* the arc CD, the *third quadrant ;* and the arc DA, the *fourth quadrant.* The point at which an arc terminates, is called its *extremity,* and an arc is said to be in that quadrant in which its extremity is situated. Thus, the arc AM is in the *first quadrant,* the arc AM' in the *second,* the arc AM'' in the *third,* and the arc AM''' in the *fourth.*

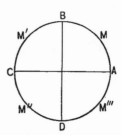

49. The *complement* of an arc has been defined to be the difference between that arc and 90° (Art. 23); geometrically considered, the *complement* of an arc is *the arc included between the extremity of the arc and the secondary origin.* Thus, MB is the complement of AM; M'B, the complement of AM'; M''B, the complement of AM'', and so on. When the arc is greater than a quadrant, the complement is negative, according to the conventional principle agreed upon (Art. 48).

The *supplement* of an arc has been defined to be the difference between that arc and 180° (Art. 24); geometrically considered, it is *the arc included between the extremity of the arc and the left-hand extremity of the initial diameter.* Thus, MC is the supplement of AM, and M''C the supplement of AM''. The supplement is negative, when the arc is greater than two quadrants.

50. *The sine of an arc is the distance from the initial diameter to the extremity of the arc.* Thus, PM is the sine of AM, and P''M'' is the sine of the arc AM''. The term *distance* is used in the sense of *shortest* or *perpendicular distance.*

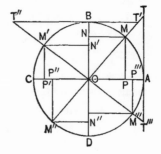

51. *The cosine of an arc is the distance from the secondary diameter to the extremity of the arc:* thus, NM is the cosine of AM, and N'M' is the cosine of AM'.

The cosine may be measured on the initial diameter: thus, OP is equal to the cosine of AM, and OP' to the cosine of AM'; that is, the cosine of an arc is equal to the distance, measured on the initial diameter, from the centre of the arc to the foot of the sine.

52. *The versed-sine of an arc is the distance from the sine to the origin of arcs:* thus, PA is the versed-sine of AM, and P'A is the versed-sine of AM'.

53. *The co-versed-sine of an arc is the distance from the cosine to the secondary origin:* thus, NB is the co-versed-sine of AM, and N"B is the co-versed-sine of AM".

54. *The tangent of an arc is that part of a perpendicular to the initial diameter, at the origin of arcs, included between the origin and the prolongation of the diameter drawn to the extremity of the arc:* thus, AT is the tangent of AM, or of AM", and AT''' is the tangent of AM', or of AM'''.

55. *The cotangent of an arc is that part of a perpendicular to the secondary diameter, at the secondary origin, included between the secondary origin and the prolongation of the diameter drawn to the extremity of the arc:* thus, BT' is the cotangent of AM, or of AM", and BT" is the cotangent of AM , or of AM'''.

56. *The secant of an arc is the distance from the centre of the arc to the extremity of the tangent:* thus, OT is the secant of AM, or of AM", and OT''' is the secant of AM'. or of AM'''.

57. *The cosecant of an arc is the distance from the centre of the arc to the extremity of the cotangent:* thus, OT′ is the cosecant of AM, or of AM″, and OT″ is the cosecant of AM′, or of AM‴.

The prefix *co*, as used here, is equivalent to *complement;* thus, the *cosine* of an arc is the "*complement sine*," that is, the *sine of the complement*, of that arc, and so on, as explained in Art. 27.

The eight *trigonometrical functions* above defined are also called *circular functions.*

RULES FOR DETERMINING THE ALGEBRAIC SIGNS OF CIRCULAR FUNCTIONS.

58. All distances estimated *upward* are regarded as *positive;* consequently, all distances estimated *downward* must be considered *negative.*

Thus, AT, PM, NB, P′M′, are positive, and AT‴, P‴M‴, P″M″, &c., are negative.

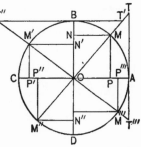

All distances estimated *toward the right* are regarded as *positive;* consequently, all distances estimated *toward the left* must be considered *negative.*

Thus, NM, BT′, PA, &c., are positive, and N′M′, BT″, &c., are negative.

These two rules are sufficient for determining the algebraic signs of all the circular functions, except the secant and cosecant. For the secant and cosecant, the following is the rule:

All distances estimated from the centre in a direction *toward the extremity* of the arc are regarded as *positive;*

consequently, all distances estimated in a direction *away from the extremity* of the arc must be considered *negative.*

Thus, OT, regarded as the secant of AM, is estimated in a direction *toward* M, and is *positive;* but OT, regarded as the secant of AM″, is estimated in a direction *away from* M″, and is *negative.*

These conventional rules enable us to give at once the proper sign to any function of an arc in any quadrant.

59. In accordance with the above rules, and the definitions of the circular functions, we have the following principles :

The sine is positive in the first and second quadrants, and negative in the third and fourth.

The cosine is positive in the first and fourth quadrants, and negative in the second and third.

The versed-sine and the co-versed-sine are always positive.

The tangent and cotangent are positive in the first and third quadrants, and negative in the second and fourth.

The secant is positive in the first and fourth quadrants, and negative in the second and third.

The cosecant is positive in the first and second quadrants, and negative in the third and fourth.

LIMITING VALUES OF THE CIRCULAR FUNCTIONS.

60. The limiting values of the circular functions are those values which they have at the beginning and the end of the different quadrants. Their numerical values are discovered by following them as the arc increases from 0° around to 360°, and so on around through 450°,

540°, &c. The signs of these values are determined by the principle, that *the sign of a varying magnitude up to the limit, is the sign at the limit.* For illustration, let us examine the limiting values of the sine and the tangent.

If we suppose the arc to be 0, the sine will be 0; as the arc increases, the sine increases until the arc becomes equal to 90°, when the sine becomes equal to + 1, which is its greatest possible value; as the arc increases from 90°, the sine diminishes until the arc becomes equal to 180°, when the sine becomes equal to + 0; as the arc increases from 180°, the sine becomes negative, and increases numerically, but *decreases algebraically,* until the arc becomes equal to 270°, when the sine becomes equal to − 1, which is its least *algebraical* value; as the arc increases from 270°, the sine decreases numerically, but *increases algebraically,* until the arc becomes 360°, when the sine becomes equal to − 0. It is − 0, for this value of the arc, in accordance with the principle of limits.

The tangent is 0 when the arc is 0, and increases till the arc becomes 90°, when the tangent is + ∞; in passing through 90°, the tangent changes from + ∞ to − ∞, and as the arc increases the tangent decreases numerically, but increases algebraically, till the arc becomes equal to 180°, when the tangent becomes equal to − 0; from 180° to 270° the tangent is again positive, and at 270° it becomes equal to + ∞; from 270° to 360°, the tangent is again negative, and at 360° it becomes equal to − 0.

If we still suppose the arc to increase after reaching 360°, the functions will again go through the same changes, that is, the functions of an arc are the same as the functions of that arc increased by 360°, 720°, &c.

By discussing the limiting values of all the circular functions we may form the following table:

TABLE I.

Arc = 0°.	Arc = 90°.	Arc = 180°.	Arc = 270°.	Arc = 360°.
sin = 0	sin = 1	sin = 0	sin = −1	sin = −0
cos = 1	cos = 0	cos = −1	cos = −0	cos = 1
v-sin = 0	v-sin = 1	v-sin = 2	v-sin = 1	v-sin = 0
co-v-sin = 1	co-v-sin = 0	co-v-sin = 1	co-v-sin = 2	co-v-sin = 1
tan = 0	tan = ∞	tan = −0	tan = ∞	tan = −0
cot = ∞	cot = 0	cot = −∞	cot = 0	cot = −∞
sec = 1	sec = ∞	sec = −1	sec = −∞	sec = 1
cosec = ∞	cosec = 1	cosec = ∞	cosec = −1	cosec − −∞

RELATIONS BETWEEN THE CIRCULAR FUNCTIONS OF ANY ARC.

61. Let AM, denoted by a, represent any arc whose radius is 1. Draw the lines as represented in the figure. Then we shall have,

OM = OA = 1;   PM = ON = sin a;

NM = OP = cos a;   PA = ver-sin a;

NB = co-ver-sin a;   AT = tan a;

BT′ = cot a;   OT = sec a;

and   OT′ = cosec a.

From the right-angled triangle OPM, we have,

$$\overline{PM}^2 + \overline{OP}^2 = \overline{OM}^2, \quad \text{or,} \quad \sin^2 a + \cos^2 a = 1. \cdot \cdot \cdot (1.)$$

The symbols $\sin^2 a$, $\cos^2 a$, &c., denote the square of the sine of a, the square of the cosine of a, &c.

From formula (1) we have, by transposition,

$$\sin^2 a = 1 - \cos^2 a; \cdot \cdot \cdot \cdot \quad (2.)$$

$$\cos^2 a = 1 - \sin^2 a. \cdot \cdot \cdot \cdot \cdot \cdot (3.)$$

We have, from the figure,

$$PA = OA - OP,$$

or, $\quad\quad\quad$ ver-sin $a = 1 - \cos a;$ · · $\quad$ · · · (4.)

and, $\quad\quad\quad\quad$ NB = OB — ON,

or, $\quad\quad\quad$ co-ver-sin $a = 1 - \sin a.$ · · · · · · (5.)

From the similar triangles OAT and OPM, we have,

OP : PM :: OA : AT, $\quad$ or, $\quad$ $\cos a : \sin a :: 1 : \tan a;$

whence, $\quad\quad\quad\quad\quad$ $\tan a = \dfrac{\sin a}{\cos a}.$ · · · · · · · (6.)

From the similar triangles ONM and OBT', we have,

ON : NM :: OB : BT', $\quad$ or, $\quad$ $\sin a : \cos a :: 1 : \cot a;$

whence, $\quad\quad\quad\quad\quad$ $\cot a = \dfrac{\cos a}{\sin a}.$ · · · · · · · (7.)

Multiplying (6) and (7), member by member, we have,

$$\tan a \cot a = 1; \quad \cdots \cdots \cdots \text{(8.)}$$

whence, by division, $\quad\quad$ $\tan a = \dfrac{1}{\cot a};$ · · · · · (9.)

and $\quad\quad\quad\quad\quad\quad$ $\cot a = \dfrac{1}{\tan a}.$ · · · · ·(10.)

From the similar triangles OPM and OAT, we have,

OP : OM :: OA : OT, $\quad$ or, $\quad$ $\cos a : 1 :: 1 : \sec a;$

whence, $\quad\quad\quad\quad\quad$ $\sec a = \dfrac{1}{\cos a}.$ · · · · · · ·(11.)

From the similar triangles ONM and OBT', we have,

ON : OM :: OB : OT',   or, $\sin a : 1 :: 1 : \operatorname{cosec} a$;

whence, $\operatorname{cosec} a = \dfrac{1}{\sin a}.$ · · · · · · (12.)

From the right-angled triangle OAT, we have,

$\overline{OT}^2 = \overline{OA}^2 + \overline{AT}^2$;   or, $\sec^2 a = 1 + \tan^2 a.$ · · · (13.)

From the right-angled triangle OBT', we have,

$\overline{OT'}^2 = \overline{OB}^2 + \overline{BT'}^2$;   or, $\operatorname{cosec}^2 a = 1 + \cot^2 a.$ · (14.)

It is to be observed that formulas (5), (7), (12), and (14), may be deduced from formulas (4), (6), (11), and (13), by substituting $90° - a$, for a, and then making the proper reductions.

Collecting the preceding formulas, we have the following table:

TABLE II.

(1.)	$\sin^2 a + \cos^2 a$	$= 1.$	(9.)	$\tan a$	$= \dfrac{1}{\cot a}.$
(2.)	$\sin^2 a$	$= 1 - \cos^2 a.$	(10.)	$\cot a$	$= \dfrac{1}{\tan a}.$
(3.)	$\cos^2 a$	$= 1 - \sin^2 a.$			
(4.)	ver-sin a	$= 1 - \cos a.$	(11.)	$\sec a$	$= \dfrac{1}{\cos a}.$
(5.)	co-ver-sin a	$= 1 - \sin a.$			
(6.)	$\tan a$	$= \dfrac{\sin a}{\cos a}.$	(12.)	$\operatorname{cosec} a$	$= \dfrac{1}{\sin a}.$
(7.)	$\cot a$	$= \dfrac{\cos a}{\sin a}.$	(13.)	$\sec^2 a$	$= 1 + \tan^2 a.$
(8.)	$\tan a \cot a$	$= 1.$	(14.)	$\operatorname{cosec}^2 a$	$= 1 + \cot^2 a.$

FUNCTIONS OF NEGATIVE ARCS.

62. Let AM''', estimated from A toward D, be numerically equal to AM; then, if we denote
the arc AM by a, the arc AM''' will
be denoted by $-a$ (Art. 48).

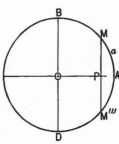

A being the middle point of the
arc M'''AM, the radius OA bisects the
chord M'''M at right angles (B. III.,
P. VI.); therefore, PM''' is numerically equal to PM, but PM''' being
measured downward from the initial
diameter is negative, while PM being
measured upward is positive, and, therefore, PM''' $= -$ PM :
OP is equal to the cosine of both AM''' and AM (Art. 61);
hence, we have,

$$\sin(-a) = -\sin a, \quad \cdots \quad \cdots \quad (1.)$$

$$\cos(-a) = \cos a. \quad \cdots \quad \cdots \quad (2.)$$

Dividing (1) by (2), member by member, and then dividing (2) by (1), member by member, we have (formulas 6
and 7, Art. 61),

$$\tan(-a) = -\tan(a); \qquad \cot(-a) = -\cot a.$$

Taking the reciprocals of the members of (2), and then
the reciprocals of the members of (1), we have (formulas
11 and 12, Art. 61),

$$\sec(-a) = \sec a; \qquad \csc(-a) = -\csc a.$$

FUNCTIONS OF ARCS

FORMED BY ADDING AN ARC TO, OR SUBTRACTING IT FROM, ANY
NUMBER OF QUADRANTS.

63. Let a denote any arc less than 90°. By definition, we have,

$$\sin (90° - a) = \cos a; \qquad \cos (90° - a) = \sin a.$$
$$\tan (90° - a) = \cot a; \qquad \cot (90° - a) = \tan a.$$
$$\sec (90° - a) = \operatorname{cosec} a; \qquad \operatorname{cosec} (90° - a) = \sec a.$$

Let the arc $BM' = AM = a$; then 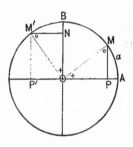 $AM' = 90° + a$. Draw lines, as in the figure. Then $PM = \sin a$; $OP = \cos a$; $ON = P'M' = \sin (90° + a)$; $NM' = \cos (90° + a)$.

The right-angled triangles ONM' and OPM have the angles NOM' and POM equal (B. III., P. XV.), the angles ONM' and OPM equal, both being right angles, and therefore (B. I., P. XXV., C. 2), the angles $OM'N$ and OMP equal; they have, also, the sides OM' and OM equal, and are, consequently (B. I., P. VI.), equal in all respects: hence, $ON = OP$, and $NM' = PM$. These are *numerical* relations; by the rules for signs, Art. 58, ON and OP are both positive, NM' is negative, and PM positive; and hence, *algebraically*, $ON = OP$, and $NM' = -PM$; therefore, we have,

$$\sin (90° + a) = \cos a; \quad \cdot \quad \cdot \quad \cdot \quad \cdot \text{(1.)}$$
$$\cos (90° + a) = -\sin a. \cdot \quad \cdot \quad \cdot \quad \cdot \text{(2.)}$$

Dividing (1) by (2), member by member, we have,

$$\frac{\sin (90° + a)}{\cos (90° + a)} = \frac{\cos a}{-\sin a};$$

or (formulas 6 and 7, Art. 61).

$$\tan (90° + a) = -\cot a.$$

In like manner, dividing (2) by (1), member by member, we have,

$$\cot (90° + a) = - \tan a.$$

Taking the reciprocals of both members of (2), we have (formulas 11 and 12, Art. 61),

$$\sec (90° + a) = - \operatorname{cosec} a.$$

In like manner, taking the reciprocals of both members of (1), we have,

$$\operatorname{cosec} (90° + a) = \sec a.$$

Again, let $M''C = AM = a$; then $AM'' = 180° - a$. As before, the right-angled triangles $OP''M''$ and OPM may be proved equal in all respects, giving the *numerical* relations, $P''M'' = PM$, and $OP'' = OP$, and, by the application of the rules for signs, Art. 58, may be obtained, $P''M'' = PM$, and $OP'' = - OP$; hence,

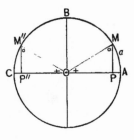

$$\sin (180° - a) = \sin a ; \cdot \ \cdot \ \cdot \ \cdot \ \cdot \ \cdot \ \text{(1.)}$$

$$\cos (180° - a) = - \cos a. \cdot \ \cdot \ \cdot \ \cdot \ \cdot \ \text{(2.)}$$

From these equations (1) and (2), and formulas (6), (7), (11), and (12), Art. 61, may be obtained, as before,

$$\tan (180° - a) = - \tan a ;$$

$$\cot (180° - a) = - \cot a ;$$

$$\sec (180° - a) = - \sec a ;$$

$$\operatorname{cosec} (180° - a) = \operatorname{cosec} a.$$

In like manner, the values of the several functions of the remaining arcs in question may be obtained in terms of functions of the arc a. Tabulating the results, we have the following

TABLE III.

Arc = 90° + a.		Arc = 270° − a.	
sin = cos a,	cos = − sin a,	sin = − cos a,	cos = − sin a,
tan = − cot a,	cot = − tan a,	tan = cot a,	cot = tan a,
sec = − cosec a,	cosec = sec a.	sec = − cosec a,	cosec = − sec a.
Arc = 180° − a.		**Arc = 270° + a.**	
sin = sin a,	cos = − cos a,	sin = − cos a,	cos = sin a,
tan = − tan a,	cot = − cot a,	tan = − cot a,	cot = − tan a,
sec = − sec a,	cosec = cosec a.	sec = cosec a,	cosec = − sec a.
Arc = 180° + a.		**Arc = 360° − a.**	
sin = − sin a,	cos = − cos a,	sin = − sin a,	cos = cos a,
tan = tan a,	cot = cot a,	tan = − tan a,	cot = − cot a,
sec = − sec a,	cosec = − cosec a.	sec = sec a,	cosec = − cosec a.

It will be observed that, when the arc is added to, or subtracted from, an *even* number of quadrants, the name of the function is the *same* in both columns; and when the arc is added to, or subtracted from, an *odd* number of quadrants, the names of the functions in the two columns are *contrary:* in all cases, the algebraic sign is determined by the rules already given (Art. 58).

By means of this table, we may find the functions of any arc in terms of the functions of an arc less than 90°. Thus,

$$\sin 115° = \sin (90° + 25°) = \cos 25°,$$

$$\sin 284° = \sin (270° + 14°) = -\cos 14°,$$

$$\sin 400° = \sin (360° + 40°) = \sin 40°,$$

$$\tan 210° = \tan (180° + 30°) = \tan 30°.$$

&c. &c. &c.

PARTICULAR VALUES OF CERTAIN FUNCTIONS.

64. Let MAM′ be any arc, denoted by $2a$, M′M its chord, and OA a radius drawn perpendicular to M′M : then will PM $= \frac{1}{2}$M′M, and AM $= \frac{1}{2}$M′AM (B. III., P. VI.). But PM is the sine of AM, or, PM $= \sin a$: hence,

$$\sin a = \tfrac{1}{2}\text{M′M} ;$$

that is, *the sine of an arc is equal to one half the chord of twice the arc.*

Let M′AM $= 60°$; then will AM $= 30°$, and M′M will equal the radius, or 1 (B. V., P. IV.): hence, we have

$$\sin 30° = \tfrac{1}{2} ;$$

that is, *the sine of 30° is equal to half the radius.*

Also, $\qquad \cos 30° = \sqrt{1 - \sin^2 30°} = \tfrac{1}{2}\sqrt{3} ;$

hence, $\qquad \tan 30° = \dfrac{\sin 30°}{\cos 30°} = \sqrt{\dfrac{1}{3}}.$

Again, let M′AM $= 90°$: then will AM $= 45°$, and M′M $= \sqrt{2}$ (B. V., P. III.): hence, we have

$$\sin 45° = \tfrac{1}{2}\sqrt{2} ;$$

Also, $\qquad \cos 45° = \sqrt{1 - \sin^2 45°} = \tfrac{1}{2}\sqrt{2} ;$

hence, $\qquad \tan 45° = \dfrac{\sin 45°}{\cos 45°} = 1.$

Many other numerical values might be deduced.

FORMULAS

EXPRESSING RELATIONS BETWEEN THE CIRCULAR FUNCTIONS OF DIFFERENT ARCS.

65. Let AB and BM represent two arcs, having the common radius 1; denote the first by a, and the second by b; then, AM $=$ $a + b$. From M draw PM perpendicular to CA, and NM perpendicular to CB; from N draw NP′ perpendicular, and NL parallel, to CA.

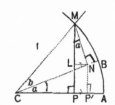

Then, by definition, we have

PM $=$ sin $(a + b)$,   NM $=$ sin b,   and   CN $=$ cos b.

From the figure, we have

$$PM \ = \ PL + LM. \ \cdot \ \cdot \ \cdot \ \cdot \ \cdot \ \cdot \ (1.)$$

From the right-angled triangle CP′N (Art. 37), we have

$$P'N \ = \ CN \sin a \, ;$$

or, since $P'N \ = \ PL,$

$$PL \ = \ \cos b \sin a \ = \ \sin a \cos b.$$

Since the triangle MLN is similar to CP′N (B. IV., P. XXI.), the angle LMN is equal to the angle P′CN; hence, from the right-angled triangle MLN, we have

$$LM \ = \ NM \cos a \ = \ \sin b \cos a \ = \ \cos a \sin b.$$

Substituting the values of PM, PL, and LM, in equation (1), we have

$$\sin (a + b) \ = \ \sin a \cos b + \cos a \sin b \, ; \ \cdot \ (A.)$$

that is, *the sine of the sum of two arcs is equal to the sine of the first into the cosine of the second, plus the cosine of the first into the sine of the second.*

Since the above formula is true for any values of a and b, we may substitute $-b$ for b; whence,

$$\sin (a - b) = \sin a \cos (- b) + \cos a \sin (- b);$$

but (Art. 62),

$$\cos (- b) = \cos b, \quad \text{and} \quad \sin (- b) = - \sin b;$$

hence, $\quad \sin (a - b) = \sin a \cos b - \cos a \sin b;$ · **(B.)**

that is, *the sine of the difference of two arcs is equal to the sine of the first into the cosine of the second, minus the cosine of the first into the sine of the second.*

If, in formula **(B)**, we substitute $(90° - a)$, for a, we have

$$\sin (90°-a-b) = \sin (90°-a) \cos b - \cos (90°-a) \sin b; \quad (2.)$$

but (Art. 63),

$$\sin (90° - a - b) = \sin [90° - (a + b)] = \cos (a + b),$$

and, $\qquad\qquad\qquad \sin (90° - a) = \cos a,$

$$\cos (90° - a) = \sin a;$$

hence, by substitution in equation (2), we have

$$\cos (a + b) = \cos a \cos b - \sin a \sin b; \quad · \quad \textbf{(C.)}$$

that is, *the cosine of the sum of two arcs is equal to the rectangle of their cosines, minus the rectangle of their sines.*

If, in formula **(C)**, we substitute $-b$, for b, we find

$$\cos (a - b) = \cos a \cos (- b) - \sin a \sin (- b),$$

or, $\qquad \cos (a - b) = \cos a \cos b + \sin a \sin b;$ · · **(D.)**

that is, *the cosine of the difference of two arcs is equal to the rectangle of their cosines, plus the rectangle of their sines.*

If we divide formula **(A)** by formula **(C)**, member by member, we have

$$\frac{\sin (a + b)}{\cos (a + b)} = \frac{\sin a \cos b + \cos a \sin b}{\cos a \cos b - \sin a \sin b}.$$

Dividing both terms of the second member by $\cos a \cos b$, recollecting that the sine divided by the cosine is equal to the tangent, we find

$$\tan (a + b) = \frac{\tan a + \tan b}{1 - \tan a \tan b}; \quad \cdots \quad \textbf{(E.)}$$

that is, *the tangent of the sum of two arcs, is equal to the sum of their tangents, divided by 1 minus the rectangle of their tangents.*

If, in formula **(E)**, we substitute $- b$ for b, recollecting that $\tan (- b) = - \tan b$, we have

$$\tan (a - b) = \frac{\tan a - \tan b}{1 + \tan a \tan b}; \quad \cdots \quad \textbf{(F.)}$$

that is, *the tangent of the difference of two arcs is equal to the difference of their tangents, divided by 1 plus the rectangle of their tangents.*

In like manner, dividing formula **(C)** by formula **(A)**, member by member, and reducing, we have

$$\cot (a + b) = \frac{\cot a \cot b - 1}{\cot a + \cot b}; \quad \cdots \quad \textbf{(G.)}$$

and thence, by the substitution of $-b$ for b,

$$\cot(a-b) = \frac{\cot a \cot b + 1}{\cot b - \cot a} \quad \cdots \quad \textbf{(H.)}$$

FUNCTIONS OF DOUBLE ARCS AND HALF ARCS.

66. If, in formulas **(A)**, **(C)**, **(E)**, and **(G)**, we make $b = a$, we find

$$\sin 2a = 2 \sin a \cos a; \quad \cdots \quad \textbf{(A'.)}$$

$$\cos 2a = \cos^2 a - \sin^2 a; \quad \cdots \quad \textbf{(C'.)}$$

$$\tan 2a = \frac{2 \tan a}{1 - \tan^2 a}; \quad \cdots \quad \textbf{(E'.)}$$

$$\cot 2a = \frac{\cot^2 a - 1}{2 \cot a}. \quad \cdots \quad \textbf{(G'.)}$$

Substituting in **(C')** for $\cos^2 a$, its value, $1 - \sin^2 a$; and afterwards for $\sin^2 a$, its value, $1 - \cos^2 a$, we have

$$\cos 2a = 1 - 2 \sin^2 a,$$

$$\cos 2a = 2 \cos^2 a - 1;$$

whence, by solving these equations,

$$\sin a = \sqrt{\frac{1 - \cos 2a}{2}}; \quad \cdots \quad \textbf{(1.)}$$

$$\cos a = \sqrt{\frac{1 + \cos 2a}{2}}. \quad \cdots \quad \textbf{(2.)}$$

We also have, from the same equations,

$$1 - \cos 2a = 2 \sin^2 a; \quad \cdots \quad \textbf{(3.)}$$

$$1 + \cos 2a = 2 \cos^2 a. \quad \cdots \quad \textbf{(4.)}$$

Dividing equation (**A′**), first by equation (4), and then by equation (3), member by member, we have

$$\frac{\sin 2a}{1 + \cos 2a} = \tan a; \quad \cdots \quad \text{(5.)}$$

$$\frac{\sin 2a}{1 - \cos 2a} = \cot a. \quad \cdots \quad \text{(6.)}$$

Substituting $\tfrac{1}{2}a$ for a, in equations (1), (2), (5), and (6), we have

$$\sin \tfrac{1}{2}a = \sqrt{\frac{1 - \cos a}{2}}; \quad \cdots \quad \text{(A″.)}$$

$$\cos \tfrac{1}{2}a = \sqrt{\frac{1 + \cos a}{2}}; \quad \cdots \quad \text{(C″.)}$$

$$\tan \tfrac{1}{2}a = \frac{\sin a}{1 + \cos a}; \quad \cdots \quad \text{(E″.)}$$

$$\cot \tfrac{1}{2}a = \frac{\sin a}{1 - \cos a}. \quad \cdots \quad \text{(G″.)}$$

Taking the reciprocals of both members of the last two formulas, we have also,

$$\cot \tfrac{1}{2}a = \frac{1 + \cos a}{\sin a}, \quad \text{and} \quad \tan \tfrac{1}{2}a = \frac{1 - \cos a}{\sin a}.$$

ADDITIONAL FORMULAS.

67. If formulas (**A**) and (**B**) are first added, member to member, and then subtracted, member from member, and the same operations are performed upon (**C**) and (**D**), we obtain

$$\sin (a + b) + \sin (a - b) = 2 \sin a \cos b;$$

$$\sin (a + b) - \sin (a - b) = 2 \cos a \sin b;$$

$$\cos (a + b) + \cos (a - b) = 2 \cos a \cos b;$$

$$\cos (a - b) - \cos (a + b) = 2 \sin a \sin b.$$

If in these we make

$$a + b = p, \qquad \text{and} \qquad a - b = q,$$

whence,

$$a = \tfrac{1}{2} (p + q), \qquad\qquad b = \tfrac{1}{2} (p - q);$$

and then substitute in the above formulas, we obtain

$$\sin p + \sin q = 2 \sin \tfrac{1}{2} (p + q) \cos \tfrac{1}{2} (p - q). \quad \cdot \quad \textbf{(K.)}$$

$$\sin p - \sin q = 2 \cos \tfrac{1}{2} (p + q) \sin \tfrac{1}{2} (p - q). \quad \cdot \quad \textbf{(L.)}$$

$$\cos p + \cos q = 2 \cos \tfrac{1}{2} (p + q) \cos \tfrac{1}{2} (p - q). \quad \cdot \quad \textbf{(M.)}$$

$$\cos q - \cos p = 2 \sin \tfrac{1}{2} (p + q) \sin \tfrac{1}{2} (p - q). \quad \cdot \quad \textbf{(N.)}$$

From formulas **(L)** and **(K)**, by division, we obtain

$$\frac{\sin p - \sin q}{\sin p + \sin q} = \frac{\cos \tfrac{1}{2} (p + q) \sin \tfrac{1}{2} (p - q)}{\sin \tfrac{1}{2} (p + q) \cos \tfrac{1}{2} (p - q)}$$

$$= \frac{\tan \tfrac{1}{2} (p - q)}{\tan \tfrac{1}{2} (p + q)}. \quad \cdot \quad \cdot \quad \cdot \quad \cdot \quad \cdot \quad \textbf{(1.)}$$

Hence, since p and q represent any arcs whatever, *the sum of the sines of two arcs is to their difference, as the tangent of one half the sum of the arcs is to the tangent of one half their difference.*

Also, in like manner, we obtain

$$\frac{\sin p + \sin q}{\cos p + \cos q} = \frac{\sin \frac{1}{2}(p+q)\cos \frac{1}{2}(p-q)}{\cos \frac{1}{2}(p+q)\cos \frac{1}{2}(p-q)} = \tan \frac{1}{2}(p+q), \quad (2.)$$

$$\frac{\sin p - \sin q}{\cos p + \cos q} = \frac{\cos \frac{1}{2}(p+q)\sin \frac{1}{2}(p-q)}{\cos \frac{1}{2}(p+q)\cos \frac{1}{2}(p-q)} = \tan \frac{1}{2}(p-q), \quad (3.)$$

$$\frac{\sin p + \sin q}{\sin (p+q)} = \frac{\sin \frac{1}{2}(p+q)\cos \frac{1}{2}(p-q)}{\sin \frac{1}{2}(p+q)\cos \frac{1}{2}(p+q)} = \frac{\cos \frac{1}{2}(p-q)}{\cos \frac{1}{2}(p+q)}, \quad (4.)$$

$$\frac{\sin p - \sin q}{\sin (p+q)} = \frac{\sin \frac{1}{2}(p-q)\cos \frac{1}{2}(p+q)}{\sin \frac{1}{2}(p+q)\cos \frac{1}{2}(p+q)} = \frac{\sin \frac{1}{2}(p-q)}{\sin \frac{1}{2}(p+q)}, \quad (5.)$$

$$\frac{\sin (p-q)}{\sin p - \sin q} = \frac{\sin \frac{1}{2}(p-q)\cos \frac{1}{2}(p-q)}{\sin \frac{1}{2}(p-q)\cos \frac{1}{2}(p+q)} = \frac{\cos \frac{1}{2}(p-q)}{\cos \frac{1}{2}(p+q)}, \quad (6.)$$

all of which give proportions analogous to that deduced from formula (1).

Since the second members of (6) and (4) are the same, we have

$$\frac{\sin p - \sin q}{\sin (p - q)} = \frac{\sin (p + q)}{\sin p + \sin q}; \quad \cdots \quad (7.)$$

that is, *the sine of the difference of two arcs is to the difference of the sines, as the sum of the sines is to the sine of the sum.*

All of the preceding formulas may be made homogeneous in terms of R, R being any radius, as explained in Art. 30; or, we may simply introduce R, as a factor, into each term as many times as may be necessary to render all of its terms of the same degree.

METHOD OF COMPUTING A TABLE OF NATURAL SINES.

68. Since the length of the semi-circumference of a circle whose radius is 1, is equal to the number 3.14159265..., if we divide this number by 10800, the number of minutes in 180°, the quotient, .0002908882..., will be the length of the arc of *one minute;* and since this arc is so small that it does not differ materially from its sine or tangent, this may be placed in the table as *the sine of one minute.*

Formula (3) of Table II., gives

$$\cos 1' = \sqrt{1 - \sin^2 1'} = .9999999577. \quad \cdot \quad (1.)$$

Having thus determined, to a near degree of approximation, the sine and cosine of one minute, we take the first formula of Art. 67, and put it under the form,

$$\sin (a + b) = 2 \sin a \cos b - \sin (a - b),$$

and make in this, $b = 1'$, and then in succession,

$$a = 1', \quad a = 2', \quad a = 3', \quad a = 4', \quad \&c.,$$

and obtain,

$$\sin 2' = 2 \sin 1' \cos 1' - \sin 0 = .0005817764 \ldots$$

$$\sin 3' = 2 \sin 2' \cos 1' - \sin 1' = .0008726646 \ldots$$

$$\sin 4' = 2 \sin 3' \cos 1' - \sin 2' = .0011635526 \ldots$$

$$\sin 5' = \qquad \&c.,$$

thus obtaining the sine of every number of degrees and minutes from $1'$ to $45°$.

The cosines of the corresponding arcs may be computed by means of equation (1).

Having found the sines and cosines of arcs less than 45°, those of the arcs between 45° and 90° may be deduced, by considering that the sine of an arc is equal to the cosine of its complement, and the cosine equal to the sine of its complement. Thus,

$$\sin 50° = \sin(90° - 40°) = \cos 40°, \qquad \cos 50° = \sin 40°,$$

in which the second members are known from the previous computations.

To find the tangent of any arc, divide its sine by its cosine. To find the cotangent, take the reciprocal of the corresponding tangent.

As the accuracy of the calculation of the sine of any arc, by the above method, depends upon the accuracy of each previous calculation, it would be well to verify the work, by calculating the sines of the degrees separately (after having found the sines of one and two degrees), by the last proportion of Art. 67. Thus,

$$\sin 1° \ : \ \sin 2° - \sin 1° \ :: \ \sin 2° + \sin 1° \ : \ \sin 3°;$$

$$\sin 2° \ : \ \sin 3° - \sin 1° \ :: \ \sin 3° + \sin 1° \ : \ \sin 4°; \quad \&c.$$

SPHERICAL TRIGONOMETRY.

69. SPHERICAL TRIGONOMETRY is that branch of Mathematics which treats of the solution of spherical triangles.

In every spherical triangle there are six parts: three sides and three angles. In general, any three of these parts being given, the remaining parts may be found.

GENERAL PRINCIPLES.

70. For the purpose of deducing the formulas required in the solution of spherical triangles, we shall suppose the triangles to be situated on spheres whose radii are equal to 1. The formulas thus deduced may be rendered applicable to triangles lying on any sphere, by making them homogeneous in terms of the radius of that sphere, as explained in Art. 30. The only cases considered will be those in which each of the sides and angles is less than 180°.

Any angle of a spherical triangle is the same as the diedral angle included by the planes of its sides, and its measure is equal to that of the angle included between two right lines, one in each plane, and both perpendicular to their common intersection at the same point (B. VI., D. 4).

The radius of the sphere being equal to 1, each side of the triangle will measure the angle, at the centre, subtended by it. Thus, in the triangle ABC, the angle at A

is the same as that included between the planes AOC and AOB; and the side a is the measure of the plane angle BOC, O being the centre of the sphere, and OB the radius, equal to 1.

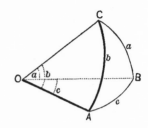

71. Spherical triangles, like plane triangles, are divided into two classes, *right-angled spherical tri-angles*, and *oblique-angled spherical triangles*. Each class will be considered in turn.

We shall, as before, denote the angles by the capital letters A, B, and C, and the sides opposite by the small letters a, b, and c.

FORMULAS
USED IN SOLVING RIGHT-ANGLED SPHERICAL TRIANGLES.

72. Let CAB be a sperical triangle, right-angled at A, and let O be the centre of the sphere on which it is situated. Denote the angles of the triangle by the letters A, B, and C, and the sides opposite by the letters a, b, and c, recollecting that B and C may change places, provided that b and c change places at the same time.

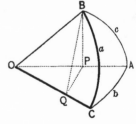

Draw OA, OB, and OC, each equal to 1. From B, draw BP perpendicular to OA, and from P draw PQ perpendicular to OC; then join the points Q and B, by the line QB. The line QB will be perpendicular to OC (B. VI., P. VI.), and the angle PQB will be equal to the inclination of the planes OCB and OCA; that is, it will be equal to the spherical angle C.

We have, from the figure,

$$PB = \sin c, \qquad OP = \cos c, \qquad QB = \sin a, \qquad OQ = \cos a.$$

From the right-angled triangles OQP and QPB, we have

$$OQ = OP \cos AOC; \qquad or, \qquad \cos a = \cos c \cos b. \quad \cdot \quad (1.)$$

$$PB = QB \sin PQB; \qquad or, \qquad \sin c = \sin a \sin C. \quad \cdot \quad (2.)$$

From the right-angled triangle QPB, we have

$$\cos PQB, \ \ or \ \ \cos C = \frac{QP}{QB};$$

but, from the right-angled triangle PQO, we have

$$QP = OQ \tan QOP = \cos a \tan b;$$

substituting for QP and QB their values, we have

$$\cos C = \frac{\cos a \tan b}{\sin a} = \cot a \tan b. \quad \cdot \quad \cdot \quad (3.)$$

From the right-angled triangle OQP, we have

$$\sin QOP, \ \ or \ \ \sin b = \frac{QP}{OP};$$

but, from the right-angled triangle QPB, we have

$$QP = PB \cot PQB = \sin c \cot C;$$

substituting for QP and OP their values, we have

$$\sin b = \frac{\sin c \cot C}{\cos c} = \tan c \cot C. \quad \cdot \quad \cdot \quad \cdot \quad (4.)$$

If, in (2), we change c and C into b and B, we have

$$\sin b = \sin a \sin B. \quad \cdot \quad \cdot \quad \cdot \quad \cdot \quad \cdot \quad (5.)$$

If, in (3), we change b and C into c and B, we have

$$\cos B = \cot a \tan c. \quad \cdot \quad \cdot \quad \cdot \quad \quad \cdot \quad (6.)$$

If, in (4), we change b, c, and C, into c, b, and B, we have

$$\sin c = \tan b \cot B. \quad \cdot \quad \cdot \quad \cdot \quad \cdot \quad \cdot \quad (7.)$$

Multiplying (4) by (7), member by member, we have

$$\sin b \sin c = \tan b \tan c \cot B \cot C.$$

Dividing both members by $\tan b \tan c$, we have

$$\cos b \cos c = \cot B \cot C ;$$

and substituting for $\cos b \cos c$, its value, $\cos a$, taken from (1), we have

$$\cos a = \cot B \cot C. \quad \cdot \quad \cdot \quad \cdot \cdot \quad \cdot \quad (8.)$$

Formula (6) may be written under the form

$$\cos B = \frac{\cos a \sin c}{\sin a \cos c}.$$

Substituting for $\cos a$, its value, $\cos b \cos c$, taken from (1), and reducing, we have

$$\cos B = \frac{\cos b \sin c}{\sin a}.$$

Again, substituting for $\sin c$, its value, $\sin a \sin C$, taken from (2), and reducing, we have

$$\cos B = \cos b \sin C. \quad \cdot \quad \cdot \quad \cdot \quad \cdot \quad \cdot \quad \textbf{(9.)}$$

Changing B, b, and C, in (9), into C, c, and B, we have

$$\cos C = \cos c \sin B. \quad \cdot \quad \cdot \quad \cdot \quad \cdot \quad \cdot \quad (10.)$$

These ten formulas are sufficient for the solution of any right-angled spherical triangle whatever. For the purpose of classifying them under two general rules, and for convenience in remembering them, these formulas are usually put under other forms by the use of

NAPIER'S CIRCULAR PARTS.

73. *The two sides about the right angle, the complements of their opposite angles, and the complement of the hypothenuse*, are called Napier's Circular Parts.

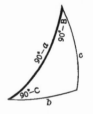

If we take *any three* of the five parts, as shown in the figure, they will either be *adjacent* to each other, or one of them will be separated from each of the two others by an intervening part. In the first case, the one lying between the two other parts is called the *middle part*, and the two others, *adjacent parts*. In the second case, the one separated from both the other parts, is called the *middle part*, and the two others, *opposite parts*. Thus, if $90°-a$ is the middle part, $90° - B$ and $90° - C$ are *adjacent parts;* and b and c are *opposite parts;* if c is the middle part, b and $90° - B$ are *adjacent parts* (the right angle not being considered), and $90° - C$ and $90° - a$ are *opposite parts:* and similarly, for each of the other parts, taken as a middle part.

74. Let us now consider, in succession, each of the five parts as a middle part, when the two other parts are opposite. Beginning with the hypothenuse, we have, from formulas (1), (2), (5), (9), and (10), Art. 72,

$$\sin (90° - a) = \cos b \cos c ; \cdot \cdot \cdot \cdot \cdot \cdot \cdot \quad (1.)$$

$$\sin c \qquad = \cos (90° - a) \cos (90° - C) ; \quad (2.)$$

$$\sin b \qquad = \cos (90° - a) \cos (90° - B) ; \quad (3.)$$

$$\sin (90° - B) = \cos b \cos (90° - C) ; \cdot \cdot \cdot \cdot \quad (4.)$$

$$\sin (90° - C) = \cos c \cos (90° - B). \cdot \cdot \cdot \cdot \quad (5.)$$

Comparing these formulas with the figure, we see that

The sine of the middle part is equal to the rectangle of the cosines of the opposite parts.

Let us now take the same middle parts, and the other parts adjacent. Formulas (8), (7), (4), (6), and (3), Art. 72, give

$$\sin (90° - a) = \tan (90° - B) \tan (90° - C) ; \quad (6.)$$

$$\sin c \qquad = \tan b \tan (90° - B) ; \quad \cdot \cdot \cdot \quad (7.)$$

$$\sin b \qquad = \tan c \tan (90° - C) ; \quad \cdot \cdot \cdot \quad (8.)$$

$$\sin (90° - B) = \tan (90° - a) \tan c ; \quad \cdot \cdot \cdot \quad (9.)$$

$$\sin (90° - C) = \tan (90° - a) \tan b. \quad \cdot \cdot \cdot (10.)$$

Comparing these formulas with the figure, we see that

The sine of the middle part is equal to the rectangle of the tangents of the adjacent parts.

These two rules are called Napier's rules for circular parts, and are sufficient to solve any right-angled spherical triangle.

75. In applying Napier's rules for circular parts, the part sought will be determined by its sine. Now, the same sine corresponds to two different arcs, or angles, supplements of each other; it is, therefore, necessary to discover such relations between the given and the required parts, as will serve to point out which of the two arcs, or angles, is to be taken.

Two parts of a spherical triangle are said to be of *the same species*, when they are each less than 90°, or each greater than 90°; and of *different species*, when one is less and the other greater than 90°.

From formulas (9) and (10), Art. 72, we have,

$$\sin C = \frac{\cos B}{\cos b}, \quad \text{and} \quad \sin B = \frac{\cos C}{\cos c};$$

since the angles B and C are each less than 180°, their sines must always be positive: hence, cos B must have the same sign as cos b, and the cos C must have the same sign as cos c. This can only be the case when B is of the same species as b, and C of the same species as c; that is, *each side about the right angle is always of the same species as its opposite angle.*

From formula (1), we see that when a is less than 90°, or when cos a is positive, the cosines of b and c will have the same sign; and hence, b and c will be of the *same species:* when a is greater than 90°, or when cos a is negative, the cosines of b and c will have contrary signs, and hence b and c will be of *different species:*

therefore, *when the hypothenuse is less than* 90°, *the two sides about the right angle, and consequently the two oblique angles, will be of the same species; when the hypothenuse is greater than* 90°, *the two sides about the right angle, and consequently the two oblique angles, will be of different species.*

These two principles enable us to determine the nature of the part sought, in every case, except when an oblique angle and the side opposite are given, to find the remaining parts. In this case, there may be *two solutions, one solution,* or *no solution.*

There may be two cases:

1°. Let there be given B and *b*, and B *acute.* Construct B and prolong its sides till they meet in B'. Then will BCB' and BAB' be semi-circumferences of great 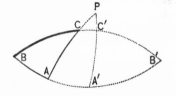 circles, and the spherical angles B and B' will be equal to each other. As B is acute, its measure is the *longest* arc of a great circle that can be drawn perpendicular to the side BA and included between the sides of the angle B (B. IX., Gen. S. 2); hence, if the given side is *greater* than the measure of the given angle opposite, that is, if *b* > B, no triangle can be constructed, that is, there can be *no solution:* if *b* = B, BC' and BA' will each be a quadrant (B. IX., P. IV.), and the triangle BA'C', or its equal B'A'C', will be birectangular (B. IX., P. XIV., C. 3), and there will be but *one solution:* if *b* < B, there will be *two solutions,* BAC and B'AC, the required parts of one being supplements of the required parts of the other.

Since B < 90°, if *b* < B, *b* differs *more* from 90° than B does; and' if *b* > B, *b* differs *less* from 90° than B.

2d. Let B be *obtuse*. Construct B as before. As B is
obtuse, its measure is the *short-
est* arc of a great circle that can
be drawn perpendicular to the
side BA and included between
the sides of the angle B (B. IX.,
Gen. S. 2); hence, if $b <$ B, there
can be *no solution:* if $b =$ B, the

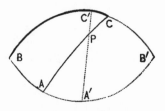

corresponding triangle, BA′C′ or B′A′C′, will be birectangular
and there will be but *one solution,* as before: and if
$b >$ B, there will be *two solutions,* BAC and B′AC.

Since B $> 90°$, if $b >$ B, b differs *more* from $90°$ than
B does; and if $b <$ B, b differs *less* from $90°$ than B.

Hence, it appears, from both cases, that

If b differs *more* from $90°$ than B, there will be *two
solutions,* the required parts in the one case being supplements of the required parts in the other case.

If $b =$ B, the triangle will be birectangular, and there
will be but *one solution.*

If b differs *less* from $90°$ than B, the triangle can not
be constructed, that is, there will be *no solution.*

SOLUTION OF RIGHT-ANGLED SPHERICAL TRI-ANGLES.

76. In a right-angled spherical triangle, the right
angle is always known. If any two of the other parts
are given, the remaining parts may be found by Napier's
rules for circular parts. Six cases may arise. There may
be given,

I. The hypothenuse and one side.

II. The hypothenuse and one oblique angle.

III. The two sides about the right angle.

IV. One side and its adjacent angle.

V. One side and its opposite angle.

VI. The two oblique angles.

In any one of these cases, we select that part which is either adjacent to, or separated from, each of the other given parts, and calling it the middle part, we employ that one of Napier's rules which is applicable. Having determined a third part, the two others may then be found in a similar manner. It is to be observed, that the formulas employed are to be rendered homogeneous, in terms of R, as explained in Art. 30. This is done by simply multiplying the radius, R, into the middle part.

Examples.

1. Given $a = 105°\ 17'\ 29''$, and $b = 38°\ 47'\ 11''$, to find C, c, and B.

Since $a > 90°$, b and c must be of different species, that is, $c > 90°$, and hence C $> 90°$.

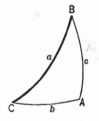

Operation.

Formula (10), Art. 74, gives for $90° -$ C, middle part,

$$\log \cos C = \log \cot a + \log \tan b - 10;$$

$\log \cot a$ ($105°\ 17'\ 29''$)	9.436811
$\log \tan b$ ($38°\ 47'\ 11''$)	9.905055
$\log \cos C$ · · ·	9.341866 $\therefore$ C $= 102°\ 41'\ 33''$.

Formula (2), Art. 74, gives for c, middle part,

$$\log \sin c = \log \sin a + \log \sin C - 10;$$

$\log \sin a$ $(105° 17' 29'')$ 9.984346
$\log \sin C$ $(102° 41' 33'')$ 9.989256
     $\log \sin c$ $\cdot$ $\cdot$ $\cdot$ 9.973602 $\therefore$ $c = 109° 46' 32''.$

Formula (4) gives for $90° - B$, middle part,

$$\log \cos B = \log \sin C + \log \cos b - 10;$$

$\log \sin C$ $(102° 41' 33'')$ 9.989256
$\log \cos b$ $(38° 47' 11'')$ 9.891808
     $\log \cos B$ $\cdot$ $\cdot$ $\cdot$ 9.881064 $\therefore$ $B = 40° 29' 50''.$

Ans. $c = 109° 46' 32''$, $B = 40° 29' 50''$, $C = 102° 41' 33''.$

It is better, in all cases, to find the required parts in terms of the two given parts. This may always be done by one of the formulas of Art. 74. Select the formula which contains the two given parts and the required part, and transform it, if necessary, so as to find the required part in terms of the given parts.

Thus, let a and B be given, to find C. Regarding $90° - a$ as a middle part, we have, from formula (6),

$$\cos a = \cot B \cot C;$$

whence, $\cot C = \dfrac{\cos a}{\cot B};$

and, by the application of logarithms,

$$\log \cot C = \log \cos a + (a. c.) \log \cot B;$$

from which C may be found. In like manner, other cases may be treated.

2. Given $b = 51° 30'$, and $B = 58° 35'$, to find a, c, and C.

Because $b < B$, there are two solutions.

Operation.

Formula (7) gives for c, middle part,

$$\log \sin c = \log \tan b + \log \cot B - 10 ;$$

log tan b	(51° 30')	10.099395
log cot B	(58° 35')	9.785900
log sin c · · ·		9.885295

$\therefore\ c = 50° 09' 51''$,

and $c' = 129° 50' 09''$.

Formula (3) gives

$$\sin b = \sin a \sin B,$$

whence,

$$\sin a = \frac{\sin b}{\sin B},$$

and hence,

$$\log \sin a = \log \sin b + (\text{a. c.}) \log \sin B ;$$

log sin b	(51° 30')	9.893544
(a. c.) log sin B	(58° 35')	0.068848
log sin a · ·		9.962392

$\therefore\ a = 66° 29' 53''$,

$a' = 113° 30' 07''$.

Formula (4) gives

$$\cos B = \cos b \sin C,$$

whence,

$$\sin C = \frac{\cos B}{\cos b},$$

and hence,

$$\log \sin C = \log \cos B + (\text{a. c.}) \log \cos b ;$$

log cos B	(58° 35')	9.717053
(a. c.) log cos b	(51° 30')	0.205850
log sin C · ·		9.922903

$\therefore\ C = 56° 51' 38''$,

$C' = 123° 08' 22''$.

As a *check*, to test the accuracy of the above work, formula (2) may be used. Thus, from that formula,

$$\log \sin c = \log \sin a + \log \sin C - 10.$$

As found above,

$$\begin{aligned}
\log \sin a \quad \cdot \quad \cdot \quad & 9.962392 \\
\log \sin C \quad \cdot \quad \cdot \quad & \underline{9.922903} \\
\log \sin c \quad \cdot \quad & 9.885295
\end{aligned}$$

As the test is satisfied, the work is probably correct. Other cases may be treated in like manner.

3. Given $a = 86° 51'$, and $B = 18° 03' 32''$, to find b, c, and C.

Ans. $b = 18° 01' 50''$, $c = 86° 41' 14''$, $C = 88° 58' 25''$.

4. Given $b = 155° 27' 54''$, and $c = 29° 46' 08''$, to find a, B, and C.

Ans. $a = 142° 09' 13''$, $B = 137° 24' 21''$, $C = 54° 01' 16''$.

5. Given $c = 73° 41' 35''$, and $B = 99° 17' 33''$, to find a, b, and C.

Ans. $a = 92° 42' 17''$, $b = 99° 40' 30''$, $C = 73° 54' 47''$.

6. Given $b = 115° 20'$, and $B = 91° 01' 47''$, to find a, c, and C.

$a = 64° 41' 11''$, $c = 177° 49' 27''$, $C = 177° 35' 36''$.
$a' = 115° 18' 49''$, $c' = 2° 10' 33''$, $C' = 2° 24' 24''$.

7. Given $B = 47° 13' 43''$, and $C = 126° 40' 24''$, to find a, b, and c.

Ans. $a = 133° 32' 26''$, $b = 32° 08' 56''$, $c = 144° 27' 03''$.

QUADRANTAL SPHERICAL TRIANGLES.

77. A QUADRANTAL SPHERICAL TRIANGLE is one in which one side is equal to 90°. To solve such a triangle, we pass to its supplemental polar triangle, by subtracting each side and each angle from 180° (B. IX., P. VI.). The resulting polar triangle will be right-angled, and may be solved by the rules already given. The supplemental polar triangle of any quadrantal triangle being solved, the parts of the given triangle may be found by subtracting each part of the supplemental triangle from 180°.

Example.

Let A′B′C′ be a quadrantal triangle, in which

$$B'C' = 90°,$$

$$B' = 75° \ 42',$$

and $$c' = 18° \ 37'.$$

Passing to the supplemental polar triangle, we have

$$A = 90°, \quad b = 104° \ 18', \quad \text{and} \quad C = 161° \ 23'.$$

Solving this triangle by previous rules, we find

$$a = 76° \ 25' \ 11'', \quad c = 161° \ 55' \ 20'', \quad B = 94° \ 31' \ 21'';$$

hence, the required parts of the given quadrantal triangle are,

$$A' = 103° \ 34' \ 49'', \quad C' = 18° \ 04' \ 40'', \quad b' = 85° \ 28' \ 39'',$$

Other quadrantal triangles may be solved in like manner.

FORMULAS

USED IN SOLVING OBLIQUE-ANGLED SPHERICAL TRIANGLES.

78. To show that, in a spherical triangle, the sines of the sides are proportional to the sines of their opposite angles.

Let ABC represent an oblique-angled spherical triangle. From any vertex, as C, draw the arc of a great circle, CB', perpendicular to the opposite side. The two triangles ACB' and BCB' will be right-angled at B'.

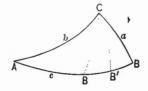

From the triangle ACB', we have, formula (2) Art. 74,

$$\sin CB' = \sin A \sin b.$$

From the triangle BCB', we have

$$\sin CB' = \sin B \sin a.$$

Equating these values of sin CB', we have

$$\sin A \sin b = \sin B \sin a;$$

from which results the proportion,

$$\sin a \ : \ \sin b \ :: \ \sin A \ : \ \sin B. \cdot \ \cdot \ \cdot \quad \textbf{(1.)}$$

In like manner, we may deduce

$$\sin a \ : \ \sin c \ :: \ \sin A \ : \ \sin C, \cdot \ \cdot \ \cdot \quad \textbf{(2.)}$$

$$\sin b \ : \ \sin c \ :: \ \sin B \ : \ \sin C. \cdot \ \cdot \ \cdot \quad \textbf{(3.)}$$

That is, in any spherical triangle, *the sines of the sides are proportional to the sines of their opposite angles.*

Had the perpendicular fallen on the prolongation of AB, the same relation would have been found.

79. To find an expression for the cosine of any side of a spherical triangle.

Let ABC represent any spherical triangle, and O the centre of the sphere on which it is situated. Draw the radii OA, OB, and OC; from C draw CP perpendicular to the plane AOB; from P, the foot of this perpendicular, draw PD and PE respectively perpendicular to OA and OB; join CD and CE, these lines will be respectively perpendicular to OA and OB 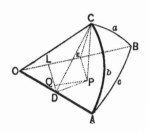 (B. VI., P. VI.), and the angles CDP and CEP will be equal to the angles A and B respectively. Draw DL and PQ, the one perpendicular, and the other parallel to OB. We then have

$$OE = \cos a, \quad DC = \sin b, \quad OD = \cos b.$$

We have from the figure,

$$OE = OL + QP. \quad \cdot \quad \cdot \quad \cdot \quad \cdot \quad \cdot \quad (1.)$$

In the right-angled triangle OLD,

$$OL = OD \cos DOL = \cos b \cos c.$$

The right-angled triangle PQD has its sides respectively perpendicular to those of OLD; it is, therefore, similar to it, and the angle QDP is equal to c, and we have

$$QP = PD \sin QDP = PD \sin c. \quad \cdot \quad \cdot \quad \cdot \quad (2.)$$

The right-angled triangle CPD gives

$$PD = CD \cos CDP = \sin b \cos A;$$

substituting this value in (2), we have

$$QP = \sin b \sin c \cos A;$$

and now substituting these values of OE, OL, and QP, in (1), we have

$$\cos a = \cos b \cos c + \sin b \sin c \cos A. \quad \cdot \quad (3.)$$

In the same way, we may deduce,

$$\cos b = \cos a \cos c + \sin a \sin c \cos B, \quad \cdot \quad \cdot \quad (4.)$$

$$\cos c = \cos a \cos b + \sin a \sin b \cos C. \quad \cdot \quad \cdot \quad (5.)$$

That is, *the cosine of any side of a spherical triangle is equal to the rectangle of the cosines of the two other sides, plus the rectangle of the sines of these sides into the cosine of their included angle.*

80. To find an expression for the cosine of any angle of a spherical triangle.

If we represent the angles of the supplemental polar triangle of ABC, by A′, B′, and C′, and the sides by a', b', and c', we have (B. IX., P. VI.),

$$a = 180° - A', \quad b = 180° - B', \quad c = 180° - C',$$

$$A = 180° - a', \quad B = 180° - b', \quad C = 180° - c'.$$

Substituting these values in equation (3), of the preceding article, and recollecting that

$$\cos (180° - A') = - \cos A',$$

$$\sin (180° - B') = \sin B', \ \&c.,$$

we have

$$- \cos A' = \cos B' \cos C' - \sin B' \sin C' \cos a' \,;$$

or, changing the signs and omitting the primes (since the preceding result is true for any triangle),

$$\cos A = \sin B \sin C \cos a - \cos B \cos C. \quad \cdot \quad \cdot \quad (1.)$$

In the same way, we may deduce,

$$\cos B = \sin A \sin C \cos b - \cos A \cos C, \quad \cdot \quad \cdot \quad (2.)$$

$$\cos C = \sin A \sin B \cos c - \cos A \cos B. \quad \cdot \quad \cdot \quad (3.)$$

That is, *the cosine of any angle of a spherical triangle is equal to the rectangle of the sines of the two other angles into the cosine of their included side, minus the rectangle of the cosines of these angles.*

The formulas deduced in Arts. 79 and 80, for cos a, cos A, etc., are not convenient for use, as logarithms can not be applied to them; other formulas are, therefore, derived from them, to which logarithms may be applied.

81. To find an expression for the cosine of one half of any angle of a spherical triangle.

From equation (3), Art. 79, we deduce,

$$\cos A = \frac{\cos a - \cos b \cos c}{\sin b \sin c}. \quad \cdot \quad \cdot \quad \cdot \quad \cdot \quad (1.)$$

If we add this equation, member by member, to the number 1, and recollect that $1 + \cos A$, in the first member, is equal to $2 \cos^2 \frac{1}{2}A$ (Art. 66), and reduce, we have

$$2 \cos^2 \tfrac{1}{2}A = \frac{\sin b \sin c + \cos a - \cos b \cos c}{\sin b \sin c};$$

or, formula (C), Art. 65,

$$2 \cos^2 \tfrac{1}{2}A = \frac{\cos a - \cos (b + c)}{\sin b \sin c}. \quad \cdot \quad \cdot \quad \cdot \quad (2.)$$

And since, formula (N), Art. 67,

$$\cos a - \cos (b + c) = 2 \sin \tfrac{1}{2} (a + b + c) \sin \tfrac{1}{2} (b + c - a),$$

equation (2) becomes, after dividing both members by 2,

$$\cos^2 \tfrac{1}{2}A = \frac{\sin \tfrac{1}{2} (a + b + c) \sin \tfrac{1}{2} (b + c - a)}{\sin b \sin c}.$$

If in this we make

$$\tfrac{1}{2} (a + b + c) = \tfrac{1}{2}s \, ;$$

whence,
$$\tfrac{1}{2} (b + c - a) = \tfrac{1}{2}s - a,$$

and extract the square root of both members, we have

$$\cos \tfrac{1}{2}A = \sqrt{\frac{\sin \tfrac{1}{2}s \sin (\tfrac{1}{2}s - a)}{\sin b \sin c}}. \quad \cdot \quad \cdot \quad \cdot \quad (3.)$$

That is, *the cosine of one half of any angle of a spherical triangle is equal to the square root of the sine of one half of the sum of the three sides, into the sine of one half this sum minus the side opposite the angle, divided by the rectangle of the sines of the adjacent sides.*

If we subtract equation (1), of this article, member by member, from the number 1, and recollect that

$$1 - \cos A = 2 \sin^2 \tfrac{1}{2}A,$$

we find, after reduction,

$$\sin \tfrac{1}{2}A = \sqrt{\frac{\sin (\tfrac{1}{2}s - b) \sin (\tfrac{1}{2}s - c)}{\sin b \sin c}}. \quad \cdot \quad \cdot \quad (4.)$$

Dividing equation (4) by equation (3), member by member, we obtain

$$\tan \tfrac{1}{2}A = \sqrt{\frac{\sin (\tfrac{1}{2}s - b) \sin (\tfrac{1}{2}s - c)}{\sin \tfrac{1}{2}s \sin (\tfrac{1}{2}s - a)}}. \quad \cdot \quad \cdot \quad (5.)$$

82. From the foregoing values of the functions of one half of any angle, may be deduced values of the functions of one half of any side of a spherical triangle.

Representing the angles and sides of the supplemental polar triangle of ABC as in Art. 80, we have

$$A = 180° - a', \quad b = 180° - B', \quad c = 180° - C',$$

$$\tfrac{1}{2}s = 270° - \tfrac{1}{2}(A' + B' + C'),$$

$$\tfrac{1}{2}s - a = 90° - \tfrac{1}{2}(B' + C' - A').$$

Substituting these values in (3), Art. 81, and reducing by the aid of the formulas in Table III., Art. 63, we find

$$\sin \tfrac{1}{2}a' = \sqrt{\frac{-\cos \tfrac{1}{2}(A' + B' + C') \cos \tfrac{1}{2}(B' + C' - A')}{\sin B' \sin C'}}.$$

Place $\qquad \tfrac{1}{2}(A' + B' + C') = \tfrac{1}{2}S;$

whence, $\qquad \tfrac{1}{2}(B' + C' - A') = \tfrac{1}{2}S - A'.$

Substituting and omitting the primes, we have

$$\sin \tfrac{1}{2}a = \sqrt{\frac{-\cos \tfrac{1}{2}S \cos (\tfrac{1}{2}S - A)}{\sin B \sin C}}. \quad \cdot \quad \cdot \quad (1.)$$

In a similar way, we may deduce from (4), Art. 81,

$$\cos \tfrac{1}{2}a = \sqrt{\frac{\cos (\tfrac{1}{2}S - B) \cos (\tfrac{1}{2}S - C)}{\sin B \sin C}}. \quad \cdot \quad (2.)$$

and thence, $\quad \tan \tfrac{1}{2}a = \sqrt{\dfrac{-\cos \tfrac{1}{2}S \cos (\tfrac{1}{2}S - A)}{\cos (\tfrac{1}{2}S - B) \cos (\tfrac{1}{2}S - C)}}. \quad \cdot \quad (3.)$

83. To deduce Napier's Analogies.

From equation (1), Art. 80, we have

$$\cos A + \cos B \cos C = \sin B \sin C \cos a$$

$$= \sin C \frac{\sin A}{\sin a} \sin b \cos a ; \quad (1.)$$

since, from proportion (1), Art. 78, we have

$$\sin B = \frac{\sin A}{\sin a} \sin b.$$

Also, from equation (2), Art. 80, we have

$$\cos B + \cos A \cos C = \sin A \sin C \cos b$$

$$= \sin C \frac{\sin A}{\sin a} \sin a \cos b. \quad (2.)$$

Adding (1) and (2), and dividing by sin C, we obtain

$$(\cos A + \cos B) \frac{1 + \cos C}{\sin C} = \frac{\sin A}{\sin a} \sin (a + b). \quad (3.)$$

The proportion,

$$\sin A \; : \; \sin B \; :: \; \sin a \; : \; \sin b,$$

taken first by composition, and then by division, gives

$$\sin A + \sin B = \frac{\sin A}{\sin a} (\sin a + \sin b), \;\; \cdot \;\; \cdot \;\; (4.)$$

$$\sin A - \sin B = \frac{\sin A}{\sin a} (\sin a - \sin b). \;\; \cdot \;\; \cdot \;\; (5.)$$

Dividing (4) and (5), in succession, by (3), we obtain

$$\frac{\sin A + \sin B}{\cos A + \cos B} \times \frac{\sin C}{1 + \cos C} = \frac{\sin a + \sin b}{\sin (a + b)}. \quad (6.)$$

$$\frac{\sin A - \sin B}{\cos A + \cos B} \times \frac{\sin C}{1 + \cos C} = \frac{\sin a - \sin b}{\sin (a + b)}. \quad (7.)$$

But, by formulas (2) and (4), Art. 67, and formula (**E''**), Art. 66, equation (6) becomes

$$\tan \tfrac{1}{2} (A + B) \tan \tfrac{1}{2}C = \frac{\cos \tfrac{1}{2}(a - b)}{\cos \tfrac{1}{2}(a + b)}; \quad \cdot \quad \cdot \quad (8.)$$

and, by the similar formulas (3) and (5), of Art. 67, equation (7) becomes

$$\tan \tfrac{1}{2} (A - B) \tan \tfrac{1}{2}C = \frac{\sin \tfrac{1}{2}(a - b)}{\sin \tfrac{1}{2}(a + b)}. \quad \cdot \quad \cdot \quad (9.)$$

As $\tan \tfrac{1}{2}C = \dfrac{1}{\cot \tfrac{1}{2}C}$, formulas (8) and (9) may be written

$$\frac{\tan \tfrac{1}{2} (A + B)}{\cot \tfrac{1}{2}C} = \frac{\cos \tfrac{1}{2}(a - b)}{\cos \tfrac{1}{2}(a + b)}, \quad \cdot \quad \cdot \quad \cdot \quad (8'.)$$

$$\frac{\tan \tfrac{1}{2} (A - B)}{\cot \tfrac{1}{2}C} = \frac{\sin \tfrac{1}{2}(a - b)}{\sin \tfrac{1}{2}(a + b)}. \quad \cdot \quad \cdot \quad \cdot \quad (9'.)$$

These last two formulas give the proportions known as *the first set of Napier's Analogies*; viz.,

$$\cos \tfrac{1}{2}(a+b) \;:\; \cos \tfrac{1}{2}(a-b) \;::\; \cot \tfrac{1}{2}C \;:\; \tan \tfrac{1}{2}(A+B). \quad (10.)$$

$$\sin \tfrac{1}{2}(a+b) \;:\; \sin \tfrac{1}{2}(a-b) \;::\; \cot \tfrac{1}{2}C \;:\; \tan \tfrac{1}{2}(A-B). \quad (11.)$$

If in these we substitute the values of a, b, C, A, and B, in terms of the corresponding parts of the supplemental polar triangle, as expressed in Art. 80, we obtain

$$\cos \tfrac{1}{2}(A+B) \;:\; \cos \tfrac{1}{2}(A-B) \;::\; \tan \tfrac{1}{2}c \;:\; \tan \tfrac{1}{2}(a+b), \quad (12.)$$

$$\sin \tfrac{1}{2}(A+B) \;:\; \sin \tfrac{1}{2}(A-B) \;::\; \tan \tfrac{1}{2}c \;:\; \tan \tfrac{1}{2}(a-b), \quad (13.)$$

the second set of Napier's Analogies.

In applying logarithms to any of the preceding formulas, they must be made homogeneous in terms of R, as explained in Art. 30.

In all the formulas, the letters may be interchanged at pleasure, provided that, when one large letter is substituted for another, the like substitution is made in the corresponding small letters, and the reverse: for example, C may be substituted for A, provided that at the same time c is substituted for a, &c.

Note.—It may be noted that, in formulas (10) and (12), whenever the sign of the first term of the proportion is *minus*, the sign of the last term must, also, be *minus*, *i. e.*, whenever $\frac{1}{2}(a+b)$ is greater than $90°$, $\frac{1}{2}(A+B)$ must, also, be greater than $90°$, and the reverse; and similarly, whenever $\frac{1}{2}(a+b)$ is less than $90°$, $\frac{1}{2}(A+B)$ must, also, be less than $90°$, and the reverse.

SOLUTION OF OBLIQUE-ANGLED SPHERICAL TRIANGLES.

84. In the solution of oblique-angled triangles six different cases may arise: viz., there may be given,

     I. Two sides and an angle opposite one of them.

     II. Two angles and a side opposite one of them.

     III. Two sides and their included angle.

     IV. Two angles and their included side.

     V. The three sides.

     VI. The three angles.

CASE I.

Given two sides and an angle opposite one of them.

85. The solution, in this case, is commenced by finding the angle opposite the second given side, for which purpose formula (1), Art. 78, is employed.

As this angle is found by means of its sine, and because the same sine corresponds to two different arcs, there would seem to be two different solutions. To ascertain when there are *two solutions*, when *one solution*, and when *no solution* at all, it becomes necessary to examine the relations which may exist between the given parts. Two cases may arise, viz., the given angle may be *acute*, or it may be *obtuse*.

We shall consider each case separately (B. IX., Gen. S. 1).

1st Case: A < 90°.

Let A be the given acute angle, and let *a* and *b* be the given sides. Prolong the arcs AC and AB till they meet at A', forming the lune AA'; and from C, draw the arc CB″ perpendicular to ABA'. From C, as a pole, and with the

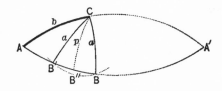

arc *a*, describe the arc of a small circle BB'. If this circle cuts ABA', in two points between A and A', there will be *two solutions*; for if C be joined with each point of intersection by the arc of a great circle, we shall have two triangles, ABC and AB'C, both of which will conform to the conditions of the problem.

If only one point of intersection lies between A and A', or if the small circle is tangent to ABA', there will be but *one solution.*

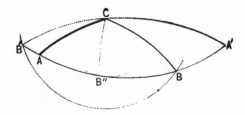

If there is no point of intersection, or if there are points of intersection which do not lie between A and A', there will be *no solution.*

From formula (2), Art. 72, we have

$$\sin CB'' = \sin b \sin A,$$

from which the perpendicular may be found. This perpendicular will be less than 90°, since it can not exceed the measure of the angle A (B. IX., Gen. S. 2, 1°); denote its value by *p*. By inspection of the figure, we find the following relations:

1. *When* a *is greater than* p, *and at the same time less than both* b *and* $180° - b$, *there will be two solutions.*

2. *When* a *is greater than* p, *and intermediate in value between* b *and* $180° - b$; *or, when* a *is equal to* p, *there will be but one solution.*

If $a = b$, and is also less than $180° - b$, one of the points of intersection will be at A, and there will be but one solution.

3. *When* a *is greater than* p, *and at the same time greater than both* b *and* $180° - b$; *or, when* a *is less than* p, *there will be no solution.*

2d Case: A > 90°.

Adopt the same construction as before. In this case, the perpendicular will be greater than 90°, because it can not be less than the measure of the angle A (B. IX., Gen. S. 2, 2°): it will, also, be greater than any other arc CA, CB, CA', that can be drawn from C to ABA'. By a course of reasoning en-tirely analogous to that in the preceding case, we have the following principles:

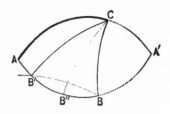

4. *When* a *is less than* p, *and at the same time greater than both* b *and* 180° − b, *there will be two solutions.*

5. *When* a *is less than* p, *and intermediate in value between* b *and* 180° − b; *or, when* a *is equal to* p, *there will be but one solution.*

6. *When* a *is less than* p, *and at the same time less than both* b *and* 180° − b; *or, when* a *is greater than* p, *there will be no solution.*

Having found the angle or angles opposite the second side, the solution may be completed by means of Napier's Analogies.

Examples.

1. Given $a = 43° \ 27' \ 36''$, $b = 82° \ 58' \ 17''$, and $A = 29° \ 32' \ 29''$, to find B, C, and c.

We see that $a > p$, since p can not exceed A (B. IX., Gen. S. 2, 1°); we see, further, that a is less than both

b and $180° - b$; hence, from the first condition there will be two solutions.

Applying logarithms to formula (1), Art. 78, we have

$$\log \sin B = (a.\ c.)\ \log \sin a + \log \sin b + \log \sin A - 10;$$

(a. c.) log sin a · ·	(43° 27′ 36″)	· ·	0.162508
log sin b · ·	(82° 58′ 17″)	· ·	9.996724
log sin A · ·	(29° 32′ 29″)	· ·	9.692893
log sin B · · · · · · ·			9.852125

$$\therefore \quad B = 45° 21′ 01″, \quad \text{and} \quad B' = 134° 38′ 59″.$$

From the first of Napier's Analogies (10), Art. 83, we find

$$\log \cot \tfrac{1}{2}C = (a.\ c.)\ \log \cos \tfrac{1}{2}(a - b) + \log \cos \tfrac{1}{2}(a + b)$$
$$+ \log \tan \tfrac{1}{2}(A + B) - 10.$$

Taking the first value of B, we have

$$\tfrac{1}{2}(A + B) = 37° 26′ 45″;$$

also, $$\tfrac{1}{2}(a + b) = 63° 12′ 56″;$$

and $$\tfrac{1}{2}(a - b) = 19° 45′ 20″.$$

(a. c.) log cos $\tfrac{1}{2}(a - b)$ · ·	(19° 45′ 20″)	·	0.026344
log cos $\tfrac{1}{2}(a + b)$ · ·	(63° 12′ 56″)	·	9.653825
log tan $\tfrac{1}{2}(A + B)$ · ·	(37° 26′ 45″)	·	9.884130
log cot $\tfrac{1}{2}C$ · · · · · · ·			9.564299

$$\therefore \quad \tfrac{1}{2}C = 69° 51′ 45″, \quad \text{and} \quad C = 139° 43′ 30″.$$

The side c may be found by means of formula (12), Art. 83, or by means of formula (2), Art. 78.

Applying logarithms to the proportion,

$$\sin A \ : \ \sin C \ :: \ \sin a \ : \ \sin c,$$

we have

$$\log \sin c = (a.\ c.) \log \sin A + \log \sin C + \log \sin a - 10;$$

(a. c.) log sin A	· · (29° 32′ 29″) ·	0.307107
log sin C	· ·(139° 43′ 30″) ·	9.810539
log sin a	· · (43° 27′ 36″) ·	9.837492
log sin c	· · · · · · ·	9.955138

$$\therefore \ \ c = 115° \ 35′ \ 48″$$

We take the greater value of c, because the angle C, being greater than the angle B, requires that the side c should be greater than the side b. By using the second value of B, we may find, in a similar manner,

$$C' = 32° \ 20′ \ 28″, \quad \text{and} \quad c' = 48° \ 16′ \ 18″.$$

2. Given $a = 97° \ 35′$, $b = 27° \ 08′ \ 22″$, and A $= 40° \ 51′$ $18″$, to find B, C, and c.

Ans. B $= 17° \ 31′ \ 09″$, C $= 144° \ 48′ \ 10″$, $c = 119° \ 08′ \ 25″$.

3. Given $a = 115° \ 20′ \ 10″$, $b = 57° \ 30′ \ 06″$, and A $= 126° \ 37′ \ 30″$, to find B, C, and c.

Ans. B $= 48° \ 29′ \ 48″$, C $= 61° \ 40′ \ 16″$, $c = 82° \ 34′ \ 04″$.

4. Given $b = 79° \ 14′$, $c = 30° \ 20′ \ 45″$, and B $= 121°$ $10′ \ 26″$, to find C, A, and a.

Ans. C $= 26° \ 06′ \ 16″$, A $= 49° \ 44′ \ 16″$, $a = 61° \ 11′ \ 06″$.

CASE II.

Given two angles and a side opposite one of them.

86. The solution, in this case, is commenced by finding the side opposite the second given angle, by means of formula (1), Art. 78. The solution is completed as in Case I.

Since the second side is found by means of its sine, there may be two solutions. To investigate this case, we pass to the supplemental polar triangle, by substituting for each part its supplement. In this triangle, there will be given two sides and an angle opposite one; it may therefore be discussed as in the preceding case. When the supplemental triangle has *two solutions, one solution,* or *no solution,* the given triangle will, in like manner, have *two solutions, one solution,* or *no solution.*

Let the given parts be A′, B′, and $a′$, and let $p′$ be the arc, C′D′, of a great circle drawn from the extremity of the given side perpendicular to the side opposite : we have

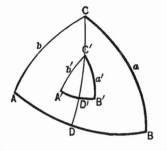

$$\sin p′ = \sin a′ \sin B′.$$

There will be two cases: $a′$ may be *less* than 90°; or, $a′$ may be *greater* than 90°.

1st Case: $a′ < 90°$.

Passing to the supplemental polar triangle, we shall have given a, b, A; and since, in the given triangle, $a′ < 90°$, in this supplemental triangle A > 90°: call the perpendicular CD, p. The conditions determining the number of solutions in this supplemental triangle are given in principles 4, 5, 6, Art. 85.

From principle 4, Art. 85, it appears that, for two solutions, a must be less than p, that is,

$$a < p:$$

subtracting each member of this inequality from 180°, we have

$$180° - a > 180° - p;$$

but, $180° - a = A'$; and (B. IX., P. VI., C. 2), $180° - p = p'$ hence

$$A' > p':$$

again, it appears from principle 4, that a must be greater than b, that is,

$$a > b;$$

subtracting each member of this inequality from 180°, we have

$$180° - a < 180° - b;$$

or, $$A' < B':$$

it further appears from the same principle, that a must be greater than $180° - b$, that is,

$$a > 180° - b;$$

subtracting each member of this inequality from 180°, we have

$$180° - a < 180° - (180° - b);$$

or. $$A' < 180° - B'$$

Collecting the results, and, for convenience, omitting the primes, we have the following principle:

Two angles and a side opposite one of them being given, and the given side less than 90°, *i. e.*, A, B, *a* given, and $a < 90°$;

1. *When* A *is greater than* p, *and at the same time less than both* B *and* 180° — B, *there will be two solutions.*

In like manner, from principle 5, Art. 85, we have

2. *When* A *is greater than* p, *and intermediate in value between* B *and* 180° — B; *or, when* A *is equal to* p, *there will be but one solution.*

And from principle 6, Art. 85, we have

3. *When* A *is greater than* p, *and at the same time greater than both* B *and* 180° — B; *or, when* A *is less than* p, *there will be no solution.*

It is to be noted that, in this case, the perpendicular is less than 90°, and less, also, than the given side; *i. e.*,

$$p \lesssim a.$$

2d Case: $a' > 90°$.

Passing to the supplemental polar triangle, we shall have given a, b, A, and $A < 90°$. The conditions determining the number of solutions in this supplemental triangle are given in principles 1, 2, 3, Art. 85.

From principle 1, Art. 85, it appears that, for two solutions, a must be greater than p, that is,

$$a > p;$$

subtracting each member of this inequality from 180°, we have

$$180° - a < 180° - p;$$

or, $$A' < p':$$

in the same manner as before, we may obtain from this principle 1,

$$A' > B';$$

and $$A' > 180° - B'.$$

As before, collecting the results and omitting the primes, we have the following principle:

Two angles and a side opposite one of them being given, the given side greater than 90°, *i. e.*, A, B, *a* given, and $a > 90°$;

4. *When* A *is less than* p, *and at the same time greater than both* B *and* 180° — B, *there will be two solutions.*

In like manner, from principle 2, Art. 85, we have

5. *When* A *is less than* p, *and intermediate in value between* B *and* 180° — B; *or, when* A *is equal to* p, *there will be but one solution.*

And from principle 3, Art. 85, we have

6. *When* A *is less than* p, *and at the same time less than both* B *and* 180° — B; *or, when* A *is greater than* p, *there will be no solution.*

It is to be noted that, in this case, the perpendicular is greater than 90°, and greater, also, than the given side; *i. e.*, $p > a.$

From the principles deduced in Articles 85 and 86, it is evident that, if the given parts of the spherical triangles considered are named as in the accompanying table, we shall have the following principles, applicable to *all* the cases:

Perpendicular.	Odd.	Adjacent.	Opposite.
p	A	b	a
	a	B	A

7. The sine of p is equal to the rectangle of the sines of the odd part and the adjacent part.

8. p is always of the *same species* as the odd part, and *differs more* from 90° than the odd part, *i. e.*, when the odd part is *less* than 90°, p is *still less;* and when the odd part is *greater* than 90°, p is *still greater.*

9. There will be *two solutions:*

1°. When (odd part being *less* than 90°) the opposite part is *greater* than p, and *less* than the adjacent part and its supplement.

2°. When (odd part being *greater* than 90°) the opposite part is *less* than p, and *greater* than the adjacent part and its supplement.

10. There will be *one solution:*

1°. When (odd part being *less* than 90°) the opposite part is *greater* than p, and *intermediate in value* between the adjacent part and its supplement.

2°. When (odd part being *greater* than 90°) the

opposite part is *less* than *p*, and *intermediate in value* between the adjacent part and its supplement.

3°. When the opposite part is *equal* to *p*.

11. There will be *no solution:*

1°. When (odd part being *less* than 90°) the opposite part is either *less* than *p*, or *greater* than *p* and *greater also* than both the adjacent part and its supplement.

2°. When (odd part being *greater* than 90°) the opposite part is either *greater* than *p*, or *less* than *p* and *less also* than both the adjacent part and its supplement.

Examples.

1. Given A = 95° 16′, B = 80° 42′ 10″, and *a* = 57° 38′, to find *c*, *b*, and C.

p might be computed from the formula,

$$\log \sin p = \log \sin B + \log \sin a - 10;$$

but it is not necessary, as *p* < *a* (see principle 8).

· Because A > *p*, and intermediate between 80° 42′ 10″ and 99° 17′ 50″, there will, from the second condition, be but one solution.

Applying logarithms to proportion (1), Art. 78, we have

$$\log \sin b = (a.\ c.)\ \log \sin A + \log \sin B + \log \sin a - 10;$$

(a. c.) log sin A	(95° 16′)	0.001837
log sin B	(80° 42′ 10″)	9.994257
log sin *a*	(57° 38′)	9.926671
log sin *b* · · · ·		9.922765 ∴ *b* = 56° 49′ 57″.

We take the smaller value of b, for the reason that A, being greater than B, requires that a should be greater than b.

Applying logarithms to proportion (12), Art. 83, we have

$$\log \tan \tfrac{1}{2}c = (\text{a. c.}) \log \cos \tfrac{1}{2}(A - B) + \log \cos \tfrac{1}{2}(A + B)$$
$$+ \log \tan \tfrac{1}{2}(a + b) - 10 ;$$

we have $\tfrac{1}{2}(A + B) = 87° \; 59' \; 05''$,

$\tfrac{1}{2}(a + b) = 57° \; 13' \; 58''$,

and $\tfrac{1}{2}(A - B) = \quad 7° \; 16' \; 55''$;

(a. c.) $\log \cos \tfrac{1}{2}(A - B)$ ·	$(7° \; 16' \; 55'')$ ·	0.003517
$\log \cos \tfrac{1}{2}(A + B)$ ·	$(87° \; 59' \; 05'')$ ·	8.546124
$\log \tan \tfrac{1}{2}(a + b)$ ·	$(57° \; 13' \; 58'')$ ·	10.191352
$\log \tan \tfrac{1}{2}c$ · · · · · ·		8.740993

$$\therefore \; \tfrac{1}{2}c = 3° \; 09' \; 09'', \;\; \text{and} \;\; c = 6° \; 18' \; 18''.$$

Applying logarithms to the proportion,

$$\sin a \; : \; \sin c \; :: \; \sin A \; : \; \sin C,$$

we have

$$\log \sin C = (\text{a. c.}) \log \sin a + \log \sin c + \log \sin A - 10 ;$$

(a. c.) $\log \sin a$	$(57° \; 38')$ · ·	0.073329
$\log \sin c$	$(6° \; 18' \; 18'')$·	9.040685
$\log \sin A$	$(95° \; 16')$ · ·	9.998163
$\log \sin C$ · · · · ·		9.112177 $\;\therefore\; C = 7° \; 26' \; 21''.$

The smaller value of C is taken, for the same reason as before.

2. Given $A = 50° 12'$, $B = 58° 08'$, and $a = 62° 42'$, to find b, c, and C.

$b = \quad 79° 12' 10''$, $c = 119° 03' 26''$, $C = 130° 54' 28''$,

$b' = 100° 47' 50''$, $c' = 152° 14' 18''$, $C' = 156° 15' 06''$.

3. Given $C = 115° 20'$, $A = 57° 30'$, and $c = 126° 38'$, to find a, b, and B.

Ans. $a = 48° 29' 13''$, $b = 118° 20' 44''$, $B = 97° 35' 06''$.

CASE III.

Given two sides and their included angle.

87. The remaining angles are found by means of Napier's Analogies, and the remaining side as in the preceding cases.

Examples.

1. Given $a = 62° 38'$, $b = 10° 13' 19''$, and $C = 150° 24' 12''$, to find c, A, and B.

Applying logarithms to proportions (10) and (11), Art. 83, we have

$$\log \tan \tfrac{1}{2} (A + B) = \text{(a. c.) } \log \cos \tfrac{1}{2} (a + b) + \log \cos \tfrac{1}{2} (a - b)$$
$$+ \log \cot \tfrac{1}{2} C - 10 ;$$

$$\log \tan \tfrac{1}{2} (A - B) = \text{(a. c.) } \log \sin \tfrac{1}{2} (a + b) + \log \sin \tfrac{1}{2} (a - b)$$
$$+ \log \cot \tfrac{1}{2} C - 10 ;$$

we have $\qquad \tfrac{1}{2} (a - b) = 26° 12' 20''$,

$$\tfrac{1}{2} C = 75° 12' 06'',$$

and $\qquad \tfrac{1}{2} (a + b) = 36° 25' 39''$.

(a. c.) log cos $\frac{1}{2}(a + b)$ · (36° 25′ 39″) · 0.094415
    log cos $\frac{1}{2}(a - b)$ · (26° 12′ 20″) · 9.952897
    log cot $\frac{1}{2}$C · · · · (72° 12′ 06″) · 9.421901
        log tan $\frac{1}{2}$(A + B) · · · · · · 9.469213

                ∴ $\frac{1}{2}$(A + B) = 16° 24′ 51″.

(a. c.) log sin $\frac{1}{2}(a + b)$ · (36° 25′ 39″) · 0.226356
    log sin $\frac{1}{2}(a - b)$ · (26° 12′ 20″) · 9.645022
    log cot $\frac{1}{2}$C · · · (75° 12′ 06″) · 9.421901
        log tan $\frac{1}{2}$(A − B) · · · · · · 9.293279

                ∴ $\frac{1}{2}$(A − B) = 11° 06′ 53″.

The greater angle is equal to the half sum plus the half difference, and the less is equal to the half sum minus the half difference. Hence, we have

A = 27° 31′ 44″,   and   B = 5° 17′ 58″.

Applying logarithms to proportion (13), Art. 83, we have

log tan $\frac{1}{2}c$ = (a. c.) log sin $\frac{1}{2}$(A − B) + log sin $\frac{1}{2}$(A + B)
                + log tan $\frac{1}{2}(a - b)$ − 10 ;

(a. c.) log sin $\frac{1}{2}$(A − B) · (11° 06′ 53″) · 0.714952
    log sin $\frac{1}{2}$(A + B) · (16° 24′ 51″) · 9.451139
    log tan $\frac{1}{2}(a - b)$ · (26° 12′ 20″) · 9.692125
        log tan $\frac{1}{2}c$ · · · · · · · · 9.858216

    ∴ $\frac{1}{2}c$ = 35° 48′ 33″,   and   c = 71° 37′ 06″.

2. Given $a =$ 68° 46′ 02″, $b =$ 37° 10′, and C = 39° 23′ 23″, to find c, A, and B.

Ans. A = 120° 59′ 21″, B = 33° 45′ 13″, c = 43° 37′ 48″.

3. Given $a = 84°\ 14'\ 29''$, $b = 44°\ 13'\ 45''$, and C = $36°\ 45'\ 28''$, to find A and B.

Ans. A = $130°\ 05'\ 22''$, B = $32°\ 26'\ 06''$.

4. Given $b = 61°\ 12'$, $c = 131°\ 44'$, and A = $88°\ 40'$, to find B, C, and a. (See Note, Art. 83.)

Ans. B = $66°\ 55'\ 59''$, C = $128°\ 25'\ 05''$, $a = 72°\ 12'\ 46''$.

CASE IV.

Given two angles and their included side.

88. The solution of this case is entirely analogous to that of Case III.

Applying logarithms to proportions (12) and (13), Art. 83, and to proportion (11), Art. 83, we have

$$\log \tan \tfrac{1}{2} (a + b) = (\text{a. c.}) \log \cos \tfrac{1}{2} (A + B) + \log \cos \tfrac{1}{2} (A - B)$$
$$+ \log \tan \tfrac{1}{2}c - 10 ;$$

$$\log \tan \tfrac{1}{2} (a - b) = (\text{a. c.}) \log \sin \tfrac{1}{2} (A + B) + \log \sin \tfrac{1}{2} (A - B)$$
$$+ \log \tan \tfrac{1}{2}c - 10 ;$$

$$\log \cot \tfrac{1}{2}C = (\text{a. c.}) \log \sin \tfrac{1}{2} (a - b) + \log \sin \tfrac{1}{2} (a + b)$$
$$+ \log \tan \tfrac{1}{2} (A - B) - 10.$$

The application of these formulas is sufficient for the solution of all cases.

Examples.

1. Given A = $81°\ 38'\ 20''$, B = $70°\ 09'\ 38''$, and $c = 59°\ 16'\ 22''$, to find C, a, and b.

Ans. C = $64°\ 46'\ 24''$, $a = 70°\ 04'\ 17''$, $b = 63°\ 21'\ 27''$.

2. Given A $= 34°$ 15' 03", B $= 42°$ 15' 13", and $c =$ 76° 35' 36", to find C, a, and b.

Ans. C $= 121°$ 36' 12", $a = 40°$ 0' 10", $b = 50°$ 10' 30".

3. Given B $= 82°$ 24', C $= 120°$ 38', and $a = 75°$ 19', to find A, b, and c.

Ans. A $= 73°$ 31' 13", $b = 90°$ 50' 50", $c = 119°$ 46' 22".

CASE V.

Given the three sides, to find the remaining parts.

89. The angles may be found by means of formula (3), Art. 81 ; or, one angle being found by that formula, the two others may be found by means of Napier's Analogies.

Examples.

1. Given $a = 74°$ 23', $b = 35°$ 46' 14", and $c = 100°$ 39', to find A, B, and C.

Applying logarithms to formula (3), Art. 81, we have

$$\log \cos \tfrac{1}{2}A \; = \; 10 + \tfrac{1}{2}\,[\log \sin \tfrac{1}{2}s + \log \sin (\tfrac{1}{2}s - a)$$
$$+ \; (a.\,c.)\,\log \sin b + (a.\,c.)\,\log \sin c - 20]\,;$$

or,

$$\log \cos \tfrac{1}{2}A \; = \; \tfrac{1}{2}\,[\log \sin \tfrac{1}{2}s + \log \sin (\tfrac{1}{2}s - a)$$
$$+ \; (a.\,c.)\,\log \sin b + (a.\,c.)\,\log \sin c]\,;$$

we have $\tfrac{1}{2}s = 105°$ 24' 07",

and $\tfrac{1}{2}s - a = 31°$ 01' 07".

$$\begin{array}{lllll}
\log \sin \tfrac{1}{2}s & \cdot\ \cdot\ \cdot & (105°\ 24'\ 07'') & \cdot & 9.984116 \\
\log \sin (\tfrac{1}{2}s - a) & \cdot & (31°\ 01'\ 07'') & \cdot & 9.712074 \\
\text{(a. c.) } \log \sin b & \cdot\ \cdot\ \cdot & (35°\ 46'\ 14'') & \cdot & 0.233185 \\
\text{(a. c.) } \log \sin c & \cdot\ \cdot\ \cdot & (100°\ 39') & \cdot\ \cdot\ \cdot & 0.007546 \\
& & & 2\ ) & \overline{19.936921} \\
\log \cos \tfrac{1}{2}A & \cdot\ \cdot\ \cdot\ \cdot\ \cdot\ \cdot\ \cdot\ \cdot & & \cdot\ \cdot\ & 9.968460
\end{array}$$

$$\therefore\ \tfrac{1}{2}A = 21°\ 34'\ 23'',\ \text{and}\ A = 43°\ 08'\ 46''.$$

Using the same formula as before, and substituting B for A, b for a, and a for b, and recollecting that $\tfrac{1}{2}s - b = 69°\ 37'\ 53''$, we have

$$\begin{array}{lllll}
\log \sin \tfrac{1}{2}s & \cdot\ \cdot\ \cdot & (105°\ 24'\ 07'') & \cdot & 9.984116 \\
\log \sin (\tfrac{1}{2}s - b) & \cdot & (69°\ 37'\ 53'') & \cdot & 9.971958 \\
\text{(a. c.) } \log \sin a & \cdot\ \cdot\ \cdot & (74°\ 23') & \cdot\ \cdot\ \cdot & 0.016336 \\
\text{(a. c.) } \log \sin c & \cdot\ \cdot\ \cdot & (100°\ 39') & \cdot\ \cdot\ \cdot & 0.007546 \\
& & & 2\ ) & \overline{19.979956} \\
\log \cos \tfrac{1}{2}B & \cdot\ \cdot\ \cdot\ \cdot\ \cdot\ \cdot\ \cdot\ \cdot & & \cdot\ \cdot\ & 9.989978
\end{array}$$

$$\therefore\ \tfrac{1}{2}B = 12°\ 15'\ 43'',\ \text{and}\ B = 24°\ 31'\ 26''.$$

Using the same formula, substituting C for A, c for a, and a for c, recollecting that $\tfrac{1}{2}s - c = 4°\ 45'\ 07''$, we have

$$\begin{array}{lllll}
\log \sin \tfrac{1}{2}s & \cdot\ \cdot\ \cdot & (105°\ 24'\ 07'') & \cdot & 9.984116 \\
\log \sin (\tfrac{1}{2}s - c) & \cdot & (4°\ 45'\ 07'') & \cdot & 8.918250 \\
\text{(a. c.) } \log \sin a & \cdot\ \cdot\ \cdot & (74°\ 23') & \cdot\ \cdot\ \cdot & 0.016336 \\
\text{(a. c.) } \log \sin b & \cdot\ \cdot\ \cdot & (25°\ 46'\ 14'') & \cdot & 9.233185 \\
& & & 2\ ) & \overline{19.151887} \\
\log \cos \tfrac{1}{2}C & \cdot\ \cdot\ \cdot\ \cdot\ \cdot\ \cdot\ \cdot\ \cdot & & \cdot\ \cdot\ & 9.575943
\end{array}$$

$$\therefore\ \tfrac{1}{2}C = 67°\ 52'\ 25'',\ \text{and}\ C = 135°\ 44'\ 50''.$$

2. Given $a = 56°\ 40'$, $b = 83°\ 13'$, and $c = 114°\ 30'$, to find A, B, and C.

Ans. A $= 48°\ 31'\ 18''$, B $= 62°\ 55'\ 44''$, C $= 125°\ 18'\ 56''$.

3. Given $a = 115°\ 15'$, $b = 125°\ 30'$, and $c = 110°\ 15'$, to find A, B, and C.

Ans. A $= 145°\ 15'\ 04''$, B $= 149°\ 07'\ 52$, C $= 143°\ 45'\ 10''$.

CASE VI.

The three angles being given, to find the sides.

90. The solution in this case is entirely analogous to the preceding one.

Applying logarithms to formula (2), Art. 82, we have

$$\log \cos \tfrac{1}{2}a = \tfrac{1}{2}\,[\log \cos (\tfrac{1}{2}S - B) + \log \cos (\tfrac{1}{2}S - C)$$
$$+ (\text{a. c.})\ \log \sin B + (\text{a. c.})\ \log \sin C].$$

In the same manner as before, we change the letters, to suit each case.

Examples.

1. Given A $= 48°\ 30'$, B $= 125°\ 20'$, and C $= 62°\ 54'$, to find a, b, and c.

Ans. $a = 56°\ 39'\ 30''$, $b = 114°\ 29'\ 58''$ $c = 83°\ 12'\ 06''$.

2. Given A $= 109°\ 55'\ 42''$, B $= 116°\ 38'\ 33''$, and C $= 120°\ 43'\ 37''$, to find a, b, and c.

Ans. $a = 98°\ 21'\ 40''$, $b = 109°\ 50'\ 22''$, $c = 115°\ 13'\ 28''$.

3. Given A $= 160°\ 20'$, B $= 135°\ 15'$, and C $= 148°\ 25'$, to find a, b, and c.

Ans. $a = 155°\ 56'\ 10''$, $b = 58°\ 32'\ 12''$, $c = 140°\ 36'\ 48''$.

MENSURATION.

91. MENSURATION is that branch of Mathematics which treats of the measurement of Geometrical Magnitudes.

92. The measurement of a quantity is the operation of finding how many times it contains another quantity of the same kind, taken as a standard. This standard is called the *unit of measure.*

93. The unit of measure for surfaces is a *square*, one of whose sides is the linear unit. The unit of measure for volumes is a *cube*, one of whose edges is the linear unit.

If the linear unit is *one foot*, the superficial unit is *one square foot*, and the unit of volume is *one cubic foot.* If the linear unit is *one yard*, the superficial unit is *one square yard*, and the unit of volume is *one cubic yard.*

94. In Mensuration, the expression *product of two lines*, is used to denote the product obtained by multiplying the number of linear units in one line by the number of linear units in the other. The expression *product of three lines*, is used to denote the continued product of the number of linear units in each of the three lines.

Thus, when we say that the area of a parallelogram is equal to the product of its base and altitude, we mean that the number of superficial units in the parallelogram is equal to the number of linear units in the base, multiplied by the number of linear units in the altitude. In like manner, the number of units of volume, in a rectangular parallelopipedon, is equal to the number of superficial units in its base multiplied by the number of linear units in its altitude, and so on.

MENSURATION OF PLANE FIGURES.

To find the area of a parallelogram.

95. From the principle demonstrated in Book IV., Prop. V., we have the following

RULE.—*Multiply the base by the altitude; the product will be the area required.*

Examples.

1. Find the area of a parallelogram, whose base is 12.25, and whose altitude is 8.5.   *Ans.* 104.125.

2. What is the area of a square, whose side is 204.3 feet?   *Ans.* 41738.49 sq. ft.

3. How many square yards are there in a rectangle whose base is 66.3 feet, and altitude 33.3 feet?
                               Ans. 245.31 sq. yds.

4. What is the area of a rectangular board, whose length is $12\frac{1}{2}$ feet, and breadth 9 inches? *Ans.* $9\frac{3}{8}$ sq. ft.

5. What is the number of square yards in a parallelogram, whose base is 37 feet, and altitude 5 feet 3 inches?
                               Ans. $21\frac{7}{12}$.

To find the area of a plane triangle.

96. *First Case.* When the base and altitude are given.

From the principle demonstrated in Book IV., Prop. VI., we may write the following

RULE. — *Multiply the base by half the altitude; the product will be the area required.*

Examples.

1. Find the area of a triangle, whose base is 625, and altitude 520 feet. *Ans.* 162500 sq. ft.

2. Find the area of a triangle, in square yards, whose base is 40, and altitude 30 feet. *Ans.* 66⅔.

3. Find the area of a triangle, in square yards, whose base is 49, and altitude 25¼ feet. *Ans.* 68.7361.

Second Case. When two sides and their included angle are given.

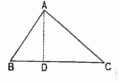

Let ABC represent a plane triangle, in which the side AB = c, BC = a, and the angle B, are given. From A draw AD perpendicular to BC; this will be the altitude of the triangle. From formula (1), Art. 37, Plane Trigonometry, we have

$$AD = c \sin B.$$

Denoting the area of the triangle by Q, and applying the rule last given, we have

$$Q = \frac{ac \sin B}{2}; \quad \text{or,} \quad 2Q = ac \sin B.$$

Substituting for sin B, $\frac{\sin B}{R}$ (Trig., Art. 30), and applying logarithms, we have

$$\log (2Q) = \log a + \log c + \log \sin B - 10;$$

hence, we may write the following

Rule.—*Add together the logarithms of the two sides and the logarithmic sine of their included angle; from this sum subtract 10; the remainder will be the logarithm of double the area of the triangle. Find, from the table, the number corresponding to this logarithm, and divide it by 2; the quotient will be the required area.*

Examples.

1. What is the area of a triangle, in which two sides, a and b, are respectively equal to 125.81, and 57.65, and whose included angle C is 57° 25'?

          Ans. $2Q = 6111.4$, and $Q = 3055.7$.

2. What is the area of a triangle, whose sides are 30 and 40, and their included angle 28° 57'?

          Ans. 290.427.

3. What is the number of square yards in a triangle, of which the sides are 25 feet and 21.25 feet, and their included angle 45°?　　　　*Ans.* 20.8694.

LEMMA.

To find half an angle, when the three sides of a plane triangle are given.

97. Let ABC be a plane triangle, the angles and sides being denoted as in the figure.

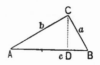

When the angle, A, is *acute*, we have (B. IV., P. XII.),

$$a^2 = b^2 + c^2 - 2c \cdot AD :$$

but (Art. 37), AD $= b \cos A$; hence,

$$a^2 = b^2 + c^2 - 2bc \cos A.$$

When the angle A is *obtuse*, we have (B. IV., P. XIII.),

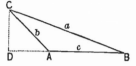

$$a^2 = b^2 + c^2 + 2c \cdot AD:$$

but (Art. 37), $\qquad AD = b \cos CAD:$

but the angle CAD is the supplement of the angle A of the given triangle, and, therefore (Art. 63),

$$\cos CAD = - \cos A ;$$

hence, $\qquad AD = - b \cos A,$

and, consequently, we have

$$a^2 = b^2 + c^2 - 2bc \cos A.$$

So that whether the angle, A, is acute or obtuse, we have

$$a^2 = b^2 + c^2 - 2bc \cos A; \quad \cdot \quad \cdot \quad \cdot \quad (1.)$$

whence, $\qquad \cos A = \dfrac{b^2 + c^2 - a^2}{2bc}. \quad \cdot \quad \cdot \quad \cdot \quad (2.)$

If we add 1 to each member, and recollect that $1 + \cos A = 2 \cos^2 \frac{1}{2}A$ (Art. 66) equation (4), we have

$$2 \cos^2 \tfrac{1}{2}A = \frac{2bc + b^2 + c^2 - a^2}{2bc}$$

$$= \frac{(b + c)^2 - a^2}{2bc}$$

$$= \frac{(b + c + a)(b + c - a)}{2bc};$$

or, $\qquad \cos^2 \tfrac{1}{2}A = \dfrac{(b + c + a)(b + c - a)}{4bc}. \cdot \quad \cdot \quad (3.)$

If we put $\qquad b + c + a = s,$

we have $\qquad \dfrac{b + c + a}{2} = \tfrac{1}{2}s,$

and $\qquad \dfrac{b + c - a}{2} = \tfrac{1}{2}s - a.$

Substituting in (3), and extracting the square root,

$$\cos \tfrac{1}{2}A = \sqrt{\frac{\tfrac{1}{2}s\,(\tfrac{1}{2}s - a)}{bc}}, \quad \cdots \quad (4.)$$

the plus sign, only, being used, since $\tfrac{1}{2}A < 90°$; hence, as A represents any angle,

The cosine of half of any angle of a plane triangle, is equal to the square root of the product of half the sum of the three sides, and half that sum minus *the side opposite the angle, divided by the rectangle of the adjacent sides.*

By applying logarithms, we have

$$\log \cos \tfrac{1}{2}A = \tfrac{1}{2}\,[\log \tfrac{1}{2}s + \log (\tfrac{1}{2}s - a) + \text{(a. c.)} \log b$$
$$+ \text{(a. c.)} \log c]. \quad \cdot \quad (A.)$$

If we subtract each member of equation (2) from 1, and recollect that $1 - \cos A = 2 \sin^2 \tfrac{1}{2}A$ (Art. 66), we have

$$2 \sin^2 \tfrac{1}{2}A = \frac{2bc - b^2 - c^2 + a^2}{2bc}$$

$$= \frac{a^2 - (b - c)^2}{2bc}$$

$$= \frac{(a + b - c)\,(a - b + c)}{2bc}. \quad \cdots \quad (5.)$$

Placing, as before, $a + b + c = s$,

we have $\dfrac{a + b - c}{2} = \tfrac{1}{2}s - c$,

and $\dfrac{a - b + c}{2} = \tfrac{1}{2}s - b$.

Substituting in (5) and reducing, we have

$$\sin \tfrac{1}{2}A = \sqrt{\frac{(\tfrac{1}{2}s - b)\,(\tfrac{1}{2}s - c)}{bc}}; \quad \cdot \quad \cdot \quad \cdot \quad (6.)$$

hence,

The sine of half an angle of a plane triangle, is equal to the square root of the product of half the sum of the three sides minus one of the adjacent sides and half that sum minus the other adjacent side, divided by the rectangle of the adjacent sides.

Applying logarithms, we have

$\log \sin \tfrac{1}{2}A = \tfrac{1}{2} \left[\log\left(\tfrac{1}{2}s - b\right) + \log\left(\tfrac{1}{2}s - c\right) + \text{(a. c.) } \log b \right.$
$\left. + \text{(a. c.) } \log c \right]. \quad \cdot \quad (\mathbf{B.})$

Third Case. To find the area of a triangle when the three sides are given.

Let ABC represent a triangle whose sides a, b, and c are given. From the principle demonstrated in the last case, we have

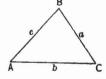

$$Q = \tfrac{1}{2}bc \sin A.$$

But, from formula (**A′**), Trig., Art. 66, we have

$$\sin A = 2 \sin \tfrac{1}{2}A \cos \tfrac{1}{2}A;$$

whence, $\qquad Q = bc \sin \tfrac{1}{2}A \cos \tfrac{1}{2}A.$

Substituting for $\sin \tfrac{1}{2}A$ and $\cos \tfrac{1}{2}A$, their values, taken from Lemma, and reducing, we have

$$Q = \sqrt{\tfrac{1}{2}s\, (\tfrac{1}{2}s - a)\, (\tfrac{1}{2}s - b)\, (\tfrac{1}{2}s - c)}\,;$$

hence, we may write the following

RULE.—*Find half the sum of the three sides, and from it subtract each side separately. Find the continued product of the half sum and the three remainders, and extract its square root; the result will be the area required.*

It is generally more convenient to employ logarithms; for this purpose, applying logarithms to the last equation, we have

$$\log Q = \tfrac{1}{2}\,[\log \tfrac{1}{2}s + \log (\tfrac{1}{2}s - a) + \log (\tfrac{1}{2}s - b) + \log (\tfrac{1}{2}s - c)]\,;$$

hence, we have the following

RULE.—*Find the half sum and the three remainders as before, then find the half sum of their logarithms; the number corresponding to the resulting logarithm will be the area required.*

Examples.

1. Find the area of a triangle, whose sides are 20, 30, and 40.

We have $\tfrac{1}{2}s = 45$, $\tfrac{1}{2}s - a = 25$, $\tfrac{1}{2}s - b = 15$, $\tfrac{1}{2}s - c = 5$. By the first rule,

$$Q = \sqrt{45 \times 25 \times 15 \times 5} = 290.4737, \textit{Ans.}$$

By the second rule,

$$
\begin{array}{lllll}
\log \tfrac{1}{2}s & \cdot \ \cdot \ \cdot & (45) & \cdot \ \cdot \ \cdot \ \cdot & 1.653213 \\
\log (\tfrac{1}{2}s - a) & \cdot \ \cdot & (25) & \cdot \ \cdot \ \cdot \ \cdot & 1.397940 \\
\log (\tfrac{1}{2}s - b) & \cdot \ \cdot & (15) & \cdot \ \cdot \ \cdot \ \cdot & 1.176091 \\
\log (\tfrac{1}{2}s - c) & \cdot \ \cdot & (5) & \cdot \ \cdot \ \cdot \ \cdot & 0.698970 \\
& & & 2\,) & \overline{4.926214} \\
& & \log Q & \cdot \ \cdot \ \cdot \ \cdot & 2.463107
\end{array}
$$

$$\therefore \ Q = 290.4737, \ Ans.$$

2. How many square yards are there in a triangle, whose sides are 30, 40, and 50 feet? *Ans.* 66⅔.

To find the area of a trapezoid.

98. From the principle demonstrated in Book IV., Prop. VII., we may write the following

RULE.—*Find half the sum of the parallel sides, and multiply it by the altitude; the product will be the area required.*

Examples.

1. In a trapezoid the parallel sides are 750 and 1225, and the perpendicular distance between them is 1540; what is the area? *Ans.* 1520750.

2. How many square feet are contained in a plank, whose length is 12 feet 6 inches, the breadth at the greater end 15 inches, and at the less end 11 inches? *Ans.* 13¹¹⁄₂₄.

3. How many square yards are there in a trapezoid, whose parallel sides are 240 feet, 320 feet, and altitude 66 feet? *Ans.* 2053⅓ sq. yd.

To find the area of any quadrilateral.

99. From what precedes, we deduce the following

RULE.—*Join the vertices of two opposite angles by a diagonal; from each of the other vertices let fall perpendiculars upon this diagonal; multiply the diagonal by half of the sum of the perpendiculars, and the product will be the area required.*

Examples.

1. What is the area of the quadrilateral ABCD, the diagonal AC being 42, and the perpendiculars Dg, Bb, equal to 18 and 16 feet?

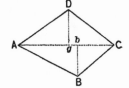

           Ans. 714 sq. ft.

2. How many square yards of paving are there in the quadrilateral, whose diagonal is 65 feet, and the two perpendiculars let fall on it 28 and $33\frac{1}{2}$ feet?   *Ans.* $222\frac{1}{18}$.

To find the area of any polygon.

100. From what precedes, we have the following

RULE.—*Draw diagonals dividing the proposed polygon into trapezoids and triangles: then find the area of these figures separately, and add them together for the area of the whole polygon.*

Example.

1. Let it be required to determine the area of the polygon ABCDE, having five sides.

Let us suppose that we have measured the diagonals and perpendiculars,

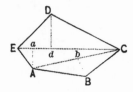

and found AC = 36.21, EC = 39.11, Bb = 4, Dd = 7.26, Aa = 4.18 : required the area. *Ans.* 296.1292.

To find the area of a regular polygon.

101. Let AB, denoted by s, repre- sent one side of a regular polygon whose centre is C. Draw CA and CB, and from C draw CD perpendicular to AB. Then will CD be the apothem, and we shall have AD = BD.

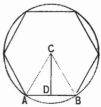

Denote the number of sides of the polygon by n; then will the angle ACB, at the centre, be equal to $\dfrac{360°}{n}$ (B. V., page 144, D. 2), and the angle ACD, which is half of ACB, will be equal to $\dfrac{180°}{n}$.

In the right-angled triangle ADC, we shall have, formula (3), Art. 37, Trig.,

$$CD = \tfrac{1}{2}s \tan CAD.$$

But CAD, being the complement of ACD, we have

$$\tan CAD = \cot ACD ;$$

hence, $$CD = \tfrac{1}{2}s \cot \dfrac{180°}{n},$$

a formula by means of which the apothem may be com- puted.

But the area is equal to the perimeter multiplied by half the apothem (Book V., Prop. VIII.): hence the fol- lowing

RULE.—*Find the apothem, by the preceding formula; multiply the perimeter by half the apothem; the product will be the area required.*

Examples.

1. What is the area of a regular hexagon, each of whose sides is 20?

We have $\qquad CD = 10 \times \cot 30°$;

or, $\qquad \log CD = \log 10 + \log \cot 30° - 10.$

$$\log \tfrac{1}{2}s \quad \cdot \quad \cdot \quad \cdot \quad (10) \quad \cdot \quad 1.000000$$

$$\log \cot \frac{180°}{n} \cdot \quad (30°) \quad \cdot \quad 10.238561$$

$$\log CD \cdot \quad \cdot \quad \cdot \quad \cdot \quad \cdot \quad \underline{1.238561} \quad \therefore \quad CD = 17.3205.$$

The perimeter is equal to 120: hence, denoting the area by Q,

$$Q = \frac{120 \times 17.3205}{2} = 1039.23, \; Ans.$$

2. What is the area of an octagon, one of whose sides is 20?   *Ans.* 1931.37.

The areas of some of the most important of the regular polygons have been computed by the preceding method, on the supposition that each side is equal to 1, and the results are given in the following

TABLE.

NAMES.	SIDES.	AREAS.	NAMES.	SIDES.	AREAS.
Triangle,	3	0.4330127	Octagon,	8	4.8284271
Square,	4	1.0000000	Nonagon,	9	6.1818242
Pentagon,	5	1.7204774	Decagon,	10	7.6942088
Hexagon,	6	2.5980762	Undecagon,	11	9.3656399
Heptagon,	7	3.6339124	Dodecagon,	12	11.1961524

The areas of similar polygons are to each other as the squares of their homologous sides (Book IV., Prop. XXVII.).

Denoting the area of a regular polygon whose side is s by Q, and that of a similar polygon whose side is 1 by T, the tabular area, we have

$$Q \ : \ T \ :: \ s^2 \ : \ 1^2;$$

$$\therefore \ Q = Ts^2;$$

hence, the following

RULE.—*Multiply the corresponding tabular area by the square of the given side; the product will be the area required.*

Examples.

1. What is the area of a regular hexagon, each of whose sides is 20?

We have $T = 2.5980762$, and $s^2 = 400$: hence,

$$Q = 2.5980762 \times 400 = 1039.23048, \ Ans.$$

2. Find the area of a pentagon, whose side is 25.

Ans. 1075.298375.

3. Find the area of a decagon, whose side is 20.

Ans. 3077.68352.

To find the circumference of a circle, when the diameter is given.

102. From the principle demonstrated in Book V., Prop. XVI., we may write the following

RULE. — *Multiply the given diameter by 3.1416; the product will be the circumference required.*

Examples.

1. What is the circumference of a circle, whose diameter is 25?   *Ans.* 78.54.

2. If the diameter of the earth is 7921 miles, what is the circumference?   *Ans.* 24884.6136.

To find the diameter of a circle, when the circumference is given.

103. From the preceding case, we may write the following

Rule.—*Divide the given circumference by* 3.1416; *the quotient will be the diameter required.*

Examples.

1. What is the diameter of a circle, whose circumference is 11652.1944?   *Ans.* 3709.

2. What is the diameter of a circle, whose circumference is 6850?   *Ans.* 2180.41.

To find the length of an arc containing any number of degrees.

104. The length of an arc of 1°, in a circle whose diameter is 1, is equal to the circumference, or 3.1416, divided by 360; that is, it is equal to 0.0087266: hence, the length of an arc of n degrees will be $n \times 0.0087266$. To find the length of an arc containing n degrees, when the diameter is d, we employ the principle demonstrated in Book V., Prop. XIII., C. 2: hence, we may write the following

RULE.—*Multiply the number of degrees in the arc by .0087266, and the product by the diameter of the circle; the result will be the length required.*

Examples.

1. What is the length of an arc of 30 degrees, the diameter being 18 feet?　　　　*Ans.* 4.712364 ft.

2. What is the length of an arc of 12° 10', or 12⅙°, the diameter being 20 feet?　　　*Ans.* 2.123472 ft.

To find the area of a circle.

105. From the principle demonstrated in Book V., Prop. XV., we may write the following

RULE.—*Multiply the square of the radius by 3.1416; the product will be the area required;*

Examples.

1. Find the area of a circle, whose diameter is 10 and circumference 31.416.　　　*Ans.* 78.54.

2. How many square yards in a circle whose diameter is 3½ feet?　　　　*Ans.* 1.069016.

3. What is the area of a circle whose circumference is 12 feet?　　　　*Ans.* 11.4595.

To find the area of a circular sector.

106. From the principle demonstrated in Book V., Prop. XIV., C. 1 and 2, we may write the following

RULE.—I. *Multiply half the length of the arc by the radius; or,*

II. *Find the area of the whole circle, by the last rule; then write the proportion, 360 is to the number of degrees in the arc of the sector, as the area of the circle is to the area of the sector.*

Examples.

1. Find the area of a circular sector, whose arc contains 18°, the diameter of the circle being 3 feet.

Ans. 0.35343 sq. ft.

2. Find the area of a sector, whose arc is 20 feet, the radius being 10.   *Ans.* 100.

3. Required the area of a sector, whose arc is 147° 29′ and radius 25 feet.   *Ans.* 804.3986 sq. ft.

To find the area of a circular segment.

107. Let AB represent the chord corresponding to the two segments ACB and AFB. Draw AE and BE. The segment ACB is equal to the sector EACB, *minus* the triangle AEB. The segment AFB is equal to the sector EAFB, *plus* the triangle AEB. Hence, we have the following

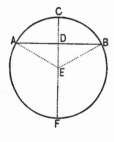

RULE.—*Find the area of the corresponding sector, and also of the triangle formed by the chord of the segment and the two extreme radii of the sector; subtract the latter from the former when the segment is less than a semicircle, and add the latter to the former when the segment is greater than a semicircle; the result will be the area required.*

Examples.

1. Find the area of a segment, whose chord is 12 and whose radius is 10.

Solving the triangle AEB, we find the angle AEB is equal to 73° 44', the area of the sector EACB equal to 64.35, and the area of the triangle AEB equal to 48; hence, the segment ACB is equal to 16.35.

2. Find the area of a segment, whose height is 18, the diameter of the circle being 50.   *Ans.* 636.4834.

3. Required the area of a segment, whose chord is 16, the diameter being 20.   *Ans.* 44.764.

To find the area of a circular ring contained between the circumferences of two concentric circles.

108. Let R and r denote the radii of the two circles, R being greater than r. The area of the outer circle is $R^2 \times 3.1416$, and that of the inner circle is $r^2 \times 3.1416$; hence, the area of the ring is equal to $(R^2 - r^2) \times 3.1416$. Hence, the following

RULE.—*Find the difference of the squares of the radii of the two circles, and multiply it by 3.1416; the product will be the area required.*

Examples.

1. The diameters of two concentric circles being 10 and 6, required the area of the ring contained between their circumferences.   *Ans.* 50.2656.

2. What is the area of the ring, when the diameters of the circles are 10 and 20?   *Ans.* 235.62.

MENSURATION OF BROKEN AND CURVED SUR-FACES.

To find the area of the entire surface of a right prism.

109. From the principle demonstrated in Book VII., Prop. I., we may write the following

RULE.—*Multiply the perimeter of the base by the altitude, the product will be the area of the convex surface; to this add the areas of the two bases; the result will be the area required.*

Examples.

1. Find the surface of a cube, the length of each side being 20 feet.   *Ans.* 2400 sq. ft.

2. Find the whole surface of a triangular prism, whose base is an equilateral triangle having each of its sides equal to 18 inches, and altitude 20 feet.

Ans. 91.949 sq. ft.

To find the area of the entire surface of a right pyramid.

110. From the principle demonstrated in Book VII., Prop. IV., we may write the following

RULE.—*Multiply the perimeter of the base by half the slant height; the product will be the area of the convex surface; to this add the area of the base; the result will be the area required.*

Examples.

1. Find the convex surface of a right triangular pyramid, the slant height being 20 feet, and each side of the base 3 feet.   *Ans.* 90 sq. ft.

2. What is the entire surface of a right pyramid, whose slant height is 27 feet, and the base a pentagon of which each side is 25 feet? *Ans.* 2762.798 sq. ft.

To find the area of the convex surface of a frustum of a right pyramid.

111. From the principle demonstrated in Book VII., Prop. IV., S., we may write the following

RULE.—*Multiply the half sum of the perimeters of the two bases by the slant height; the product will be the area required.*

Examples.

1. How many square feet are there in the convex surface of the frustum of a square pyramid, whose slant height is 10 feet, each side of the lower base 3 feet 4 inches, and each side of the upper base 2 feet 2 inches? *Ans.* 110 sq. ft.

2. What is the convex surface of the frustum of a heptagonal pyramid, whose slant height is 55 feet, each side of the lower base 8 feet, and each side of the upper base 4 feet? *Ans.* 2310 sq. ft.

112. Since a cylinder may be regarded as a prism whose base has an infinite number of sides, and a cone as a pyramid whose base has an infinite number of sides, the rules just given may be applied to find the areas of the surfaces of right cylinders, cones, and frustums of cones, by simply changing the term *perimeter* to circumference.

Examples.

1. What is the convex surface of a cylinder, the diameter of whose base is 20, and whose altitude 50?

Ans. 3141.6.

2. What is the entire surface of a cylinder, the altitude being 20, and diameter of the base 2 feet?

Ans. 131.9472 sq. ft.

3. Required the convex surface of a cone, whose slant height is 50 feet, and the diameter of its base 8½ feet.

Ans. 667.59 sq. ft.

4. Required the entire surface of a cone, whose slant height is 36, and the diameter of its base 18 feet.

Ans. 1272.348 sq. ft.

5. Find the convex surface of the frustum of a cone, the slant height of the frustum being 12½ feet, and the circumferences of the bases 8.4 feet and 6 feet.

Ans. 90 sq. ft.

6. Find the entire surface of the frustum of a cone, the slant height being 16 feet, and the radii of the bases 3 feet and 2 feet.   *Ans.* 292.1688 sq. ft.

To find the area of the surface of a sphere.

113. From the principle demonstrated in Book VIII., Prop. X., C. 1, we may write the following

RULE.—*Find the area of one of its great circles, and multiply it by 4; the product will be the area required.*

Examples.

1. What is the area of the surface of a sphere, whose radius is 16?   *Ans.* 3216.9984.

2. What is the area of the surface of a sphere, whose radius is 27.25? *Ans.* 9331.3374.

To find the area of a zone.

114. From the principle demonstrated in Book VIII., Prop. X., C. 2, we may write the following

RULE.—*Find the circumference of a great circle of the sphere, and multiply it by the altitude of the zone; the product will be the area required.*

Examples.

1. The diameter of a sphere being 42 inches, what is the area of the surface of a zone whose altitude is 9 inches? *Ans.* 1187.5248 sq. in.

2. If the diameter of a sphere is $12\frac{1}{2}$ feet, what will be the surface of a zone whose altitude is 2 feet? *Ans.* 78.54 sq. ft.

To find the area of a spherical polygon.

115. From the principle demonstrated in Book IX., Prop. XIX., we may write the following

RULE.—*From the sum of the angles of the polygon, subtract 180° taken as many times, less two, as the polygon has sides, and divide the remainder by 90°; the quotient will be the spherical excess. Find the area of a great circle of the sphere, and divide it by 2; the quotient will be the area of a tri-rectangular triangle. Multiply the area of the tri-rectangular triangle by the spherical excess, and the product will be the area required.*

This rule applies to the spherical triangle, as well as to any other spherical polygon.

Examples.

1. Required the area of a triangle, described on a sphere whose diameter is 30 feet, the angles being 140°, 92°, and 68°.   *Ans.* 471.24 sq. ft.

2. What is the area of a polygon of seven sides, described on a sphere whose diameter is 17 feet, the sum of the angles being 1080°?   *Ans.* 226.98.

3. What is the area of a regular polygon of eight sides, described on a sphere whose diameter is 30 yards, each angle of the polygon being 140°?

Ans. 157.08 sq. yds.

MENSURATION OF VOLUMES.

To find the volume of a prism.

116. From the principle demonstrated in Book VII., Prop. XIV., we may write the following

RULE.—*Multiply the area of the base by the altitude; the product will be the volume required.*

Examples.

1. What is the volume of a cube, whose side is 24 inches?   *Ans.* 13824 cu. in.

2. How many cubic feet in a block of marble, of which the length is 3 feet 2 inches, breadth 2 feet 8 inches, and height or thickness 2 feet 6 inches?

Ans. 21¼ cu. ft.

3. Required the volume of a triangular prism, whose height is 10 feet, and the three sides of its triangular base 3, 4, and 5 feet. *Ans.* 60.

To find the volume of a pyramid.

117. From the principle demonstrated in Book VII., Prop. XVII., we may write the following

RULE.—*Multiply the area of the base by one third of the altitude; the product will be the volume required.*

Examples.

1. Required the volume of a square pyramid, each side of its base being 30, and the altitude 25. *Ans.* 7500.

2. Find the volume of a triangular pyramid, whose altitude is 30, and each side of the base 3 feet.
Ans. 38.9711 cu. ft.

3. What is the volume of a pentagonal pyramid, its altitude being 12 feet, and each side of its base 2 feet?
Ans. 27.5276 cu. ft.

4. What is the volume of a hexagonal pyramid, whose altitude is 6.4 feet, and each side of its base 6 inches?
Ans. 1.38564 cu. ft.

To find the volume of a frustum of a pyramid.

118. From the principle demonstrated in Book VII., Prop. XVIII., C., we may write the following

RULE.—*Find the sum of the upper base, the lower base, and a mean proportional between them; multiply the result by one third of the altitude; the product will be the volume required.*

Examples.

1. Find the number of cubic feet in a piece of timber, whose bases are squares, each side of the lower base being 15 inches, and each side of the upper base 6 inches, the altitude being 24 feet.   *Ans.* 19.5.

2. Required the volume of a pentagonal frustum, whose altitude is 5 feet, each side of the lower base 18 inches, and each side of the upper base 6 inches.

Ans. 9.31925 cu. ft.

119. Since cylinders and cones are limiting cases of prisms and pyramids, the three preceding rules are equally applicable to them.

Examples.

1. Required the volume of a cylinder whose altitude is 12 feet, and the diameter of its base 15 feet.

Ans. 2120.58 cu. ft.

2. Required the volume of a cylinder whose altitude is 20 feet, and the circumference of whose base is 5 feet 6 inches.   *Ans.* 48.144 cu. ft.

3. Required the volume of a cone whose altitude is 27 feet, and the diameter of the base 10 feet.

Ans. 706.86 cu. ft.

4. Required the volume of a cone whose altitude is 10½ feet, and the circumference of its base 9 feet.

Ans. 22.56 cu. ft.

5. Find the volume of the frustum of a cone, the altitude being 18, the diameter of the lower base 8, and that of the upper base 4.   *Ans.* 527.7888.

6. What is the volume of the frustum of a cone, the altitude being 25, the circumference of the lower base 20, and that of the upper base 10? *Ans.* 464.216.

7. If a cask, which is composed of two equal conic frustums joined together at their larger bases, have its bung diameter 28 inches, the head diameter 20 inches, and the length 40 inches, how many gallons of wine will it contain, there being 231 cubic inches in a gallon?

Ans. 79.0613.

To find the volume of a sphere.

120. From the principle demonstrated in Book VIII., Prop. XIV., we may write the following

RULE.—*Cube the diameter of the sphere, and multiply the result by $\frac{1}{6}\pi$, that is, by 0.5236; the product will be the volume required.*

Examples.

1. What is the volume of a sphere, whose diameter is 12? *Ans.* 904.7808.

2. What is the volume of the earth, if the mean diameter is taken equal to 7918.7 miles?

Ans. 259992792082 cu. miles.

To find the volume of a wedge.

121. A WEDGE is a volume bounded by a rectangle ABCD, called the *back*, two trapezoids ABHG, DCHG, called *faces*, and two triangles ADG, CBH, called *ends*. The line GH, in which the faces meet, is called the *edge*.

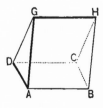

There are three cases ;

1st, When the length of the edge is equal to the length of the back ;

2d, When it is less ; and

3d, When it is greater.

In the first case, the wedge is equal in volume to a right prism, whose base is the triangle ADG, and altitude GH or AB: hence, its volume is equal to ADG multiplied by AB.

In the second case, through H, a point of the edge, pass a plane HCB perpendicular to the back, and intersecting it in the line BC parallel to AD. This plane will divide the wedge into two parts, one of which is represented by the figure.

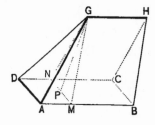

Through G, draw the plane GNM parallel to HCB, and it will divide the part of the wedge represented by the figure into the right triangular prism GNM–B, and the quadrangular pyramid ADNM–G. Draw GP perpendicular to NM: it will also be perpendicular to the back of the wedge (B. VI., P. XVII.), and hence, will be equal to the altitude of the wedge.

Denote AB by L, the breadth AD by b, the edge GH by l, the altitude by h, and the volume by V; then,

$$AM = L - l,$$

$$MB = GH = l,$$

and $\qquad$ area NGM $= \tfrac{1}{2}bh :$

then $$\text{Prism} = \tfrac{1}{2}bhl;$$

$$\text{Pyramid} = b\,(\mathsf{L} - l)\,\tfrac{1}{3}h = \tfrac{1}{3}bh\,(\mathsf{L} - l),$$

and
$$V = \tfrac{1}{2}bhl + \tfrac{1}{3}bh\,(\mathsf{L} - l)$$
$$= \tfrac{1}{2}bhl + \tfrac{1}{3}bh\mathsf{L} - \tfrac{1}{3}bhl$$
$$= \tfrac{1}{6}bh\,(l + 2\mathsf{L}).$$

We can find a similar expression for the remaining part of the wedge, and by adding, the factor within the parenthesis becomes the entire length of the edge plus twice the length of the back.

In the third case, l is greater than L; the volume of each part is equal to *the difference* of the prism and pyramid, and is of the same form as before. Hence, in either case, we have the following

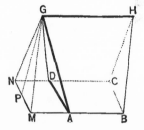

RULE.—*Add twice the length of the back to the length of the edge; multiply the sum by the breadth of the back, and that result by one sixth of the altitude; the final product will be the volume required.*

Examples.

1. If the back of a wedge is 40 by 20 feet, the edge 35 feet, and the altitude 10 feet, what is the volume?

Ans. 3833.33 cu. ft.

2. What is the volume of a wedge, whose back is 18 feet by 9, edge 20 feet, and altitude 6 feet?

Ans. 504 cu. ft.

To find the volume of a prismoid.

122. A PRISMOID is a frustum of a wedge.

Let L and B denote the length and breadth of the lower base, l and b the length and breadth of the upper base, M and m the length and breadth of the section equidistant from the bases, and h the altitude of the prismoid.

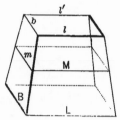

Through the edges L and l', let a plane be passed, and it will divide the prismoid into two wedges, having for bases the bases of the prismoid, and for edges the lines L and l'.

The volume of the prismoid, denoted by V, will be equal to the sum of the volumes of the two wedges; hence,

$$V = \tfrac{1}{6}Bh\,(l + 2L) + \tfrac{1}{6}bh\,(L + 2l)\,;$$

or,

$$V = \tfrac{1}{6}h\,(2BL + 2bl + Bl + bL)\,;$$

which may be written under the form,

$$V = \tfrac{1}{6}h\,[(BL + bl + Bl + bL) + BL + bl]. \quad \cdot \quad \textbf{(A.)}$$

Because the auxiliary section is midway between the bases, we have

$$2M = L + l, \quad \text{and} \quad 2m = B + b\,;$$

hence, $4Mm = (L + l)\,(B + b) = BL + Bl + bL + bl.$

Substituting in (**A**), we have

$$V = \tfrac{1}{6}h\,(BL + bl + 4Mm).$$

But BL is the area of the lower base, or lower section, *bl* is the area of the upper base, or upper section, and M*m* is the area of the middle section; hence, the following

RULE.—*To find the volume of a prismoid, find the sum of the areas of the extreme sections and four times the middle section; multiply the result by one sixth of the distance between the extreme sections; the result will be the volume required.*

This rule is used in computing volumes of earth-work in railroad cutting and embankment, and is of very extensive application. It may be shown that the same rule holds for every one of the volumes heretofore discussed in this work. Thus, in a pyramid, we may regard the base as one extreme section, and the vertex (whose area is 0), as the other extreme; their sum is equal to the area of the base. The area of a section midway between them is equal to one fourth of the base: hence, four times the middle section is equal to the base. Multiplying the sum of these by one sixth of the altitude, gives the same result as that already found. The application of the rule to the case of cylinders, frustums of cones, spheres, &c., is left as an exercise for the student.

Examples.

1. One of the bases of a rectangular prismoid is 25 feet by 20, the other 15 feet by 10, and the altitude 12 feet: required the volume. *Ans.* 3700 cu. ft.

2. What is the volume of a stick of hewn timber, whose ends are 30 inches by 27, and 24 inches by 18, its length being 24 feet? *Ans.* 102 cu. ft.

MENSURATION OF REGULAR POLYEDRONS.

123. A REGULAR POLYEDRON is a polyedron bounded by equal regular polygons.

The polyedral angles of any regular polyedron are all equal.

124. There are five regular polyedrons (Book VII., page 219).

To find the diedral angle contained between two consecutive faces of a regular polyedron.

125. As in the figure, let the vertex, O, of a polyedral angle of a tetraedron be taken as the centre of a sphere whose radius is 1: then will the three faces of this polyedral angle, by their intersections with the surface of the sphere, determine the spherical

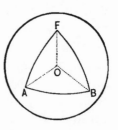

triangle FAB. The plane angles FOA, FOB, and AOB, being equal to each other, the arcs FA, FB, and AB, which measure these angles, are also equal to each other, and the spherical triangle FAB is equilateral. The angle FAB of the triangle is equal to the diedral angle of the planes FOA and AOB, that is, to the diedral angle between the faces of the tetraedron.

In like manner, if the vertex of a polyedral angle of any one of the regular polyedrons be taken as the centre of a sphere whose radius is 1, the faces of this polyedral angle will, by their intersections with the surface of the sphere, determine a regular spherical polygon; the *number of sides* of this spherical polygon will be equal to the number of faces of the polyedral angle; *each side* of the polygon will be the measure of one of the plane angles formed by the edges of the polyedral angle; and *each angle* of the polygon will be equal to the diedral angle contained between two consecutive faces of the regular polyedron.

To find the required diedral angle, therefore, it only remains to deduce a formula for finding one angle of a regular spherical polygon when the sides are given.

Let ABCDE represent a regular spherical polygon, and let P be the pole of a small circle passing through its vertices. Suppose P to be connected with each of the vertices by arcs of great circles; there will thus be formed as many equal isosceles triangles as the polygon has sides, the vertical angle in each being equal to 360° divided by the number of sides. Through P draw the arc of a great circle, PQ, perpendicular to AB: then will AQ be equal to BQ, and the angle APQ to the angle QPB (B. IX., P. XI., C.). If we denote the number of sides of the spherical polygon by n', the angle APQ will be equal to $\frac{360°}{2n'}$, or $\frac{180°}{n'}$.

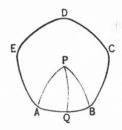

In the right-angled spherical triangle AQP, we know the base AQ, and the vertical angle APQ; hence, by Napier's rules for circular parts, we have

$$\sin (90° - APQ) = \cos (90° - PAQ) \cos AQ,$$

or,
$$\cos APQ = \sin PAQ \cos AQ;$$

denoting the side AB of the polygon by s', and the angle PAQ, which is half the angle EAB of the polygon, by $\frac{1}{2}A$, we have

$$\cos \frac{180°}{n'} = \sin \tfrac{1}{2}A \cos \tfrac{1}{2}s';$$

whence,
$$\sin \tfrac{1}{2}A = \frac{\cos \dfrac{180°}{n'}}{\cos \tfrac{1}{2}s'}.$$

Examples.

In the Tetraedron,

$$\frac{180°}{n'} = 60°, \text{ and } \tfrac{1}{2}s' = 30°; \quad \therefore A = 70° 31' 42''.$$

In the Hexaedron,

$$\frac{180°}{n'} = 60°, \text{ and } \tfrac{1}{2}s' = 45°; \quad \therefore A = 90°.$$

In the Octaedron,

$$\frac{180°}{n'} = 45°, \text{ and } \tfrac{1}{2}s' = 30°; \quad \therefore A = 109° 28' 19''.$$

In the Dodecaedron,

$$\frac{180°}{n'} = 60°, \text{ and } \tfrac{1}{2}s' = 54°; \quad \therefore A = 116° 63' 54''.$$

In the Icosaedron,

$$\frac{180°}{n'} = 36°, \text{ and } \tfrac{1}{2}s' = 30°; \quad \therefore A = 138° 11' 23''.$$

To find the volume of a regular polyedron.

126. If planes be passed through the centre of the polyedron and each of the edges, they will divide the polyedron into as many equal right pyramids as the polyedron has faces. The common vertex of these pyramids will be at the centre of the polyedron, their bases will be the faces of the polyedron, and their lateral faces will bisect the diedral angles of the polyedron. The volume of each pyramid will be equal to the product of its base and one third of its altitude, and this product multiplied by the number of faces, will be the volume of the polyedron.

It only remains to deduce a formula for finding the altitude of the several pyramids, *i. e.*, the distance from the centre to one face of the polyedron.

Conceive a perpendicular OC to be drawn from O, the centre of the polyedron, to one face; the foot of this perpendicular will be the centre of the face. From C, the foot of this perpendicular, draw a perpendicular to one side of the face in which it lies, and connect the point D with the centre of the polyedron. There will thus be formed a right-angled triangle, OCD, whose base, CD, is the apothem of the face, whose angle ODC is half the angle CDL contained between two consecutive faces of the polyedron, and whose altitude OC is the required altitude of the pyramid, or, in other words, the radius of the inscribed sphere. This will be true for any one of the regular polyedrons—the hexaedron is taken here for simplicity of illustration.

Denote the line CD by p, the angle ODC by $\frac{1}{2}$A, and the perpendicular OC by R. p may be found by the formula, given in Art. 101, for finding the apothem of a regular polygon; $\frac{1}{2}$A may be found from the formula for sin $\frac{1}{2}$A, given in Art. 125; then, in the right-angled triangle OCD, we have, formula (3), Art. 37,

$$R = p \tan \tfrac{1}{2}A.$$

Compute the area of one of the faces of the given polyedron and multiply it by $\frac{1}{3}$R, as determined by the formula just given, and multiply the result thus obtained by the number of faces of the polyedron; the final product will be the volume of the given regular polyedron.

The volumes of all the regular polyedrons have been computed on the supposition that their edges are each equal to 1, and the results are given in the following

TABLE.

NAMES.	NO. OF FACES.	VOLUMES.
Tetraedron,	4	0.1178513
Hexaedron,	6	1.0000000
Octaedron,	8	0.4714045
Dodecaedron,	12	7.6631189
Icosaedron,	20	2.1816950

From the principles demonstrated in Book VII., we may write the following

RULE.—*To find the volume of any regular polyedron, multiply the cube of its edge by the corresponding tabular volume; the product will be the volume required.*

Examples.

1. What is the volume of a tetraedron, whose edge is 15?   *Ans.* 397.75.

2. What is the volume of a hexaedron, whose edge is 12?   *Ans.* 1728.

3. What is the volume of an octaedron, whose edge is 20?   *Ans.* 3771.236.

4. What is the volume of a dodecaedron, whose edge is 25?   *Ans.* 119736.2328.

5. What is the volume of an icosaedron, whose edge is 20?   *Ans.* 17453.56.

A TABLE

LOGARITHMS OF NUMBERS

FROM 1 TO 10,000.

N.	Log.	N.	Log.	N.	Log.	N.	Log.
1	0·000000	26	1·414973	51	1·707570	76	1·880814
2	0·301030	27	1·431364	52	1·716003	77	1·886491
3	0·477121	28	1·447158	53	1·724276	78	1·892095
4	0·602060	29	1·462398	54	1·732394	79	1·897627
5	0·698970	30	1·477121	55	1·740363	80	1·903090
6	0·778151	31	1·491362	56	1·748188	81	1·908485
7	0·845098	32	1·505150	57	1·755875	82	1·913814
8	0·903090	33	1·518514	58	1·763428	83	1·919078
9	0·954243	34	1·531479	59	1·770852	84	1·924279
10	1·000000	35	1·544068	60	1·778151	85	1·929419
11	1·041393	36	1·556303	61	1·785330	86	1·934498
12	1·079181	37	1·568202	62	1·792392	87	1·939519
13	1·113943	38	1·579784	63	1·799341	88	1·944483
14	1·146128	39	1·591065	64	1·806181	89	1·949390
15	1·176091	40	1·602060	65	1·812913	90	1·954243
16	1·204120	41	1·612784	66	1·819544	91	1·959041
17	1·230449	42	1·623249	67	1·826075	92	1·963788
18	1·255273	43	1·633468	68	1·832509	93	1·968483
19	1·278754	44	1·643453	69	1·838849	94	1·973128
20	1·301030	45	1·653213	70	1·845098	95	1·977724
21	1·322219	46	1·662758	71	1·851258	96	1·982271
22	1·342423	47	1·672098	72	1·857333	97	1·986772
23	1·361728	48	1·681241	73	1·863323	98	1·991226
24	1·380211	49	1·690196	74	1·869232	99	1·995635
25	1·397940	50	1·698970	75	1·875061	100	2·000000

REMARKS. In the following table, in the nine right-hand columns of each page, where the first or leading figures change from 9's to 0's, points or dots are introduced instead of the 0's, to catch the eye, and to indicate that from thence the two figures of the Logarithm to be taken from the second column, stand in the next line below.

N.	0	1	2	3	4	5	6	7	8	9	D.
100	000000	0434	0868	1301	1784	2166	2598	3029	3461	3891	432
101	4321	4751	5181	5609	6088	6466	6894	7321	7748	8174	428
102	8600	9026	9451	9876	♦300	♦724	1147	1570	1993	2415	424
103	012837	3259	3680	4100	4521	4940	5360	5779	6197	6616	419
104	7033	7451	7868	8284	8700	9116	9532	9947	♦361	♦775	416
105	021189	1603	2016	2428	2841	3252	3664	4075	4486	4896	412
106	5306	5715	6125	6533	6942	7350	7757	8164	8571	8978	408
107	9384	9789	♦195	♦600	1004	1408	1812	2216	2619	3021	404
108	033424	3826	4227	4628	5029	5430	5830	6230	6629	7028	400
109	7426	7825	8223	8620	9017	9414	9811	♦207	♦602	♦998	396
110	041393	1787	2182	2576	2969	3362	3755	4148	4540	4932	393
111	5323	5714	6105	6495	6885	7275	7664	8053	8442	8830	389
112	9218	9606	9993	♦380	♦766	1153	1538	1924	2309	2694	386
113	053078	3463	3846	4230	4618	4996	5378	5760	6142	6524	382
114	6905	7286	7666	8046	8426	8805	9185	9568	9942	♦320	379
115	060698	1075	1452	1829	2206	2582	2958	3333	3709	4083	376
116	4458	4832	5206	5580	5953	6326	6699	7071	7443	7815	372
117	8186	8557	8928	9298	9668	♦♦38	♦407	♦776	1145	1514	369
118	071882	2250	2617	2985	3352	3718	4085	4451	4816	5182	366
119	5547	5912	6276	6640	7004	7368	7731	8094	8457	8819	363
120	079181	9543	9904	♦266	♦626	♦987	1347	1707	2067	2426	360
121	082785	3144	3503	3861	4219	4576	4934	5291	5647	6004	357
122	6360	6716	7071	7426	7781	8136	8490	8845	9198	9552	355
123	9905	♦258	♦611	♦963	1315	1667	2018	2370	2721	3071	351
124	093422	3772	4122	4471	4820	5169	5518	5866	6215	6562	349
125	6910	7257	7604	7951	8298	8644	8990	9335	9681	♦♦26	346
126	100371	0715	1059	1403	1747	2091	2434	2777	3119	3462	343
127	3804	4146	4487	4828	5169	5510	5851	6191	6531	6871	340
128	7210	7549	7888	8227	8565	8903	9241	9579	9916	♦253	338
129	110590	0926	1263	1599	1934	2270	2605	2940	3275	3609	335
130	113943	4277	4611	4944	5278	5611	5943	6276	6608	6940	333
131	7271	7603	7934	8265	8595	8926	9256	9586	9915	♦245	330
132	120574	0903	1231	1560	1888	2216	2544	2871	3198	3525	328
133	3852	4178	4504	4830	5156	5481	5806	6131	6456	6781	325
134	7105	7429	7753	8076	8399	8722	9045	9368	9690	♦♦12	323
135	130334	0655	0977	1298	1619	1939	2260	2580	2900	3219	321
136	3539	3858	4177	4496	4814	5133	5451	5769	6086	6403	318
137	6721	7037	7354	7671	7987	8303	8618	8934	9249	9564	315
138	9879	♦194	♦508	♦822	1136	1450	1763	2076	2389	2702	314
139	143015	3327	3639	3951	4263	4574	4885	5196	5507	5818	311
140	146128	6438	6748	7058	7367	7676	7985	8294	8603	8911	309
141	9219	9527	9835	♦142	♦449	♦756	1063	1370	1676	1982	307
142	152288	2594	2900	3205	3510	3815	4120	4424	4728	5032	305
143	5336	5640	5943	6246	6549	6852	7154	7457	7759	8061	303
144	8362	8664	8965	9266	9567	9868	♦168	♦469	♦769	1068	301
145	161368	1667	1967	2266	2564	2863	3161	3460	3758	4055	299
146	4353	4650	4947	5244	5541	5838	6134	6430	6726	7022	297
147	7317	7613	7908	8203	8497	8792	9086	9380	9674	9968	295
148	170262	0555	0848	1141	1434	1726	2019	2311	2603	2895	293
149	3186	3478	3769	4060	4351	4641	4932	5222	5512	5802	291
150	176091	6381	6670	6959	7248	7536	7825	8113	8401	8689	289
151	8977	9264	9552	9839	♦126	♦413	♦699	♦985	1272	1558	287
152	181844	2129	2415	2700	2985	3270	3555	3839	4123	4407	285
153	4691	4975	5259	5542	5825	6108	6391	6674	6956	7239	283
154	7521	7803	8084	8366	8647	8928	9209	9490	9771	♦♦51	281
155	190332	0612	0892	1171	1451	1730	2010	2289	2567	2846	279
156	3125	3403	3681	3959	4237	4514	4792	5069	5346	5623	278
157	5899	6176	6453	6729	7005	7281	7556	7832	8107	8382	276
158	8657	8932	9206	9481	9755	♦♦29	♦303	♦577	♦850	1124	274
159	201397	1670	1943	2216	2488	2761	3033	3305	3577	3848	272
N.	0	1	2	3	4	5	6	7	8	9	D.

N.	0	1	2	3	4	5	6	7	8	9	D.
160	204120	4391	4663	4934	5204	5475	5746	6016	6286	6556	271
161	6826	7096	7365	7634	7904	8173	8441	8710	8979	9247	269
162	9515	9783	♦♦51	♦319	♦586	♦853	1121	1388	1654	1921	267
163	212188	2454	2720	2986	3252	3518	3783	4049	4314	4579	266
164	4844	5109	5373	5638	5902	6166	6430	6694	6957	7221	264
165	7484	7747	8010	8273	8536	8798	9060	9323	9585	9846	262
166	220108	0370	0631	0892	1158	1414	1675	1986	2196	2456	261
167	2716	2976	3236	3496	3755	4015	4274	4533	4792	5051	259
168	5309	5568	5826	6084	6342	6600	6858	7115	7372	7630	258
169	7887	8144	8400	8657	8913	9170	9426	9682	9938	♦193	256
170	230449	0704	0960	1215	1470	1724	1979	2234	2488	2742	254
171	2996	3250	3504	3757	4011	4264	4517	4770	5023	5276	253
172	5528	5781	6033	6285	6537	6789	7041	7292	7544	7795	252
173	8046	8297	8548	8799	9049	9299	9550	9800	♦♦50	♦300	250
174	240549	0799	1048	1297	1546	1795	2044	2293	2541	2790	249
175	3038	3286	3534	3782	4080	4277	4525	4772	5019	5266	248
176	5513	5759	6006	6252	6499	6745	6991	7237	7482	7728	246
177	7973	8219	8464	8709	8954	9198	9443	9687	9932	♦176	245
178	250420	0664	0908	1151	1395	1638	1881	2125	2368	2610	243
179	2853	3096	3338	3580	3822	4064	4306	4548	4790	5031	242
180	255273	5514	5755	5996	6237	6477	6718	6958	7198	7439	241
181	7679	7918	8158	8398	8637	8877	9116	9355	9594	9833	239
182	260071	0310	0548	0787	1025	1263	1501	1739	1976	2214	238
183	2451	2688	2925	3162	3399	3636	3873	4109	4346	4582	237
184	4818	5054	5290	5525	5761	5996	6232	6467	6702	6937	235
185	7172	7406	7641	7875	8110	8344	8578	8812	9046	9279	234
186	9513	9746	9980	♦213	♦446	♦679	♦912	1144	1377	1609	233
187	271842	2074	2306	2538	2770	3001	3233	3464	3696	3927	232
188	4158	4389	4620	4850	5081	5311	5542	5772	6002	6232	230
189	6462	6692	6921	7151	7380	7609	7838	8067	8296	8525	229
190	278754	8982	9211	9439	9667	9895	♦123	♦351	♦578	♦806	228
191	281033	1261	1488	1715	1942	2169	2396	2622	2849	3075	227
192	3301	3527	3753	3979	4205	4431	4656	4882	5107	5332	226
193	5557	5782	6007	6232	6456	6681	6905	7130	7354	7578	225
194	7802	8026	8249	8473	8696	8920	9143	9366	9589	9812	223
195	290035	0257	0480	0702	0925	1147	1369	1591	1813	2034	222
196	2256	2478	2699	2920	3141	3363	3584	3804	4025	4246	221
197	4466	4687	4907	5127	5347	5567	5787	6007	6226	6446	220
198	6665	6884	7104	7323	7542	7761	7979	8198	8416	8635	219
199	8853	9071	9289	9507	9725	9943	♦161	♦378	♦595	♦813	218
200	301030	1247	1464	1681	1898	2114	2331	2547	2764	2980	217
201	3196	3412	3628	3844	4059	4275	4491	4706	4921	5136	216
202	5351	5566	5781	5996	6211	6425	6639	6854	7068	7282	215
203	7496	7710	7924	8137	8351	8564	8778	8991	9204	9417	213
204	9630	9843	♦♦56	♦268	♦481	♦693	♦906	1118	1330	1542	212
205	311754	1966	2177	2389	2600	2812	3023	3234	3445	3656	211
206	3867	4078	4289	4499	4710	4920	5130	5340	5551	5760	210
207	5970	6180	6390	6599	6809	7018	7227	7436	7646	7854	209
208	8063	8272	8481	8689	8898	9106	9314	9522	9730	9938	208
209	320146	0354	0562	0769	0977	1184	1391	1598	1805	2012	207
210	322219	2426	2633	2839	3046	3252	3458	3665	3871	4077	206
211	4282	4488	4694	4899	5105	5310	5516	5721	5926	6131	205
212	6336	6541	6745	6950	7155	7359	7563	7767	7972	8176	204
213	8380	8583	8787	8991	9194	9398	9601	9805	♦♦♦8	♦211	203
214	330414	0617	0819	1022	1225	1427	1630	1832	2034	2236	202
215	2438	2640	2842	3044	3246	3447	3649	3850	4051	4253	202
216	4454	4655	4856	5057	5257	5458	5658	5859	6059	6260	201
217	6460	6660	6860	7060	7260	7459	7659	7858	8058	8257	200
218	8456	8656	8855	9054	9253	9451	9650	9849	♦♦47	♦246	199
219	340444	0642	0841	1039	1237	1435	1632	1830	2028	2225	198
N	0	1	2	3	4	5	6	7	8	9	D.

N.	0	1	2	3	4	5	6	7	8	9	D.
220	342423	2620	2817	3014	3212	3409	3606	3802	3999	4196	197
221	4392	4589	4785	4981	5178	5374	5570	5766	5962	6157	196
222	6353	6549	6744	6939	7135	7330	7525	7720	7915	8110	195
223	8305	8500	8694	8889	9083	9278	9472	9666	9860	♦♦54	194
224	350248	0442	0636	0829	1023	1216	1410	1603	1796	1989	193
225	2183	2375	2568	2761	2954	3147	3339	3532	3724	3916	193
226	4108	4301	4493	4685	4876	5068	5260	5452	5643	5834	192
227	6026	6217	6408	6599	6790	6981	7172	7363	7554	7744	191
228	7935	8125	8316	8506	8696	8886	9076	9266	9456	9646	190
229	9835	♦♦25	♦215	♦404	♦593	♦783	♦972	1161	1350	1589	189
230	361728	1917	2105	2294	2482	2671	2859	3048	3236	3424	188
231	3612	3800	3988	4176	4363	4551	4739	4926	5113	5301	188
232	5488	5675	5862	6049	6236	6423	6610	6796	6983	7169	187
233	7356	7542	7729	7915	8101	8287	8473	8659	8845	9030	186
234	9216	9401	9587	9772	9958	♦143	♦328	♦513	♦698	♦883	185
235	371068	1253	1437	1622	1806	1991	2175	2360	2544	2728	184
236	2912	3096	3280	3464	3647	3831	4015	4198	4382	4565	184
237	4748	4932	5115	5298	5481	5664	5846	6029	6212	6394	183
238	6577	6759	6942	7124	7306	7488	7670	7852	8034	8216	182
239	8398	8580	8761	8943	9124	9306	9487	9668	9849	♦♦30	181
240	380211	0392	0573	0754	0934	1115	1296	1476	1656	1837	181
241	2017	2197	2377	2557	2737	2917	3097	3277	3456	3636	180
242	3815	3995	4174	4353	4533	4712	4891	5070	5249	5428	179
243	5606	5785	5964	6142	6321	6499	6677	6856	7034	7212	178
244	7390	7568	7746	7923	8101	8279	8456	8634	8811	8989	178
245	9166	9343	9520	9698	9875	♦♦51	♦228	♦405	♦582	♦759	177
246	390935	1112	1288	1464	1641	1817	1993	2169	2345	2521	176
247	2697	2873	3048	3224	3400	3575	3751	3926	4101	4277	176
248	4452	4627	4802	4977	5152	5326	5501	5676	5850	6025	175
249	6199	6374	6548	6722	6896	7071	7245	7419	7592	7766	174
250	397940	8114	8287	8461	8634	8808	8981	9154	9328	9501	173
251	9674	9847	♦♦20	♦192	♦365	♦538	♦711	♦883	1056	1228	173
252	401401	1573	1745	1917	2089	2261	2433	2605	2777	2949	172
253	3121	3292	3464	3635	3807	3978	4149	4320	4492	4663	171
254	4834	5005	5176	5346	5517	5688	5858	6029	6199	6370	171
255	6540	6710	6881	7051	7221	7391	7561	7731	7901	8070	170
256	8240	8410	8579	8749	8918	9087	9257	9426	9595	9764	169
257	9933	♦102	♦271	♦440	♦609	♦777	♦946	1114	1283	1451	169
258	411620	1788	1956	2124	2293	2461	2629	2796	2964	3132	168
259	3300	3467	3635	3803	3970	4137	4305	4472	4639	4806	167
260	414973	5140	5307	5474	5641	5808	5974	6141	6308	6474	167
261	6641	6807	6973	7139	7306	7472	7638	7804	7970	8135	166
262	8301	8467	8633	8798	8964	9129	9295	9460	9625	9791	165
263	9956	♦121	♦286	♦451	♦616	♦781	♦945	1110	1275	1439	165
264	421604	1788	1933	2097	2261	2426	2590	2754	2918	3082	164
265	3246	3410	3574	3737	3901	4065	4228	4392	4555	4718	164
266	4882	5045	5208	5371	5534	5697	5860	6023	6186	6349	163
267	6511	6674	6836	6999	7161	7324	7486	7648	7811	7973	162
268	8135	8297	8459	8621	8783	8944	9106	9268	9429	9591	162
269	9752	9914	♦♦75	♦236	♦398	♦559	♦720	♦881	1042	1203	161
270	431364	1525	1685	1846	2007	2167	2328	2488	2649	2809	161
271	2969	3130	3290	3450	3610	3770	3930	4090	4249	4409	160
272	4569	4729	4888	5048	5207	5367	5526	5685	5844	6004	159
273	6163	6322	6481	6640	6798	6957	7116	7275	7433	7592	159
274	7751	7909	8067	8226	8384	8542	8701	8859	9017	9175	158
275	9333	9491	9648	9806	9964	♦122	♦279	♦437	♦594	♦752	158
276	440909	1066	1224	1381	1538	1695	1852	2009	2166	2323	157
277	2480	2637	2793	2950	3106	3263	3419	3576	3732	3889	157
278	4045	4201	4357	4513	4669	4825	4981	5137	5293	5449	156
279	5604	5760	5915	6071	6226	6382	6537	6692	6848	7003	155
N.	0	1	2	3	4	5	6	7	8	9	D.

N.	0	1	2	3	4	5	6	7	8	9	D.
280	447158	7313	7468	7623	7778	7933	8088	8242	8397	8552	155
281	8706	8861	9015	9170	9324	9478	9633	9787	9941	♦♦95	154
282	450249	0403	0557	0711	0865	1018	1172	1326	1479	1633	154
283	1786	1940	2093	2247	2400	2553	2706	2859	3012	3165	153
284	3318	3471	3624	3777	3930	4082	4235	4387	4540	4692	153
285	4845	4997	5150	5302	5454	5606	5758	5910	6062	6214	152
286	6366	6518	6670	6821	6973	7125	7276	7428	7579	7731	152
287	7882	8033	8184	8336	8487	8638	8789	8940	9091	9242	151
288	9392	9543	9694	9845	9995	♦146	♦296	♦447	♦597	♦748	151
289	460898	1048	1198	1348	1499	1649	1799	1948	2098	2248	150
290	462398	2548	2697	2847	2997	3146	3296	3445	3594	3744	150
291	3893	4042	4191	4340	4490	4639	4788	4936	5085	5234	149
292	5383	5532	5680	5829	5977	6126	6274	6423	6571	6719	149
293	6868	7016	7164	7312	7460	7608	7756	7904	8052	8200	148
294	8347	8495	8643	8790	8938	9085	9233	9380	9527	9675	148
295	9822	9969	♦116	♦263	♦410	♦557	♦704	♦851	♦998	1145	147
296	471292	1438	1585	1732	1878	2025	2171	2318	2464	2610	146
297	2756	2903	3049	3195	3341	3487	3633	3779	3925	4071	146
298	4216	4362	4508	4653	4799	4944	5090	5235	5381	5526	146
299	5671	5816	5962	6107	6252	6397	6542	6687	6832	6976	145
300	477121	7266	7411	7555	7700	7844	7989	8133	8278	8422	145
301	8566	8711	8855	8999	9143	9287	9431	9575	9719	9863	144
302	480007	0151	0294	0438	0582	0725	0869	1012	1156	1299	144
303	1443	1586	1729	1872	2016	2159	2302	2445	2588	2731	143
304	2874	3016	3159	3302	3445	3587	3730	3872	4015	4157	143
305	4300	4442	4585	4727	4869	5011	5153	5295	5437	5579	142
306	5721	5863	6005	6147	6289	6430	6572	6714	6855	6997	142
307	7138	7280	7421	7563	7704	7845	7986	8127	8269	8410	141
308	8551	8692	8833	8974	9114	9255	9396	9537	9677	9818	141
309	9958	♦♦99	♦239	♦380	♦520	♦661	♦801	♦941	1081	1222	140
310	491362	1502	1642	1782	1922	2062	2201	2341	2481	2621	140
311	2760	2900	3040	3179	3319	3458	3597	3737	3876	4015	139
312	4155	4294	4433	4572	4711	4850	4989	5128	5267	5406	139
313	5544	5683	5822	5960	6099	6238	6376	6515	6653	6791	139
314	6930	7068	7206	7344	7483	7621	7759	7897	8035	8173	138
315	8311	8448	8586	8724	8862	8999	9137	9275	9412	9550	138
316	9687	9824	9962	♦♦99	♦236	♦374	♦511	♦648	♦785	♦922	137
317	501059	1196	1333	1470	1607	1744	1880	2017	2154	2291	137
318	2427	2564	2700	2837	2973	3109	3246	3382	3518	3655	136
319	3791	3927	4063	4199	4335	4471	4607	4743	4878	5014	136
320	505150	5286	5421	5557	5693	5828	5964	6099	6234	6370	136
321	6505	6640	6776	6911	7046	7181	7316	7451	7586	7721	135
322	7856	7991	8126	8260	8395	8530	8664	8799	8934	9068	135
323	9203	9337	9471	9606	9740	9874	♦♦9	♦143	♦277	♦411	134
324	510545	0679	0813	0947	1081	1215	1349	1482	1616	1750	134
325	1883	2017	2151	2284	2418	2551	2684	2818	2951	3084	133
326	3218	3351	3484	3617	3750	3883	4016	4149	4282	4414	133
327	4548	4681	4813	4946	5079	5211	5344	5476	5609	5741	133
328	5874	6006	6139	6271	6403	6535	6668	6800	6932	7064	132
329	7196	7328	7460	7592	7724	7855	7987	8119	8251	8382	132
330	518514	8646	8777	8909	9040	9171	9303	9434	9566	9697	131
331	9828	9959	♦♦90	♦221	♦353	♦484	♦615	♦745	♦876	1007	131
332	521138	1269	1400	1530	1661	1792	1922	2053	2183	2314	131
333	2444	2575	2705	2835	2966	3096	3226	3356	3486	3616	130
334	3746	3876	4006	4136	4266	4396	4526	4656	4785	4915	130
335	5045	5174	5304	5434	5563	5693	5822	5951	6081	6210	129
336	6339	6469	6598	6727	6856	6985	7114	7243	7372	7501	129
337	7630	7759	7888	8016	8145	8274	8402	8531	8660	8788	129
338	8917	9045	9174	9302	9430	9559	9687	9815	9943	♦♦72	128
339	530200	0328	0456	0584	0712	0840	0968	1096	1223	1351	128
N.	0	1	2	3	4	5	6	7	8	9	D.

N.	0	1	2	3	4	5	6	7	8	9	D.
340	531479	1607	1734	1862	1990	2117	2245	2372	2500	2627	128
341	2754	2882	3009	3136	3264	3391	3518	3645	3772	3899	127
342	4026	4153	4280	4407	4534	4661	4787	4914	5041	5167	127
343	5294	5421	5547	5674	5800	5927	6053	6180	6306	6432	126
344	6558	6685	6811	6937	7063	7189	7315	7441	7567	7693	126
345	7819	7945	8071	8197	8322	8448	8574	8699	8825	8951	126
346	9076	9202	9327	9452	9578	9703	9829	9954	♦♦79	♦204	125
347	540329	0455	0580	0705	0830	0955	1080	1205	1330	1454	125
348	1579	1704	1829	1953	2078	2203	2327	2452	2576	2701	125
349	2825	2950	3074	3199	3323	3447	3571	3696	3820	3944	124
350	544068	4192	4316	4440	4564	4688	4812	4936	5060	5183	124
351	5307	5431	5555	5678	5802	5925	6049	6172	6296	6419	124
352	6543	6666	6789	6913	7036	7159	7282	7405	7529	7652	123
353	7775	7898	8021	8144	8267	8389	8512	8635	8758	8881	123
354	9003	9126	9249	9371	9494	9616	9739	9861	9984	♦106	123
355	550228	0351	0473	0595	0717	0840	0962	1084	1206	1328	122
356	1450	1572	1694	1816	1938	2060	2181	2303	2425	2547	122
357	2668	2790	2911	3033	3155	3276	3398	3519	3640	3762	121
358	3883	4004	4126	4247	4368	4489	4610	4731	4852	4973	121
359	5094	5215	5336	5457	5578	5699	5820	5940	6061	6182	121
360	556303	6423	6544	6664	6785	6905	7026	7146	7267	7387	120
361	7507	7627	7748	7868	7988	8108	8228	8349	8469	8589	120
362	8709	8829	8948	9068	9188	9308	9428	9548	9667	9787	120
363	9907	♦♦26	♦146	♦265	♦385	♦504	♦624	♦743	♦863	♦982	119
364	561101	1221	1340	1459	1578	1698	1817	1936	2055	2174	119
365	2293	2412	2531	2650	2769	2887	3006	3125	3244	3362	119
366	3481	3600	3718	3837	3955	4074	4192	4311	4429	4548	119
367	4666	4784	4903	5021	5139	5257	5376	5494	5612	5730	118
368	5848	5966	6084	6202	6320	6437	6555	6673	6791	6909	118
369	7026	7144	7262	7379	7497	7614	7732	7849	7967	8084	118
370	568202	8319	8436	8554	8671	8788	8905	9023	9140	9257	117
371	9374	9491	9608	9725	9842	9959	♦♦76	♦193	♦309	♦426	117
372	570543	0660	0776	0893	1010	1126	1243	1359	1476	1592	117
373	1709	1825	1942	2058	2174	2291	2407	2523	2639	2755	116
374	2872	2988	3104	3220	3336	3452	3568	3684	3800	3915	116
375	4031	4147	4263	4379	4494	4610	4726	4841	4957	5072	116
376	5188	5303	5419	5534	5650	5765	5880	5996	6111	6226	115
377	6341	6457	6572	6687	6802	6917	7032	7147	7262	7377	115
378	7492	7607	7722	7836	7951	8066	8181	8295	8410	8525	115
379	8639	8754	8868	8983	9097	9212	9326	9441	9555	9669	114
380	579784	9898	♦♦12	♦126	♦241	♦355	♦469	♦583	♦697	♦811	114
381	580925	1039	1153	1267	1381	1495	1608	1722	1836	1950	114
382	2063	2177	2291	2404	2518	2631	2745	2858	2972	3085	114
383	3199	3312	3426	3539	3652	3765	3879	3992	4105	4218	113
384	4331	4444	4557	4670	4783	4896	5009	5122	5235	5348	113
385	5461	5574	5686	5799	5912	6024	6137	6250	6362	6475	113
386	6587	6700	6812	6925	7037	7149	7262	7374	7486	7599	112
387	7711	7823	7935	8047	8160	8272	8384	8496	8608	8720	112
388	8832	8944	9056	9167	9279	9391	9503	9615	9726	9838	112
389	9950	♦♦61	♦173	♦284	♦396	♦507	♦619	♦730	♦842	♦953	112
390	591065	1176	1287	1399	1510	1621	1732	1843	1955	2066	111
391	2177	2288	2399	2510	2621	2732	2843	2954	3064	3175	111
392	3286	3397	3508	3618	3729	3840	3950	4061	4171	4282	111
393	4393	4503	4614	4724	4834	4945	5055	5165	5276	5386	110
394	5496	5606	5717	5827	5937	6047	6157	6267	6377	6487	110
395	6597	6707	6817	6927	7037	7146	7256	7366	7476	7586	110
396	7695	7805	7914	8024	8134	8243	8353	8462	8572	8681	110
397	8791	8900	9009	9119	9228	9337	9446	9556	9665	9774	109
398	9883	9992	♦101	♦210	♦319	♦428	♦537	♦646	♦755	♦864	109
399	600973	1082	1191	1299	1408	1517	1625	1734	1843	1951	109
N.	0	1	2	3	4	5	6	7	8	9	D.

N.	0	1	2	3	4	5	6	7	8	9	D.
400	602060	2169	2277	2386	2494	2603	2711	2819	2928	3036	108
401	3144	3253	3361	3469	3577	3686	3794	3902	4010	4118	108
402	4226	4334	4442	4550	4658	4766	4874	4982	5089	5197	108
403	5305	5413	5521	5628	5736	5844	5951	6059	6166	6274	108
404	6381	6489	6596	6704	6811	6919	7026	7133	7241	7348	107
405	7455	7562	7669	7777	7884	7991	8098	8205	8312	8419	107
406	8526	8633	8740	8847	8954	9061	9167	9274	9381	9488	107
407	9594	9701	9808	9914	**21	*128	*234	*341	*447	*554	107
408	610660	0767	0873	0979	1086	1192	1298	1405	1511	1617	106
409	1723	1829	1936	2042	2148	2254	2360	2466	2572	2678	106
410	612784	2890	2996	3102	3207	3313	3419	3525	3630	3736	106
411	3842	3947	4053	4159	4264	4370	4475	4581	4686	4792	106
412	4897	5003	5108	5213	5319	5424	5529	5634	5740	5845	105
413	5950	6055	6160	6265	6370	6476	6581	6686	6790	6895	105
414	7000	7105	7210	7315	7420	7525	7629	7734	7839	7943	105
415	8048	8153	8257	8362	8466	8571	8676	8780	8884	8989	105
416	9093	9198	9302	9406	9511	9615	9719	9824	9928	**32	104
417	620136	0240	0344	0448	0552	0656	0760	0864	0968	1072	104
418	1176	1280	1384	1488	1592	1695	1799	1903	2007	2110	104
419	2214	2318	2421	2525	2628	2732	2835	2939	3042	3146	104
420	623249	3353	3456	3559	3663	3766	3869	3973	4076	4179	103
421	4282	4385	4488	4591	4695	4798	4901	5004	5107	5210	103
422	5312	5415	5518	5621	5724	5827	5929	6032	6135	6238	103
423	6340	6443	6546	6648	6751	6853	6956	7058	7161	7263	103
424	7366	7468	7571	7673	7775	7878	7980	8082	8185	8287	102
425	8389	8491	8593	8695	8797	8900	9002	9104	9206	9308	102
426	9410	9512	9613	9715	9817	9919	**21	*123	*224	*326	102
427	630428	0530	0631	0733	0835	0936	1038	1139	1241	1342	102
428	1444	1545	1647	1748	1849	1951	2052	2153	2255	2356	101
429	2457	2559	2660	2761	2862	2963	3064	3165	3266	3367	101
430	633468	3569	3670	3771	3872	3973	4074	4175	4276	4376	100
431	4477	4578	4679	4779	4880	4981	5081	5182	5283	5383	100
432	5484	5584	5685	5785	5886	5986	6087	6187	6287	6388	100
433	6488	6588	6688	6789	6889	6989	7089	7189	7290	7390	100
434	7490	7590	7690	7790	7890	7990	8090	8190	8290	8389	99
435	8489	8589	8689	8789	8888	8988	9088	9188	9287	9387	99
436	9486	9586	9686	9785	9885	9984	**84	*183	*283	*382	99
437	640481	0581	0680	0779	0879	0978	1077	1177	1276	1375	99
438	1474	1573	1672	1771	1871	1970	2069	2168	2267	2366	99
439	2465	2563	2662	2761	2860	2959	3058	3156	3255	3354	99
440	643453	3551	3650	3749	3847	3946	4044	4143	4242	4340	98
441	4439	4537	4636	4734	4832	4931	5029	5127	5226	5324	98
442	5422	5521	5619	5717	5815	5913	6011	6110	6208	6306	98
443	6404	6502	6600	6698	6796	6894	6992	7089	7187	7285	98
444	7383	7481	7579	7676	7774	7872	7969	8067	8165	8262	98
445	8360	8458	8555	8653	8750	8848	8945	9043	9140	9237	97
446	9335	9432	9530	9627	9724	9821	9919	**16	*113	*210	97
447	650308	0405	0502	0599	0696	0793	0890	0987	1084	1181	97
448	1278	1375	1472	1569	1666	1762	1859	1956	2053	2150	97
449	2246	2343	2440	2536	2633	2730	2826	2923	3019	3116	97
450	658213	3309	3405	3502	3598	3695	3791	3888	3984	4080	96
451	4177	4273	4369	4465	4562	4658	4754	4850	4946	5042	96
452	5138	5235	5331	5427	5523	5619	5715	5810	5906	6002	96
453	6098	6194	6290	6386	6482	6577	6673	6769	6864	6960	96
454	7056	7152	7247	7343	7438	7534	7629	7725	7820	7916	96
455	8011	8107	8202	8298	8393	8488	8584	8679	8774	8870	95
456	8965	9060	9155	9250	9346	9441	9536	9631	9726	9821	95
457	9916	**11	*106	*201	*296	*391	*486	*581	*676	*771	95
458	660865	0960	1055	1150	1245	1339	1434	1529	1623	1718	95
459	1813	1907	2002	2096	2191	2286	2380	2475	2569	2663	95
N.	0	1	2	3	4	5	6	7	8	9	D.

N.	0	1	2	3	4	5	6	7	8	9	D.
460	662758	2852	2947	3041	3135	3230	3324	3418	3512	3607	94
461	3701	3795	3889	3983	4078	4172	4266	4360	4454	4548	94
462	4642	4786	4830	4924	5018	5112	5206	5299	5393	5487	94
463	5581	5675	5769	5862	5956	6050	6143	6237	6331	6424	94
464	6518	6612	6705	6799	6892	6986	7079	7173	7266	7360	94
465	7453	7546	7640	7733	7826	7920	8013	8106	8199	8293	93
466	8386	8479	8572	8665	8759	8852	8945	9038	9131	9224	93
467	9317	9410	9503	9596	9689	9782	9875	9967	**60	*153	93
468	670246	0339	0431	0524	0617	0710	0802	0895	0988	1080	93
469	1173	1265	1358	1451	1543	1636	1728	1821	1913	2005	93
470	672098	2190	2283	2375	2467	2560	2652	2744	2836	2929	92
471	3021	3113	3205	3297	3390	3482	3574	3666	3758	3850	92
472	3942	4034	4126	4218	4310	4402	4494	4586	4677	4769	92
473	4861	4953	5045	5137	5228	5320	5412	5503	5595	5687	92
474	5778	5870	5962	6053	6145	6236	6328	6419	6511	6602	92
475	6694	6785	6876	6968	7059	7151	7242	7333	7424	7516	91
476	7607	7698	7789	7881	7972	8063	8154	8245	8336	8427	91
477	8518	8609	8700	8791	8882	8973	9064	9155	9246	9337	91
478	9428	9519	9610	9700	9791	9882	9973	**63	*154	*245	91
479	680336	0426	0517	0607	0698	0789	0879	0970	1060	1151	91
480	681241	1332	1422	1513	1603	1693	1784	1874	1964	2055	90
481	2145	2235	2326	2416	2506	2596	2686	2777	2867	2957	90
482	3047	3137	3227	3317	3407	3497	3587	3677	3767	3857	90
483	3947	4037	4127	4217	4307	4396	4486	4576	4666	4756	90
484	4845	4935	5025	5114	5204	5294	5383	5473	5563	5652	90
485	5742	5831	5921	6010	6100	6189	6279	6368	6458	6547	89
486	6636	6726	6815	6904	6994	7083	7172	7261	7351	7440	89
487	7529	7618	7707	7796	7886	7975	8064	8153	8242	8331	89
488	8420	8509	8598	8687	8776	8865	8953	9042	9131	9220	89
489	9309	9398	9486	9575	9664	9753	9841	9930	**19	*107	89
490	690196	0285	0373	0462	0550	0639	0728	0816	0905	0993	89
491	1081	1170	1258	1347	1435	1524	1612	1700	1789	1877	88
492	1965	2053	2142	2230	2318	2406	2494	2583	2671	2759	88
493	2847	2935	3023	3111	3199	3287	3375	3463	3551	3639	88
494	3727	3815	3903	3991	4078	4166	4254	4342	4430	4517	88
495	4605	4693	4781	4868	4956	5044	5131	5219	5307	5394	88
496	5482	5569	5657	5744	5832	5919	6007	6094	6182	6269	87
497	6356	6444	6531	6618	6706	6793	6880	6968	7055	7142	87
498	7229	7317	7404	7491	7578	7665	7752	7839	7926	8014	87
499	8101	8188	8275	8362	8449	8535	8622	8709	8796	8883	87
500	698970	9057	9144	9231	9317	9404	9491	9578	9664	9751	87
501	9838	9924	**11	**98	*184	*271	*358	*444	*531	*617	87
502	700704	0790	0877	0963	1050	1136	1222	1309	1395	1482	86
503	1568	1654	1741	1827	1913	1999	2086	2172	2258	2344	86
504	2431	2517	2603	2689	2775	2861	2947	3033	3119	3205	86
505	3291	3377	3463	3549	3635	3721	3807	3893	3979	4065	86
506	4151	4236	4322	4408	4494	4579	4665	4751	4837	4922	86
507	5008	5094	5179	5265	5350	5436	5522	5607	5693	5778	86
508	5864	5949	6035	6120	6206	6291	6376	6462	6547	6632	85
509	6718	6803	6888	6974	7059	7144	7229	7315	7400	7485	85
510	707570	7655	7740	7826	7911	7996	8081	8166	8251	8336	85
511	8421	8506	8591	8676	8761	8846	8931	9015	9100	9185	85
512	9270	9355	9440	9524	9609	9694	9779	9863	9948	**33	85
513	710117	0202	0287	0371	0456	0540	0625	0710	0794	0879	85
514	0963	1048	1132	1217	1301	1385	1470	1554	1639	1723	84
515	1807	1892	1976	2060	2144	2229	2313	2397	2481	2566	84
516	2650	2734	2818	2902	2986	3070	3154	3238	3323	3407	84
517	3491	3575	3659	3742	3826	3910	3994	4078	4162	4246	84
518	4330	4414	4497	4581	4665	4749	4833	4916	5000	5084	84
519	5167	5251	5335	5418	5502	5586	5669	5753	5836	5920	84
N.	0	1	2	3	4	5	6	7	8	9	D.

N.	0	1	2	3	4	5	6	7	8	9	D.
520	716003	6087	6170	6254	6337	6421	6504	6588	6671	6754	88
521	6838	6921	7004	7088	7171	7254	7338	7421	7504	7587	88
522	7671	7754	7837	7920	8003	8086	8169	8253	8336	8419	83
523	8502	8585	8668	8751	8834	8917	9000	9083	9165	9248	88
524	9331	9414	9497	9580	9663	9745	9828	9911	9994	••77	88
525	720159	0242	0325	0407	0490	0573	0655	0738	0821	0903	83
526	·0986	1068	1151	1233	1316	1398	1481	1563	1646	1728	82
527	1811	1893	1975	2058	2140	2222	2305	2387	2469	2552	82
528	2634	2716	2798	2881	2963	3045	3127	3209	3291	3374	82
529	3456	3538	3620	3702	3784	3866	3948	4030	4112	4194	82
530	724276	4358	4440	4522	4604	4685	4767	4849	4931	5013	82
531	5095	5176	5258	5340	5422	5503	5585	5667	5748	5830	82
532	5912	5993	6075	6156	6238	6320	6401	6483	6564	6646	82
533	6727	6809	6890	6972	7053	7134	7216	7297	7379	7460	81
534	7541	7623	7704	7785	7866	7948	8029	8110	8191	8273	81
535	8354	8435	8516	8597	8678	8759	8841	8922	9003	9084	81
536	9165	9246	9327	9408	9489	9570	9651	9732	9813	9893	81
537	9974	••55	•136	•217	•298	•378	•459	•540	•621	•702	81
538	730782	0863	0944	1024	1105	1186	1266	1347	1428	1508	81
539	1589	1669	1750	1830	1911	1991	2072	2152	2283	2313	81
540	732394	2474	2555	2635	2715	2796	2876	2956	3037	3117	80
541	3197	3278	3358	3438	3518	3598	3679	3759	3839	3919	80
542	3999	4079	4160	4240	4320	4400	4480	4560	4640	4720	80
543	4800	4880	4960	5040	5120	5200	5279	5359	5439	5519	80
544	5599	5679	5759	5838	5918	5998	6078	6157	6237	6317	80
545	6397	6476	6556	6635	6715	6795	6874	6954	7034	7113	80
546	7193	7272	7352	7431	7511	7590	7670	7749	7829	7908	79
547	7987	8067	8146	8225	8305	8384	8463	8543	8622	8701	79
548	8781	8860	8939	9018	9097	9177	9256	9335	9414	9493	79
549	9572	9651	9731	9810	9889	9968	••47	•126	•205	•284	79
550	740363	0442	0521	0600	0678	0757	0836	0915	0994	1073	79
551	1152	1230	1309	1388	1467	1546	1624	1703	1782	1860	79
552	1939	2018	2096	2175	2254	2332	2411	2489	2568	2647	79
553	2725	2804	2882	2961	3039	3118	3196	3275	3353	3431	78
554	3510	3588	3667	3745	3823	3902	3980	4058	4136	4215	78
555	4293	4371	4449	4528	4606	4684	4762	4840	4919	4997	78
556	5075	5153	5231	5309	5387	5465	5543	5621	5699	5777	78
557	5855	5933	6011	6089	6167	6245	6323	6401	6479	6556	78
558	6634	6712	6790	6868	6945	7023	7101	7179	7256	7334	78
559	7412	7489	7567	7645	7722	7800	7878	7955	8033	8110	78
560	748188	8266	8343	8421	8498	8576	8653	8731	8808	8885	77
561	8963	9040	9118	9195	9272	9350	9427	9504	9582	9659	77
562	9736	9814	9891	9968	••45	•123	•200	•277	•354	•431	77
563	750508	0586	0663	0740	0817	0894	0971	1048	1125	1202	77
564	1279	1356	1433	1510	1587	1664	1741	1818	1895	1972	77
565	2048	2125	2202	2279	2356	2433	2509	2586	2663	2740	77
566	2816	2893	2970	3047	3123	3200	3277	3353	3430	3506	77
567	3583	3660	3736	3813	3889	3966	4042	4119	4195	4272	77
568	4348	4425	4501	4578	4654	4730	4807	4883	4960	5036	76
569	5112	5189	5265	5341	5417	5494	5570	5646	5722	5799	76
570	755875	5951	6027	6103	6180	6256	6332	6408	6484	6560	76
571	6636	6712	6788	6864	6940	7016	7092	7168	7244	7320	76
572	7396	7472	7548	7624	7700	7775	7851	7927	8003	8079	76
573	8155	8230	8306	8382	8458	8533	8609	8685	8761	8836	76
574	8912	8988	9063	9139	9214	9290	9366	9441	9517	9592	76
575	9668	9743	9819	9894	.9970	••45	•121	•196	•272	•347	75
576	760422	0498	0573	0649	0724	0799	0875	0950	1025	1101	75
577	1176	1251	1326	1402	1477	1552	1627	1702	1778	1853	75
578	1928	2003	2078	2153	2228	2303	2378	2453	2529	2604	75
579	2679	2754	2829	2904	2978	3053	3128	3203	3278	3353	75
N.	0	1	2	3	4	5	6	7	8	9	D.

N.	0	1	2	3	4	5	6	7	8	9	D.
580	763428	3503	3578	3653	3727	3802	3877	3952	4027	4101	75
581	4176	4251	4326	4400	4475	4550	4624	4699	4774	4848	75
582	4923	4998	5072	5147	5221	5296	5370	5445	5520	5594	75
583	5669	5743	5818	5892	5966	6041	6115	6190	6264	6338	74
584	6413	6487	6562	6636	6710	6785	6859	6933	7007	7082	74
585	7156	7230	7304	7379	7453	7527	7601	7675	7749	7823	74
586	7898	7972	8046	8120	8194	8268	8342	8416	8490	8564	74
587	8638	8712	8786	8860	8934	9008	9082	9156	9230	9303	74
588	9377	9451	9525	9599	9673	9746	9820	9894	9968	••42	74
589	770115	0189	0263	0336	0410	0484	0557	0631	0705	0778	74
590	770852	0926	0999	1073	1146	1220	1293	1367	1440	1514	74
591	1587	1661	1734	1808	1881	1955	2028	2102	2175	2248	73
592	2322	2395	2468	2542	2615	2688	2762	2835	2908	2981	73
593	3055	3128	3201	3274	3348	3421	3494	3567	3640	3713	73
594	3786	3860	3933	4006	4079	4152	4225	4298	4371	4444	73
595	4517	4590	4663	4736	4809	4882	4955	5028	5100	5173	73
596	5246	5319	5392	5465	5538	5610	5683	5756	5829	5902	73
597	5974	6047	6120	6193	6265	6338	6411	6483	6556	6629	73
598	6701	6774	6846	6919	6992	7064	7137	7209	7282	7354	73
599	7427	7499	7572	7644	7717	7789	7862	7934	8006	8079	72
600	778151	8224	8296	8368	8441	8513	8585	8658	8730	8802	72
601	8874	8947	9019	9091	9163	9236	9308	9380	9452	9524	72
602	9596	9669	9741	9813	9885	9957	••29	•101	•173	•245	72
603	780317	0389	0461	0533	0605	0677	0749	0821	0893	0965	72
604	1037	1109	1181	1253	1324	1396	1468	1540	1612	1684	72
605	1755	1827	1899	1971	2042	2114	2186	2258	2329	2401	72
606	2473	2544	2616	2688	2759	2831	2902	2974	3046	3117	72
607	3189	3260	3332	3403	3475	3546	3618	3689	3761	3832	71
608	3904	3975	4046	4118	4189	4261	4332	4403	4475	4546	71
609	4617	4689	4760	4831	4902	4974	5045	5116	5187	5259	71
610	785330	5401	5472	5543	5615	5686	5757	5828	5899	5970	71
611	6041	6112	6183	6254	6325	6396	6467	6538	6609	6680	71
612	6751	6822	6893	6964	7035	7106	7177	7248	7319	7390	71
613	7460	7531	7602	7673	7744	7815	7885	7956	8027	8098	71
614	8168	8239	8310	8381	8451	8522	8593	8663	8734	8804	71
615	8875	8946	9016	9087	9157	9228	9299	9369	9440	9510	71
616	9581	9651	9722	9792	9863	9933	•••4	••74	•144	•215	70
617	790285	0356	0426	0496	0567	0637	0707	0778	0848	0918	70
618	0988	1059	1129	1199	1269	1340	1410	1480	1550	1620	70
619	1691	1761	1831	1901	1971	2041	2111	2181	2252	2322	70
620	792392	2462	2532	2602	2672	2742	2812	2882	2952	3022	70
621	3092	3162	3231	3301	3371	3441	3511	3581	3651	3721	70
622	3790	3860	3930	4000	4070	4139	4209	4279	4349	4418	70
623	4488	4558	4627	4697	4767	4836	4906	4976	5045	5115	70
624	5185	5254	5324	5393	5463	5532	5602	5672	5741	5811	70
625	5880	5949	6019	6088	6158	6227	6297	6366	6436	6505	69
626	6574	6644	6713	6782	6852	6921	6990	7060	7129	7198	69
627	7268	7337	7406	7475	7545	7614	7683	7752	7821	7890	69
628	7960	8029	8098	8167	8236	8305	8374	8443	8513	8582	69
629	8651	8720	8789	8858	8927	8996	9065	9134	9203	9272	69
630	799341	9409	9478	9547	9616	9685	9754	9823	9892	9961	69
631	800029	0098	0167	0236	0305	0373	0442	0511	0580	0648	69
632	0717	0786	0854	0923	0992	1061	1129	1198	1266	1335	69
633	1404	1472	1541	1609	1678	1747	1815	1884	1952	2021	69
634	2089	2158	2226	2295	2363	2432	2500	2568	2637	2705	69
635	2774	2842	2910	2979	3047	3116	3184	3252	3321	3389	68
636	3457	3525	3594	3662	3730	3798	3867	3935	4003	4071	68
637	4139	4208	4276	4344	4412	4480	4548	4616	4685	4753	68
638	4821	4889	4957	5025	5093	5161	5229	5297	5365	5433	68
639	5501	5569	5637	5705	5773	5841	5908	5976	6044	6112	68
N.	0	1	2	3	4	5	6	7	8	9	D.

N.	0	1	2	3	4	5	6	7	8	9	D.
640	806180	6248	6316	6384	6451	6519	6587	6655	6723	6790	68
641	6858	6926	6994	7061	7129	7197	7264	7332	7400	7467	68
642	7535	7603	7670	7738	7806	7873	7941	8008	8076	8143	68
643	8211	8279	8346	8414	8481	8549	8616	8684	8751	8818	67
644	8886	8953	9021	9088	9156	9223	9290	9358	9425	9492	67
645	9560	9627	9694	9762	9829	9896	9964	**31	**98	*165	67
646	810233	0300	0367	0434	0501	0569	0636	0703	0770	0837	67
647	0904	0971	1039	1106	1173	1240	1307	1374	1441	1508	67
648	1575	1642	1709	1776	1843	1910	1977	2044	2111	2178	67
649	2245	2312	2379	2445	2512	2579	2646	2713	2780	2847	67
650	812913	2980	3047	3114	3181	3247	3314	3381	3448	3514	67
651	3581	3648	3714	3781	3848	3914	3981	4048	4114	4181	67
652	4248	4314	4381	4447	4514	4581	4647	4714	4780	4847	67
653	4913	4980	5046	5113	5179	5246	5312	5378	5445	5511	66
654	5578	5644	5711	5777	5843	5910	5976	6042	6109	6175	66
655	6241	6308	6374	6440	6506	6573	6639	6705	6771	6838	66
656	6904	6970	7036	7102	7169	7235	7301	7367	7433	7499	66
657	7565	7631	7698	7764	7830	7896	7962	8028	8094	8160	66
658	8226	8292	8358	8424	8490	8556	8622	8688	8754	8820	66
659	8885	8951	9017	9083	9149	9215	9281	9346	9412	9478	66
660	819544	9610	9676	9741	9807	9873	9939	***4	**70	*136	66
661	820201	0267	0333	0399	0464	0530	0595	0661	0727	0792	66
662	0858	0924	0989	1055	1120	1186	1251	1317	1382	1448	66
663	1514	1579	1645	1710	1775	1841	1906	1972	2037	2103	65
664	2168	2233	2299	2364	2430	2495	2560	2626	2691	2756	65
665	2822	2887	2952	3018	3083	3148	3213	3279	3344	3409	65
666	3474	3539	3605	3670	3735	3800	3865	3930	3996	4061	65
667	4126	4191	4256	4321	4386	4451	4516	4581	4646	4711	65
668	4776	4841	4906	4971	5036	5101	5166	5231	5296	5361	65
669	5426	5491	5556	5621	5686	5751	5815	5880	5945	6010	65
670	826075	6140	6204	6269	6334	6399	6464	6528	6593	6658	65
671	6723	6787	6852	6917	6981	7046	7111	7175	7240	7305	65
672	7369	7434	7499	7563	7628	7692	7757	7821	7886	7951	65
673	8015	8080	8144	8209	8273	8338	8402	8467	8531	8595	64
674	8660	8724	8789	8853	8918	8982	9046	9111	9175	9239	64
675	9304	9368	9432	9497	9561	9625	9690	9754	9818	9882	64
676	9947	**11	**75	*139	*204	*268	*332	*396	*460	*525	64
677	830589	0653	0717	0781	0845	0909	0973	1037	1102	1166	64
678	1230	1294	1358	1422	1486	1550	1614	1678	1742	1806	64
679	1870	1934	1998	2062	2126	2189	2253	2317	2381	2445	64
680	832509	2573	2637	2700	2764	2828	2892	2956	3020	3083	64
681	3147	3211	3275	3338	3402	3466	3530	3593	3657	3721	64
682	3784	3848	3912	3975	4039	4103	4166	4230	4294	4357	64
683	4421	4484	4548	4611	4675	4739	4802	4866	4929	4993	64
684	5056	5120	5183	5247	5310	5373	5437	5500	5564	5627	63
685	5691	5754	5817	5881	5944	6007	6071	6134	6197	6261	63
686	6324	6387	6451	6514	6577	6641	6704	6767	6830	6894	63
687	6957	7020	7083	7146	7210	7273	7336	7399	7462	7525	63
688	7588	7652	7715	7778	7841	7904	7967	8030	8093	8156	63
689	8219	8282	8345	8408	8471	8534	8597	8660	8723	8786	63
690	838849	8912	8975	9038	9101	9164	9227	9289	9352	9415	63
691	9478	9541	9604	9667	9729	9792	9855	9918	9981	**43	63
692	840106	0169	0232	0294	0357	0420	0482	0545	0608	0671	63
693	0733	0796	0859	0921	0984	1046	1109	1172	1234	1297	63
694	1359	1422	1485	1547	1610	1672	1735	1797	1860	1922	63
695	1985	2047	2110	2172	2235	2297	2360	2422	2484	2547	62
696	2609	2672	2734	2796	2859	2921	2983	3046	3108	3170	62
697	3233	3295	3357	3420	3482	3544	3606	3669	3731	3793	62
698	3855	3918	3980	4042	4104	4166	4229	4291	4353	4415	62
699	4477	4539	4601	4664	4726	4788	4850	4912	4974	5036	62
.	0	1	2	3	4	5	6	7	8	9	D.

N.	0	1	2	3	4	5	6	7	8	9	D.
700	845098	5160	5222	5284	5346	5408	5470	5532	5594	5656	62
701	5718	5780	5842	5904	5966	6028	6090	6151	6213	6275	62
702	6337	6399	6461	6523	6585	6646	6708	6770	6832	6894	62
703	6955	7017	7079	7141	7202	7264	7326	7388	7449	7511	62
704	7573	7634	7696	7758	7819	7881	7943	8004	8066	8128	62
705	8189	8251	8312	8374	8435	8497	8559	8620	8682	8743	62
706	8805	8866	8928	8989	9051	9112	9174	9235	9297	9358	61
707	9419	9481	9542	9604	9665	9726	9788	9849	9911	9972	61
708	850033	0095	0156	0217	0279	0340	0401	0462	0524	0585	61
709	0646	0707	0769	0830	0891	0952	1014	1075	1136	1197	61
710	851258	1320	1381	1442	1503	1564	1625	1686	1747	1809	61
711	1870	1931	1992	2053	2114	2175	2236	2297	2358	2419	61
712	2480	2541	2602	2663	2724	2785	2846	2907	2968	3029	61
713	3090	3150	3211	3272	3333	3394	3455	3516	3577	3637	61
714	3698	3759	3820	3881	3941	4002	4063	4124	4185	4245	61
715	4306	4367	4428	4488	4549	4610	4670	4731	4792	4852	61
716	4913	4974	5034	5095	5156	5216	5277	5337	5398	5459	61
717	5519	5580	5640	5701	5761	5822	5882	5943	6003	6064	61
718	6124	6185	6245	6306	6366	6427	6487	6548	6608	6668	60
719	6729	6789	6850	6910	6970	7031	7091	7152	7212	7272	60
720	857332	7393	7453	7513	7574	7634	7694	7755	7815	7875	60
721	7935	7995	8056	8116	8176	8236	8297	8357	8417	8477	60
722	8537	8597	8657	8718	8778	8838	8898	8958	9018	9078	60
723	9138	9198	9258	9318	9379	9439	9499	9559	9619	9679	60
724	9739	9799	9859	9918	9978	**38	**98	*158	*218	*278	60
725	860338	0398	0458	0518	0578	0637	0697	0757	0817	0877	60
726	0937	0996	1056	1116	1176	1236	1295	1355	1415	1475	60
727	1534	1594	1654	1714	1773	1833	1893	1952	2012	2072	60
728	2131	2191	2251	2310	2370	2430	2489	2549	2608	2668	60
729	2728	2787	2847	2906	2966	3025	3085	3144	3204	3263	60
730	863323	3382	3442	3501	3561	3620	3680	3739	3799	3858	59
731	3917	3977	4036	4096	4155	4214	4274	4333	4392	4452	59
732	4511	4570	4630	4689	4748	4808	4867	4926	4985	5045	59
733	5104	5163	5222	5282	5341	5400	5459	5519	5578	5637	59
734	5696	5755	5814	5874	5933	5992	6051	6110	6169	6228	59
735	6287	6346	6405	6465	6524	6583	6642	6701	6760	6819	59
736	6878	6937	6996	7055	7114	7173	7232	7291	7350	7409	59
737	7467	7526	7585	7644	7703	7762	7821	7880	7939	7998	59
738	8056	8115	8174	8233	8292	8350	8409	8468	8527	8586	59
739	8644	8703	8762	8821	8879	8938	8997	9056	9114	9173	59
740	869232	9290	9349	9408	9466	9525	9584	9642	9701	9760	59
741	9818	9877	9935	9994	**53	*111	*170	*228	*287	*345	59
742	870404	0462	0521	0579	0638	0696	0755	0813	0872	0930	58
743	0989	1047	1106	1164	1223	1281	1339	1398	1456	1515	58
744	1573	1631	1690	1748	1806	1865	1923	1981	2040	2098	58
745	2156	2215	2273	2331	2389	2448	2506	2564	2622	2681	58
746	2739	2797	2855	2913	2972	3030	3088	3146	3204	3262	58
747	3321	3379	3437	3495	3553	3611	3669	3727	3785	3844	58
748	3902	3960	4018	4076	4134	4192	4250	4308	4366	4424	58
749	4482	4540	4598	4656	4714	4772	4830	4888	4945	5003	58
750	875061	5119	5177	5235	5293	5351	5409	5466	5524	5582	58
751	5640	5698	5756	5813	5871	5929	5987	6045	6102	6160	58
752	6218	6276	6333	6391	6449	6507	6564	6622	6680	6737	58
753	6795	6853	6910	6968	7026	7083	7141	7199	7256	7314	58
754	7871	7429	7487	7544	7602	7659	7717	7774	7832	7889	58
755	7947	8004	8062	8119	8177	8234	8292	8349	8407	8464	57
756	8522	8579	8637	8694	8752	8809	8866	8924	8981	9039	57
757	9096	9153	9211	9268	9325	9383	9440	9497	9555	9612	57
758	9669	9726	9784	9841	9898	9956	**13	*70	*127	*185	57
759	880242	0299	0356	0413	0471	0528	0585	0642	0699	0756	57
N.	0	1	2	3	4	5	6	7	8	9	D.

N.	0	1	2	3	4	5	6	7	8	9	D.
760	880814	0871	0928	0985	1042	1099	1156	1213	1271	1328	57
761	1385	1442	1499	1556	1613	1670	1727	1784	1841	1898	57
762	1955	2012	2069	2126	2183	2240	2297	2354	2411	2468	57
763	2525	2581	2638	2695	2752	2809	2866	2923	2980	3037	57
764	3093	3150	3207	3264	3321	3377	3434	3491	3548	3605	57
765	3661	3718	3775	3832	3888	3945	4002	4059	4115	4172	57
766	4229	4285	4342	4399	4455	4512	4569	4625	4682	4739	57
767	4795	4852	4909	4965	5022	5078	5135	5192	5248	5305	57
768	5361	5418	5474	5531	5587	5644	5700	5757	5813	5870	57
769	5926	5983	6039	6096	6152	6209	6265	6321	6378	6434	56
770	886491	6547	6604	6660	6716	6773	6829	6885	6942	6998	56
771	7054	7111	7167	7223	7280	7336	7392	7449	7505	7561	56
772	7617	7674	7730	7786	7842	7898	7955	8011	8067	8123	56
773	8179	8236	8292	8348	8404	8460	8516	8573	8629	8685	56
774	8741	8797	8853	8909	8965	9021	9077	9134	9190	9246	56
775	9302	9358	9414	9470	9526	9582	9638	9694	9750	9806	56
776	9862	9918	9974	♦♦80	♦♦86	♦141	♦197	♦253	♦309	♦365	56
777	890421	0477	0533	0589	0645	0700	0756	0812	0868	0924	56
778	0980	1035	1091	1147	1203	1259	1314	1370	1426	1482	56
779	1537	1593	1649	1705	1760	1816	1872	1928	1983	2039	56
780	892095	2150	2206	2262	2317	2373	2429	2484	2540	2595	56
781	2651	2707	2762	2818	2873	2929	2985	3040	3096	3151	56
782	3207	3262	3318	3373	3429	3484	3540	3595	3651	3706	56
783	3762	3817	3873	3928	3984	4039	4094	4150	4205	4261	55
784	4316	4371	4427	4482	4538	4593	4648	4704	4759	4814	55
785	4870	4925	4980	5036	5091	5146	5201	5257	5312	5367	55
786	5423	5478	5533	5588	5644	5699	5754	5809	5864	5920	55
787	5975	6030	6085	6140	6195	6251	6306	6361	6416	6471	55
788	6526	6581	6636	6692	6747	6802	6857	6912	6967	7022	55
789	7077	7132	7187	7242	7297	7352	7407	7462	7517	7572	55
790	897627	7682	7737	7792	7847	7902	7957	8012	8067	8122	55
791	8176	8231	8286	8341	8396	8451	8506	8561	8615	8670	55
792	8725	8780	8835	8890	8944	8999	9054	9109	9164	9218	55
793	9273	9328	9383	9437	9492	9547	9602	9656	9711	9766	55
794	9821	9875	9930	9985	♦♦39	♦♦94	♦149	♦203	♦258	♦312	55
795	900367	0422	0476	0531	0586	0640	0695	0749	0804	0859	55
796	0913	0968	1022	1077	1131	1186	1240	1295	1349	1404	55
797	1458	1513	1567	1622	1676	1731	1785	1840	1894	1948	54
798	2003	2057	2112	2166	2221	2275	2329	2384	2438	2492	54
799	2547	2601	2655	2710	2764	2818	2873	2927	2981	3036	54
800	903090	3144	3199	3253	3307	3361	3416	3470	3524	3578	54
801	3633	3687	3741	3795	3849	3904	3958	4012	4066	4120	54
802	4174	4229	4283	4337	4391	4445	4499	4553	4607	4661	54
803	4716	4770	4824	4878	4932	4986	5040	5094	5148	5202	54
804	5256	5310	5364	5418	5472	5526	5580	5634	5688	5742	54
805	5796	5850	5904	5958	6012	6066	6119	6173	6227	6281	54
806	6335	6389	6443	6497	6551	6604	6658	6712	6766	6820	54
807	6874	6927	6981	7035	7089	7143	7196	7250	7304	7358	54
808	7411	7465	7519	7573	7626	7680	7734	7787	7841	7895	54
809	7949	8002	8056	8110	8163	8217	8270	8324	8378	8431	54
810	908485	8539	8592	8646	8699	8753	8807	8860	8914	8967	54
811	9021	9074	9128	9181	9235	9289	9342	9396	9449	9503	54
812	9556	9610	9663	9716	9770	9823	9877	9930	9984	♦♦37	53
813	910091	0144	0197	0251	0304	0358	0411	0464	0518	0571	53
814	0624	0678	0731	0784	0838	0891	0944	0998	1051	1104	53
815	1158	1211	1264	1317	1371	1424	1477	1530	1584	1637	53
816	1690	1743	1797	1850	1903	1956	2009	2063	2116	2169	53
817	2222	2275	2328	2381	2435	2488	2541	2594	2647	2700	53
818	2753	2806	2859	2913	2966	3019	3072	3125	3178	3231	53
819	3284	3337	3390	3443	3496	3549	3602	3655	3708	3761	53
N.	0	1	2	3	4	5	6	7	8	9	D.

N.	0	1	2	3	4	5	6	7	8	9	D.
820	913814	3867	3920	3973	4026	4079	4132	4184	4237	4290	53
821	4343	4396	4449	4502	4555	4608	4660	4713	4766	4819	53
822	4872	4925	4977	5030	5083	5136	5189	5241	5294	5347	53
823	5400	5453	5505	5558	5611	5664	5716	5769	5822	5875	53
824	5927	5980	6033	6085	6138	6191	6243	6296	6349	6401	53
825	6454	6507	6559	6612	6664	6717	6770	6822	6875	6927	53
826	6980	7033	7085	7138	7190	7243	7295	7348	7400	7453	53
827	7506	7558	7611	7663	7716	7768	7820	7873	7925	7978	52
828	8030	8083	8135	8188	8240	8293	8345	8397	8450	8502	52
829	8555	8607	8659	8712	8764	8816	8869	8921	8973	9026	52
830	919078	9130	9183	9235	9287	9340	9392	9444	9496	9549	52
831	9601	9653	9706	9758	9810	9862	9914	9967	**19	**71	52
832	920123	0176	0228	0280	0332	0384	0436	0489	0541	0593	52
833	0645	0697	0749	0801	0853	0906	0958	1010	1062	1114	52
834	1166	1218	1270	1322	1374	1426	1478	1530	1582	1634	52
835	1686	1738	1790	1842	1894	1946	1998	2050	2102	2154	52
836	2206	2258	2310	2362	2414	2466	2518	2570	2622	2674	52
837	2725	2777	2829	2881	2933	2985	3037	3089	3140	3192	52
838	3244	3296	3348	3399	3451	3503	3555	3607	3658	3710	52
839	3762	3814	3865	3917	3969	4021	4072	4124	4176	4228	52
840	924279	4331	4383	4434	4486	4538	4589	4641	4693	4744	52
841	4796	4848	4899	4951	5003	5054	5106	5157	5209	5261	52
842	5312	5364	5415	5467	5518	5570	5621	5673	5725	5776	52
843	5828	5879	5931	5982	6034	6085	6137	6188	6240	6291	51
844	6342	6394	6445	6497	6548	6600	6651	6702	6754	6805	51
845	6857	6908	6959	7011	7062	7114	7165	7216	7268	7319	51
846	7370	7422	7473	7524	7576	7627	7678	7730	7781	7832	51
847	7883	7985	7986	8037	8088	8140	9191	8242	8293	8345	51
848	8396	8447	8498	8549	8601	8652	8703	8754	8805	8857	51
849	8908	8959	9010	9061	9112	9163	9215	9266	9317	9368	51
850	929419	9470	9521	9572	9623	9674	9725	9776	9827	9879	51
851	9930	9981	**32	**83	*134	*185	*236	*287	*338	*389	51
852	930440	0491	0542	0592	0643	0694	0745	0796	0847	0898	51
853	0949	1000	1051	1102	1153	1204	1254	1305	1356	1407	51
854	1458	1509	1560	1610	1661	1712	1763	1814	1865	1915	51
855	1966	2017	2068	2118	2169	2220	2271	2322	2372	2423	51
856	2474	2524	2575	2626	2677	2727	2778	2829	2879	2930	51
857	2981	3031	3082	3133	3183	3234	3285	3335	3386	3437	51
858	3487	3538	3589	3639	3690	3740	3791	3841	3892	3943	51
859	3993	4044	4094	4145	4195	4246	4296	4347	4397	4448	51
860	934498	4549	4599	4650	4700	4751	4801	4852	4902	4953	50
861	5003	5054	5104	5154	5205	5255	5306	5356	5406	5457	50
862	5507	5558	5608	5658	5709	5759	5860	5860	5910	5960	50
863	6011	6061	6111	6162	6212	6262	6313	6363	6413	6463	50
864	6514	6564	6614	6665	6715	6765	6815	6865	6916	6966	50
865	7016	7066	7117	7167	7217	7267	7317	7367	7418	7468	50
866	7518	7568	7618	7668	7718	7769	7819	7869	7919	7969	50
867	8019	8069	8119	8169	8219	8269	8320	8370	8420	8470	50
868	8520	8570	8620	8670	8720	8770	8820	8870	8920	8970	50
869	9020	9070	9120	9170	9220	9270	9320	9369	9419	9469	50
870	939519	9569	9619	9669	9719	9769	9819	9869	9918	9968	50
871	940018	0068	0118	0168	0218	0267	0317	0367	0417	0467	50
872	0516	0566	0616	0666	0716	0765	0815	0865	0915	0964	50
873	1014	1064	1114	1163	1213	1263	1313	1362	1412	1462	50
874	1511	1561	1611	1660	1710	1760	1809	1859	1909	1958	50
875	2008	2058	2107	2157	2207	2256	2306	2355	2405	2455	50
876	2504	2554	2603	2653	2702	2752	2801	2851	2901	2950	50
877	3000	3049	3099	3148	3198	3247	3297	3346	3396	3445	49
878	3495	3544	3593	3643	3692	3742	3791	3841	3890	3939	49
879	3989	4038	4088	4137	4186	4236	4285	4335	4384	4433	49
N.	0	1	2	3	4	5	6	7	8	9	D.

N.	0	1	2	3	4	5	6	7	8	9	D.
880	944488	4532	4581	4631	4680	4729	4779	4828	4877	4927	49
881	4976	5025	5074	5124	5173	5222	5272	5321	5370	5419	49
882	5469	5518	5567	5616	5665	5715	5764	5813	5862	5912	49
883	5961	6010	6059	6108	6157	6207	6256	6305	6354	6403	49
884	6452	6501	6551	6600	6649	6698	6747	6796	6845	6894	49
885	6943	6992	7041	7090	7140	7189	7238	7287	7336	7385	49
886	7434	7483	7532	7581	7630	7679	7728	7777	7826	7875	49
887	7924	7973	8022	8070	8119	8168	8217	8266	8315	8364	49
888	8413	8462	8511	8560	8609	8657	8706	8755	8804	8853	49
889	8902	8951	8999	9048	9097	9146	9195	9244	9292	9341	49
890	949390	9439	9488	9536	9585	9634	9683	9731	9780	9829	49
891	9878	9926	9975	**24	**73	*121	*170	*219	*267	*316	49
892	950365	0414	0462	0511	0560	0608	0657	0706	0754	0803	49
893	0851	0900	0949	0997	1046	1095	1148	1192	1240	1289	49
894	1338	1386	1435	1483	1532	1580	1629	1677	1726	1775	49
895	1823	1872	1920	1969	2017	2066	2114	2163	2211	2260	48
896	2308	2356	2405	2453	2502	2550	2599	2647	2696	2744	48
897	2792	2841	2889	2938	2986	3034	3083	3131	3180	3228	48
898	3276	3325	3373	3421	3470	3518	3566	3615	3663	3711	48
899	3760	3808	3856	3905	3953	4001	4049	4098	4146	4194	48
900	954243	4291	4339	4387	4435	4484	4532	4580	4628	4677	48
901	4725	4773	4821	4869	4918	4966	5014	5062	5110	5158	48
902	5207	5255	5303	5351	5399	5447	5495	5543	5592	5640	48
903	5688	5736	5784	5832	5880	5928	5976	6024	6072	6120	48
904	6168	6216	6265	6313	6361	6409	6457	6505	6553	6601	48
905	6649	6697	6745	6793	6840	6888	6936	6984	7032	7080	48
906	7128	7176	7224	7272	7320	7368	7416	7464	7512	7559	48
907	7607	7655	7703	7751	7799	7847	7894	7942	7990	8038	48
908	8086	8134	8181	8229	8277	8325	8373	8421	8468	8516	48
909	8564	8612	8659	8707	8755	8803	8850	8898	8946	8994	48
910	959041	9089	9137	9185	9232	9280	9328	9375	9423	9471	48
911	9518	9566	9614	9661	9709	9757	9804	9852	9900	9947	48
912	9995	**42	**90	*138	*185	*233	*280	*328	*376	*423	48
913	960471	0518	0566	0613	0661	0709	0756	0804	0851	0899	48
914	0946	0994	1041	1089	1136	1184	1231	1279	1326	1374	47
915	1421	1469	1516	1563	1611	1658	1706	1753	1801	1848	47
916	1895	1943	1990	2038	2085	2132	2180	2227	2275	2322	47
917	2369	2417	2464	2511	2559	2606	2653	2701	2748	2795	47
918	2843	2890	2937	2985	3032	3079	3126	3174	3221	3268	47
919	3316	3363	3410	3457	3504	3552	3599	3646	3693	3741	47
920	963788	3835	3882	3929	3977	4024	4071	4118	4165	4212	47
921	4260	4307	4354	4401	4448	4495	4542	4590	4637	4684	47
922	4731	4778	4825	4872	4919	4966	5013	5061	5108	5155	47
923	5202	5249	5296	5343	5390	5437	5484	5531	5578	5625	47
924	5672	5719	5766	5813	5860	5907	5954	6001	6048	6095	47
925	6142	6189	6236	6283	6329	6376	6423	6470	6517	6564	47
926	6611	6658	6705	6752	6799	6845	6892	6939	6986	7033	47
927	7080	7127	7173	7220	7267	7314	7361	7408	7454	7501	47
928	7548	7595	7642	7688	7735	7782	7829	7875	7922	7969	47
929	8016	8062	8109	8156	8203	8249	8296	8343	8390	8436	47
930	968483	8530	8576	8623	8670	8716	8763	8810	8856	8903	47
931	8950	8996	9043	9090	9136	9183	9229	9276	9323	9369	47
932	9416	9463	9509	9556	9602	9649	9695	9742	9789	9835	47
933	9882	9928	9975	**21	**68	*114	*161	*207	*254	*300	47
934	970347	0393	0440	0486	0533	0579	0626	0672	0719	0765	46
935	0812	0858	0904	0951	0997	1044	1090	1137	1183	1229	46
936	1276	1322	1369	1415	1461	1508	1554	1601	1647	1693	46
937	1740	1786	1832	1879	1925	1971	2018	2064	2110	2157	46
938	2203	2249	2295	2342	2388	2434	2481	2527	2573	2619	46
939	2666	2712	2758	2804	2851	2897	2943	2989	3035	3082	46
N.	0	1	2	3	4	5	6	7	8	9	D.

N.	0	1	2	3	4	5	6	7	8	9	D.
940	973128	3174	3220	3266	3313	3359	3405	3451	3497	3543	46
941	3590	3636	3682	3728	3774	3820	3866	3913	3959	4005	46
942	4051	4097	4143	4189	4235	4281	4327	4374	4420	4466	46
943	4512	4558	4604	4650	4696	4742	4788	4834	4880	4926	46
944	4972	5018	5064	5110	5156	5202	5248	5294	5340	5386	46
945	5432	5478	5524	5570	5616	5662	5707	5753	5799	5845	46
946	5891	5937	5983	6029	6075	6121	6167	6212	6258	6304	46
947	6350	6396	6442	6488	6533	6579	6625	6671	6717	6763	46
948	6808	6854	6900	6946	6992	7037	7083	7129	7175	7220	46
949	7266	7312	7358	7403	7449	7495	7541	7586	7632	7678	46
950	977724	7769	7815	7861	7906	7952	7998	8043	8089	8135	46
951	8181	8226	8272	8317	8363	8409	8454	8500	8546	8591	46
952	8637	8683	8728	8774	8819	8865	8911	8956	9002	9047	46
953	9093	9138	9184	9230	9275	9321	9366	9412	9457	9503	46
954	9548	9594	9639	9685	9730	9776	9821	9867	9912	9958	46
955	980003	0049	0094	0140	0185	0231	0276	0322	0367	0412	45
956	0458	0503	0549	0594	0640	0685	0730	0776	0821	0867	45
957	0912	0957	1003	1048	1093	1139	1184	1229	1275	1320	45
958	1366	1411	1456	1501	1547	1592	1637	1683	1728	1773	45
959	1819	1864	1909	1954	2000	2045	2090	2135	2181	2226	45
960	982271	2316	2362	2407	2452	2497	2543	2588	2633	2678	45
961	2723	2769	2814	2859	2904	2949	2994	3040	3085	3130	45
962	3175	3220	3265	3310	3356	3401	3446	3491	3536	3581	45
963	3626	3671	3716	3762	3807	3852	3897	3942	3987	4032	45
964	4077	4122	4167	4212	4257	4302	4347	4392	4437	4482	45
965	4527	4572	4617	4662	4707	4752	4797	4842	4887	4932	45
966	4977	5022	5067	5112	5157	5202	5247	5292	5337	5382	45
967	5426	5471	5516	5561	5606	5651	5696	5741	5786	5830	45
968	5875	5920	5965	6010	6055	6100	6144	6189	6234	6279	45
969	6324	6369	6413	6458	6503	6548	6593	6637	6682	6727	45
970	986772	6817	6861	6906	6951	6996	7040	7085	7130	7175	45
971	7219	7264	7309	7353	7398	7443	7488	7532	7577	7622	45
972	7666	7711	7756	7800	7845	7890	7934	7979	8024	8068	45
973	8113	8157	8202	8247	8291	8336	8381	8425	8470	8514	45
974	8559	8604	8648	8693	8737	8782	8826	8871	8916	8960	45
975	9005	9049	9094	9138	9183	9227	9272	9316	9361	9405	45
976	9450	9494	9539	9583	9628	9672	9717	9761	9806	9850	44
977	9895	9939	9983	**28	**72	*117	*161	*206	*250	*294	44
978	990339	0383	0428	0472	0516	0561	0605	0650	0694	0738	44
979	0783	0827	0871	0916	0960	1004	1049	1098	1137	1182	44
980	991226	1270	1315	1359	1403	1448	1492	1586	1580	1625	44
981	1669	1713	1758	1802	1846	1890	1935	1979	2023	2067	44
982	2111	2156	2200	2244	2288	2333	2377	2421	2465	2509	44
983	2554	2598	2642	2686	2730	2774	2819	2863	2907	2951	44
984	2995	3039	3083	3127	3172	3216	3260	3304	3348	3392	44
985	3436	3480	3524	3568	3613	3657	3701	3745	3789	3833	44
986	3877	3921	3965	4009	4053	4097	4141	4185	4229	4273	44
987	4317	4361	4405	4449	4493	4537	4581	4625	4669	4713	44
988	4757	4801	4845	4889	4933	4977	5021	5065	5108	5152	44
989	5196	5240	5284	5328	5372	5416	5460	5504	5547	5591	44
990	995635	5679	5723	5767	5811	5854	5898	5942	5986	6030	44
991	6074	6117	6161	6205	6249	6293	6337	6380	6424	6468	44
992	6512	6555	6599	6643	6687	6731	6774	6818	6862	6906	44
993	6949	6993	7037	7080	7124	7168	7212	7255	7299	7343	44
994	7386	7430	7474	7517	7561	7605	7648	7692	7736	7779	44
995	7823	7867	7910	7954	7998	8041	8085	8129	8172	8216	44
996	8259	8303	8347	8390	8434	8477	8521	8564	8608	8652	44
997	8695	8739	8782	8826	8869	8913	8956	9000	9043	9087	44
998	9131	9174	9218	9261	9305	9348	9392	9435	9479	9522	44
999	9565	9609	9652	9696	9739	9783	9826	9870	9913	9957	43
N.	0 .	1	2	3	4	5	6	7	8	9	D.

A TABLE

OF

LOGARITHMIC

SINES AND TANGENTS

FOR EVERY

DEGREE AND MINUTE

OF THE QUADRANT.

REMARK. The minutes in the left-hand column of each page, increasing downward, belong to the degrees at the top; and those increasing upward, in the right-hand column, belong to the degrees below.

M.	Sine.	D.	Cosine.	D.	Tang.	D.	Cotang.	
0	0·000000		10·000000		0·000000		Infinite.	60
1	6·463726	5017·17	000000	·00	6·463726	5017·17	13·536274	59
2	764756	2934·85	000000	·00	764756	2934·83	235244	58
3	940847	2082·31	000000	·00	940847	2082·31	059153	57
4	7·065786	1615·17	000000	·00	7·065786	1615·17	12·984214	56
5	162696	1319·68	000000	·00	162696	1319·69	837304	55
6	241877	1115·75	9·999999	·01	241878	1115·78	758122	54
7	308824	966·53	999999	·01	308825	996·53	691175	53
8	366816	852·54	999999	·01	366817	852·54	633183	52
9	417968	762·63	999999	·01	417970	762·63	582030	51
10	463725	689·88	999998	·01	463727	689·88	536273	50
11	7·505118	629·81	9·999998	·01	7·505120	629·81	12·494880	49
12	542906	579·86	999997	·01	542909	579·83	457091	48
13	577668	536·41	999997	·01	577672	536·42	422328	47
14	609853	499·38	999996	·01	609857	499·39	390143	46
15	639816	467·14	999996	·01	639820	467·15	360180	45
16	667845	438·81	999995	·01	667849	438·82	332151	44
17	694173	413·72	999995	·01	694179	413·73	305821	43
18	718997	391·35	999994	·01	719004	391·36	280997	42
19	742477	371·27	999993	·01	742484	371·28	257516	41
20	764754	353·15	999993	·01	764761	351·36	235239	40
21	7·785943	336·72	9·999992	·01	7·785951	336·73	12·214049	39
22	806146	321·75	999991	·01	806155	321·76	193845	38
23	825451	308·05	999990	·01	825460	308·06	174540	37
24	843934	295·47	999989	·02	843944	295·49	156056	36
25	861662	283·88	999988	·02	861674	283·90	138326	35
26	878695	273·17	999988	·02	878708	273·18	121292	34
27	895085	263·23	999987	·02	895099	263·25	104901	33
28	910879	253·99	999986	·02	910894	254·01	089106	32
29	926119	245·38	999985	·02	926134	245·40	073866	31
30	940842	237·33	999983	·02	940858	237·35	059142	30
31	7·955082	229·80	9·999982	·02	7·955100	229·81	12·044900	29
32	968870	222·73	999981	·02	968889	222·75	031111	28
33	982233	216·08	999980	·02	982253	216·10	017747	27
34	995198	209·81	999979	·02	995219	209·83	004781	26
35	8·007787	203·90	999977	·02	8·007809	203·92	11·992191	25
36	020021	198·31	999976	·02	020045	198·33	979955	24
37	031919	193·02	999975	·02	031945	193·05	968055	23
38	043501	188·01	999973	·02	043527	188·03	956473	22
39	054781	183·25	999972	·02	054809	183·27	945191	21
40	065776	178·72	999971	·02	065806	178·74	934194	20
41	8·076500	174·41	9·999969	·02	8·076531	174·44	11·923469	19
42	086965	170·31	999968	·02	086997	170·34	913003	18
43	097183	166·39	999966	·02	097217	166·42	902783	17
44	107167	162·65	999964	·03	107202	162·68	892797	16
45	116926	159·08	999963	·03	116963	159·10	883037	15
46	126471	155·66	999961	·03	126510	155·68	873490	14
47	135810	152·38	999959	·03	135851	152·41	864149	13
48	144953	149·24	999958	·03	144996	149·27	855004	12
49	153907	146·22	999956	·03	153952	146·27	846048	11
50	162681	143·33	999954	·03	162727	143·36	837273	10
51	8·171280	140·54	9·999952	·03	8·171328	140·57	11·828672	9
52	179713	137·86	999950	·03	179763	137·90	820237	8
53	187985	135·29	999948	·03	188036	135·32	811964	7
54	196102	132·80	999946	·03	196156	132·84	803844	6
55	204070	130·41	999944	·03	204126	130·44	795874	5
56	211895	128·10	999942	·04	211953	128·14	788047	4
57	219581	125·87	999940	·04	219641	125·90	780359	3
58	227134	123·72	999938	·04	227195	123·76	772805	2
59	234557	121·64	999936	·04	234621	121·68	765379	1
60	241855	119·63	999934	·04	241921	119·67	758079	0
	Cosine.	D.	Sine.		Cotang.	D.	Tang.	M.

M.	Sine.	D.	Cosine.	D.	Tang.	D.	Cotang.	
0	8·241855	119·68	9·999934	·04	8·241921	119·67	11·758079	60
1	249033	117·68	999932	·04	249102	117·72	750898	59
2	256094	115·80	999929	·04	256165	115·84	743835	58
3	263042	113·98	999927	·04	263115	114·02	736885	57
4	269881	112·21	999925	·04	269956	112·25	730044	56
5	276614	110·50	999922	·04	276691	110·54	723309	55
6	283243	108·83	999920	·04	283323	108·87	716677	54
7	289773	107·21	999918	·04	289856	107·26	710144	53
8	296207	105·65	999915	·04	296292	105·70	703708	52
9	302546	104·13	999913	·04	302634	104·18	697366	51
10	308794	102·66	999910	·04	308884	102·70	691116	50
11	8·314954	101·22	9·999907	·04	8·315046	101·26	11·684954	49
12	321027	99·82	999905	·04	321122	99·87	678878	48
13	327016	98·47	999902	·04	327114	98·51	672886	47
14	332924	97·14	999899	·05	333025	97·19	666975	46
15	338753	95·86	999897	·05	338856	95·90	661144	45
16	344504	94·60	999894	·05	344610	94·65	655390	44
17	350181	93·38	999891	·05	350289	93·43	649711	43
18	355783	92·19	999888	·05	355895	92·24	644105	42
19	361315	91·08	999885	·05	361430	91·08	638570	41
20	366777	89·90	999882	·05	366895	89·95	633105	40
21	8·372171	88·80	9·999879	·05	8·372292	88·85	11·627708	39
22	377499	87·72	999876	·05	377622	87·77	622378	38
23	382762	86·67	999873	·05	382889	86·72	617111	37
24	387962	85·64	999870	·05	388092	85·70	611908	36
25	393101	84·64	999867	·05	393234	84·70	606766	35
26	398179	83·66	999864	·05	398315	83·71	601685	34
27	403199	82·71	999861	·05	403338	82·76	596662	33
28	408161	81·77	999858	·05	408304	81·82	591696	32
29	413068	80·86	999854	·05	413213	80·91	586787	31
30	417919	79·96	999851	·06	418068	80·02	581932	30
31	8·422717	79·09	9·999848	·06	8·422869	79·14	11·577131	29
32	427462	78·23	999844	·06	427618	78·30	572382	28
33	432156	77·40	999841	·06	432315	77·45	567685	27
34	436800	76·57	999838	·06	436962	76·63	563038	26
35	441394	75·77	999834	·06	441560	75·83	558440	25
36	445941	74·99	999831	·06	446110	75·05	553890	24
37	450440	74·22	999827	·06	450613	74·28	549387	23
38	454893	73·46	999823	·06	455070	73·52	544930	22
39	459301	72·73	999820	·06	459481	72·79	540519	21
40	463665	72·00	999816	·06	463849	72·06	536151	20
41	8·467985	71·29	9·999812	·06	8·468172	71·35	11·531828	19
42	472263	70·60	999809	·06	472454	70·66	527546	18
43	476498	69·91	999805	·06	476693	69·98	523307	17
44	480693	69·24	999801	·06	480892	69·31	519108	16
45	484848	68·59	999797	·07	485050	68·65	514950	15
46	488963	67·94	999793	·07	489170	68·01	510830	14
47	493040	67·31	999790	·07	493250	67·38	506750	13
48	497078	66·69	999786	·07	497293	66·76	502707	12
49	501080	66·08	999782	·07	501298	66·15	498702	11
50	505045	65·48	999778	·07	505267	65·55	494733	10
51	8·508974	64·89	9·999774	·07	8·509200	64·96	11·490800	9
52	512867	64·31	999769	·07	513098	64·39	486902	8
53	516726	63·75	999765	·07	516961	63·82	483039	7
54	520551	63·16	999761	·07	520790	63·26	479210	6
55	524343	62·64	999757	·07	524586	62·72	475414	5
56	528102	62·11	999753	·07	528349	62·18	471651	4
57	531828	61·58	999748	·07	532080	61·65	467920	3
58	535523	61·06	999744	·07	535779	61·13	464221	2
59	539186	60·55	999740	·07	539447	60·62	460553	1
60	542819	60·04	999735	·07	543084	60·12	456916	0
	Cosine.	D.	Sine.		Cotang.	D.	Tang.	M.

(88 DEGREES.)

M.	Sine.	D.	Cosine.	D.	Tang.	D.	Cotang.	
0	8·542819	60·04	9·999735	·07	8·543084	60·12	11·456916	60
1	546422	59·55	999731	·07	546691	59·62	453309	59
2	549995	59·06	999726	·07	550268	59·14	449732	58
3	553539	58·58	999722	·08	553817	58·66	446183	57
4	557054	58·11	999717	·08	557336	58·19	442664	56
5	560540	57·65	999713	·08	560828	57·73	439172	55
6	563999	57·19	999708	·08	564291	57·27	435709	54
7	567431	56·74	999704	·08	567727	56·82	432273	53
8	570836	56·30	999699	·08	571137	56·38	428863	52
9	574214	55·87	999694	·08	574520	55·95	425480	51
10	577566	55·44	999689	·08	577877	55·52	422123	50
11	8·580892	55·02	9·999685	·08	8·581208	55·10	11·418792	49
12	584193	54·60	999680	·08	584514	54·68	415486	48
13	587469	54·19	999675	·08	587795	54·27	412205	47
14	590721	53·79	999670	·08	591051	53·87	408949	46
15	593948	53·39	999665	·08	594283	53·47	405717	45
16	597152	53·00	999660	·08	597492	53·08	402508	44
17	600332	52·61	999655	·08	600677	52·70	399323	43
18	603489	52·23	999650	·08	603839	52·32	396161	42
19	606623	51·86	999645	·09	606978	51·94	393022	41
20	609734	51·49	999640	·09	610094	51·58	389906	40
21	8·612823	51·12	9·999635	·09	8·613189	51·21	11·386811	39
22	615891	50·76	999629	·09	616262	50·85	383738	38
23	618937	50·41	999624	·09	619313	50·50	380687	37
24	621962	50·06	999619	·09	622343	50·15	377657	36
25	624965	49·72	999614	·09	625352	49·81	374648	35
26	627948	49·38	999608	·09	628340	49·47	371660	34
27	630911	49·04	999603	·09	631308	49·13	368692	33
28	633854	48·71	999597	·09	634256	48·80	365744	32
29	636776	48·39	999592	·09	637184	48·48	362816	31
30	639680	48·06	999586	·09	640093	48·16	359907	30
31	8·642563	47·75	9·999581	·09	8·642982	47·84	11·357018	29
32	645428	47·43	999575	·09	645853	47·53	354147	28
33	648274	47·12	999570	·09	648704	47·22	351296	27
34	651102	46·82	999564	·09	651537	46·91	348463	26
35	653911	46·52	999558	·10	654352	46·61	345648	25
36	656702	46·22	999553	·10	657149	46·31	342851	24
37	659475	45·92	999547	·10	659928	46·02	340072	23
38	662230	45·63	999541	·10	662689	45·73	337311	22
39	664968	45·35	999535	·10	665433	45·44	334567	21
40	667689	45·06	999529	·10	668160	45·26	331840	20
41	8·670393	44·79	9·999524	·10	8·670870	44·88	11·329130	19
42	673080	44·51	999518	·10	673563	44·61	326437	18
43	675751	44·24	999512	·10	676239	44·34	323761	17
44	678405	43·97	999506	·10	678900	44·17	321100	16
45	681043	43·70	999500	·10	681544	43·80	318456	15
46	683665	43·44	999493	·10	684172	43·54	315828	14
47	686272	43·18	999487	·10	686784	43·28	313216	13
48	688863	42·92	999481	·10	689381	43·03	310619	12
49	691438	42·67	999475	·10	691963	42·77	308037	11
50	693998	42·42	999469	·10	694529	42·52	305471	10
51	8·696543	42·17	9·999463	·11	8·697081	42·28	11·302919	9
52	699073	41·92	999456	·11	699617	42·03	300383	8
53	701589	41·68	999450	·11	702139	41·79	297861	7
54	704090	41·44	999443	·11	704646	41·55	295354	6
55	706577	41·21	999437	·11	707140	41·32	292860	5
56	709049	40·97	999431	·11	709618	41·08	290382	4
57	711507	40·74	999424	·11	712083	40·85	287917	3
58	713952	40·51	999418	·11	714534	40·62	285465	2
59	716383	40·29	999411	·11	716972	40·40	283028	1
60	718800	40·06	999404	·11	719396	40·17	280604	0
	Cosine.	D.	Sine.		Cotang.	D.	Tang.	M.

M.	Sine.	D.	Cosine.	D.	Tang.	D.	Cotang.	
0	8·718800	40·06	9·999404	·11	8·719396	40·17	11·280604	60
1	721204	39·84	999398	·11	721806	39·95	278194	59
2	723595	39·62	999391	·11	724204	39·74	275796	58
3	725972	39·41	999384	·11	726588	39·52	273412	57
4	728337	39·19	999378	·11	728959	39·30	271041	56
5	730688	38·98	999371	·11	731317	39·09	268683	55
6	733027	38·77	999364	·12	733663	38·89	266337	54
7	735354	38·57	999357	·12	735996	38·68	264004	53
8	737667	38·36	999350	·12	738317	38·48	261683	52
9	739969	38·16	999343	·12	740626	38·27	259374	51
10	742259	37·96	999336	·12	742922	38·07	257078	50
11	8·744536	37·76	9·999329	·12	8·745207	37·87	11·254793	49
12	746802	37·56	999322	·12	747479	37·68	252521	48
13	749055	37·37	999315	·12	749740	37·49	250260	47
14	751297	37·17	999308	·12	751989	37·29	248011	46
15	753528	36·98	999301	·12	754227	37·10	245773	45
16	755747	36·79	999294	·12	756453	36·92	243547	44
17	757955	36·61	999286	·12	758668	36·73	241332	43
18	· 760151	36·42	999279	·12	760872	36·55	239128	42
19	762337	36·24	999272	·12	763065	36·36	236935	41
20	764511	36·06	999265	·12	765246	36·18	234754	40
21	8·766675	35·88	9·999257	·12	8·767417	36·00	11·232583	39
22	768828	35·70	999250	·13	769578	35·83	230422	38
23	770970	35·53	999242	·13	771727	35·65	228278	37
24	773101	35·35	999235	·13	773866	35·48	226134	36
25	775223	35·18	999227	·13	775995	35·31	224005	35
26	· 777333	35·01	999220	·13	778114	35·14	221885	34
27	779434	34·84	999212	·13	780222	34·97	219778	33
28	781524	34·67	999205	·13	782320	34·80	217680	32
29	783605	34·51	999197	·13	784408	34·64	215592	31
30	785675	34·31	999189	·13	786486	34·47	213514	30
31	8·787736	34·18	9·999181	·13	8·788554	34·31	11·211446	29
32	789787	34·02	999174	·13	790613	34·15	209887	28
33	791828	33·86	999166	·13	792662	33·99	207338	27
34	793859	33·70	999158	·13	794701	33·83	205299	26
35	795881	33·54	999150	·13	796731	33·68	203269	25
36	797894	33·39	999142	·13	798752	33·52	201248	24
37	799897	33·23	999134	·13	800763	33·37	199237	23
38	801892	33·08	999126	·13	802765	33·22	197235	22
39	803876	32·93	999118	·13	804758	33·07	195242	21
40	805852	32·78	999110	·13	806742	32·92	193258	20
41	8·807819	32·63	9·999102	·13	8·808717	32·78	11·191283	19
42	809777	32·49	999094	·14	810683	32·62	189817	18
43	811726	32·34	999086	·14	812641	32·48	187359	17
44	813667	32·19	999077	·14	814589	32·33	185411	16
45	· 815599	32·05	999069	·14	816529	32·19	183471	15
46	817522	31·91	999061	·14	818461	32·05	181539	14
47	819436	31·77	999053	·14	820384	31·91	179616	13
48	821343	31·63	999044	·14	822298	31·77	177702	12
49	823240	31·49	999036	·14	824205	31·63	175795	11
50	825130	31·35	999027	·14	826103	31·50	173897	10
51	8·827011	31·22	9·999019	·14	8·827992	31·36	11·172008	9
52	828884	31·08	999010	·14	829874	31·23	170126	8
53	830749	30·95	999002	·14	831748	31·10	168252	7
54	832607	30·82	998993	·14	833613	30·96	166387	6
55	834456	30·69	998984	·14	835471	30·83	164529	5
56	836297	30·56	998976	·14	837321	30·70	162679	4
57	838130	30·43	998967	·15	839163	30·57	160837	3
58	839956	30·30	998958	·15	840998	30·45	159002	2
59	841774	30·17	998950	·15	842825	30·32	157175	1
60	843585	30·00	998941	·15	844644	30·19	155356	0
	Cosine.	D.	Sine.		Cotang.	D.	Tang.	M.

M.	Sine.	D.	Cosine.	D.	Tang.	D.	Cotang.	
0	8·843585	30·05	9·998941	·15	8·844644	30·19	11·155356	60
1	845387	29·92	998932	·15	846455	30·07	153545	59
2	847183	29·80	998923	·15	848260	29·95	151740	58
3	848971	29·67	998914	·15	850057	29·82	149943	57
4	850751	29·55	998905	·15	851846	29·70	148154	56
5	852525	29·43	998896	·15	853628	29·58	146372	55
6	854291	29·31	998887	·15	855403	29·46	144597	54
7	856049	29·19	998878	·15	857171	29·35	142829	53
8	857801	29·07	998869	·15	858932	29·23	141068	52
9	859546	28·96	998860	·15	860686	29·11	139314	51
10	861283	28·84	998851	·15	862433	29·00	137567	50
11	8·863014	28·73	9·998841	·15	8·864178	28·88	11·135827	49
12	864738	28·61	998832	·15	865906	28·77	134094	48
13	866455	28·50	998823	·16	867632	28·66	132368	47
14	868165	28·39	998813	·16	869351	28·54	130649	46
15	869868	28·28	998804	·16	871064	28·43	128936	45
16	871565	28·17	998795	·16	872770	28·32	127230	44
17	873255	28·06	998785	·16	874469	28·21	125531	43
18	874938	27·95	998776	·16	876162	28·11	123838	42
19	876615	27·84	998766	·16	877849	28·00	122151	41
20	878285	27·73	998757	·16	879529	27·89	120471	40
21	8·879949	27·63	9·998747	·16	8·881202	27·79	11·118798	39
22	881607	27·52	998738	·16	882869	27·68	117131	38
23	883258	27·42	998728	·16	884530	27·58	115470	37
24	884903	27·31	998718	·16	886185	27·47	113815	36
25	886542	27·21	998708	·16	887833	27·37	112167	35
26	888174	27·11	998699	·16	889476	27·27	110524	34
27	889801	27·00	998689	·16	891112	27·17	108888	33
28	891421	26·90	998679	·16	892742	27·07	107258	32
29	893035	26·80	998669	·17	894366	26·97	105634	31
30	894643	26·70	998659	·17	895984	26·87	104016	30
31	8·896246	26·60	9·998649	·17	8·897596	26·77	11·102404	29
32	897842	26·51	998639	·17	899203	26·67	100797	28
33	899432	26·41	998629	·17	900803	26·58	099197	27
34	901017	26·31	998619	·17	902398	26·48	097602	26
35	902596	26·22	998609	·17	903987	26·38	096013	25
36	904169	26·12	998599	·17	905570	26·29	094430	24
37	905736	26·03	998589	·17	907147	26·20	092853	23
38	907297	25·93	998578	·17	908719	26·10	091281	22
39	908853	25·84	998568	·17	910285	26·01	089715	21
40	910404	25·75	998558	·17	911846	25·92	088154	20
41	8·911949	25·66	9·998548	·17	8·913401	25·83	11·086599	19
42	913488	25·56	998537	·17	914951	25·74	085049	18
43	915022	25·47	998527	·17	916495	25·65	083505	17
44	916550	25·38	998516	·18	918034	25·56	081966	16
45	918073	25·29	998506	·18	919568	25·47	080432	15
46	919591	25·20	998495	·18	921096	25·38	078904	14
47	921103	25·12	998485	·18	922619	25·30	077381	13
48	922610	25·03	998474	·18	924136	25·21	075864	12
49	924112	24·94	998464	·18	925649	25·12	074351	11
50	925609	24·86	998453	·18	927156	25·03	072844	10
51	8·927100	24·77	9·998442	·18	8·928658	24·95	11·071342	9
52	928587	24·69	998431	·18	930155	24·86	069845	8
53	930068	24·60	998421	·18	931647	24·78	068353	7
54	931544	24·52	998410	·18	933134	24·70	066866	6
55	933015	24·43	998399	·18	934616	24·61	065384	5
56	934481	24·35	998388	·18	936093	24·53	063907	4
57	935942	24·27	998377	·18	937565	24·45	062435	3
58	937398	24·19	998366	·18	939032	24·37	060968	2
59	938850	24·11	998355	·18	940494	24·30	059506	1
60	940296	24·03	998344	·18	941952	24·21	058048	0
	Cosine.	D.	Sine.		Cotang.	D.	Tang.	M.

M.	Sine.	D.	Cosine.	D.	Tang.	D.	Cotang.	
0	8·940296	24·03	9·998344	·19	8·941952	24·21	11·058048	60
1	941738	23·94	998333	·19	943404	24·13	056596	59
2	943174	23·87	998322	·19	944852	24·05	055148	58
3	944606	23·79	998311	·19	946295	23·97	053705	57
4	946034	23·71	998300	·19	947734	23·90	052266	56
5	947456	23·63	998289	·19	949168	23·82	050832	55
6	948874	23·55	998277	·19	950597	23·74	049403	54
7	950287	23·48	998266	·19	952021	23·66	047979	53
8	951696	23·40	998255	·19	953441	23·60	046559	52
9	953100	23·32	998243	·19	954856	23·51	045144	51
10	954499	23·25	998232	·19	956267	23·44	043733	50
11	8·955894	23·17	9·998220	·19	8·957674	23·37	11·042326	49
12	957284	23·10	998209	·19	959075	23·29	040925	48
13	958670	23·02	998197	·19	960473	23·23	039527	47
14	960052	22·95	998186	·19	961866	23·14	038134	46
15	961429	22·88	998174	·19	963255	23·07	036745	45
16	962801	22·80	998163	·19	964639	23·00	035361	44
17	964170	22·73	998151	·19	966019	22·93	033981	43
18	965534	22·66	998139	·20	967394	22·86	032606	42
19	966893	22·59	998128	·20	968766	22·79	031234	41
20	968249	22·52	998116	·20	970133	22·71	029867	40
21	8·969600	22·44	9·998104	·20	8·971496	22·65	11·028504	39
22	970947	22·38	998092	·20	972855	22·57	027145	38
23	972289	22·31	998080	·20	974209	22·51	025791	37
24	973628	22·24	998068	·20	975560	22·44	024440	36
25	974962	22·17	998056	·20	976906	22·37	023094	35
26	976293	22·10	998044	·20	978248	22·30	021752	34
27	977619	22·03	998032	·20	979586	22·23	020414	33
28	978941	21·97	998020	·20	980921	22·17	019079	32
29	980259	21·90	998008	·20	982251	22·10	017749	31
30	981573	21·83	997996	·20	983577	22·04	016423	30
31	8·982883	21·77	9·997985	·20	8·984899	21·97	11·015101	29
32	984189	21·70	997972	·20	986217	21·91	013783	28
33	985491	21·63	997959	·20	987532	21·84	012468	27
34	986789	21·57	997947	·20	988842	21·78	011158	26
35	988083	21·50	997935	·21	990149	21·71	009851	25
36	989374	21·44	997922	·21	991451	21·65	008549	24
37	990660	21·38	997910	·21	992750	21·58	007250	23
38	991943	21·31	997897	·21	994045	21·52	005955	22
39	993222	21·25	997885	·21	995337	21·46	004663	21
40	994497	21·19	997872	·21	996624	21·40	003376	20
41	8·995768	21·12	9·997860	·21	8·997908	21·34	11·002092	19
42	997036	21·06	997847	·21	999188	21·27	000812	18
43	998299	21·00	997835	·21	9·000465	21·21	10·999535	17
44	999560	20·94	997822	·21	001738	21·15	998262	16
45	9·000816	20·87	997809	·21	003007	21·09	996993	15
46	002069	20·82	997797	·21	004272	21·03	995728	14
47	003318	20·76	997784	·21	005534	20·97	994466	13
48	004563	20·70	997771	·21	006792	20·91	993208	12
49	005805	20·64	997758	·21	008047	20·85	991953	11
50	007044	20·58	997745	·21	009298	20·80	990702	10
51	9·008278	20·52	9·997732	·21	9·010546	20·74	10·989454	9
52	009510	20·46	997719	·21	011790	20·68	988210	8
53	010787	20·40	997706	·21	013031	20·62	986969	7
54	011962	20·34	997693	·22	014268	20·56	985732	6
55	013182	20·29	997680	·22	015502	20·51	984498	5
56	014400	20·23	997667	·22	016732	20·45	983268	4
57	015613	20·17	997654	·22	017959	20·40	982041	3
58	016824	20·12	997641	·22	019183	20·33	980817	2
59	018031	20·06	997628	·22	020403	20·28	979597	1
60	019235	20·00	997614	·22	021620	20·23	978380	0
	Cosine.	D.	Sine.		Cotang.	D.	Tang.	M.

M.	Sine.	D.	Cosine.	D.	Tang.	D.	Cotang.	
0	9·019235	20·00	9·997614	·22	9·021620	20·23	10·978880	60
1	020435	19·95	997601	·22	022834	20·17	977166	59
2	021632	19·89	997588	·22	024044	20·11	975956	58
3	022825	19·84	997574	·22	025251	20·06	974749	57
4	024016	19·78	997561	·22	026455	20·00	973545	56
5	025208	19·73	997547	·22	027655	19·95	972345	55
6	026386	19·67	997534	·23	028852	19·90	971148	54
7	027567	19·62	997520	·23	030046	19·85	969954	53
8	028744	19·57	997507	·23	031237	19·79	968763	52
9	029918	19·51	997493	·23	032425	19·74	967575	51
10	031089	19·47	997480	·23	033609	19·69	966391	50
11	9·032257	19·41	9·997466	·23	9·034791	19·64	10·965209	49
12	033421	19·36	997452	·23	035969	19·58	964081	48
13	034582	19·30	997439	·23	037144	19·53	962856	47
14	035741	19·25	997425	·23	038316	19·48	961684	46
15	036896	19·20	997411	·23	039485	19·43	960515	45
16	038048	19·15	997397	·23	040651	19·38	959349	44
17	039197	19·10	997383	·23	041813	19·33	958187	43
18	040342	19·05	997369	·23	042973	19·28	957027	42
19	041485	18·99	997355	·23	044130	19·23	955870	41
20	042625	18·94	997341	·23	045284	19·18	954716	40
21	9·043762	18·89	9·997327	·24	9·046434	19·13	10·953566	39
22	044895	18·84	997313	·24	047582	19·08	952418	38
23	046026	18·79	997299	·24	048727	19·03	951273	37
24	047154	18·75	997285	·24	049869	18·98	950131	36
25	048279	18·70	997271	·24	051008	18·93	948992	35
26	049400	18·65	997257	·24	052144	18·89	947856	34
27	050519	18·60	997242	·24	053277	18·84	946723	33
28	051635	18·55	997228	·24	054407	18·79	945593	32
29	052749	18·50	997214	·24	055535	18·74	944465	31
30	053859	18·45	997199	·24	056659	18·70	943341	30
31	9·054966	18·41	9·997185	·24	9·057781	18·65	10·942219	29
32	056071	18·36	997170	·24	058900	18·60	941100	28
33	057172	18·31	997156	·24	060016	18·55	939984	27
34	058271	18·27	997141	·24	061130	18·51	938870	26
35	059367	18·22	997127	·24	062240	18·46	937760	25
36	060460	18·17	997112	·24	063348	18·42	936652	24
37	061551	18·13	997098	·24	064453	18·37	935547	23
38	062639	18·08	997083	·25	065556	18·33	934444	22
39	063724	18·04	997068	·25	066655	18·28	933345	21
40	064806	17·99	997053	·25	067752	18·24	932248	20
41	9·065885	17·94	9·997039	·25	9·068846	18·19	10·931154	19
42	066962	17·90	997024	·25	069938	18·15	930062	18
43	068036	17·86	997009	·25	071027	18·10	928973	17
44	069107	17·81	996994	·25	072113	18·06	927887	16
45	070176	17·77	996979	·25	073197	18·02	926803	15
46	071242	17·72	996964	·25	074278	17·97	925722	14
47	072306	17·68	996949	·25	075356	17·93	924644	13
48	073366	17·63	996934	·25	076432	17·89	923568	12
49	074424	17·59	996919	·25	077505	17·84	922495	11
50	075480	17·55	996904	·25	078576	17·80	921424	10
51	9·076533	17·50	9·996889	·25	9·079644	17·76	10·920356	9
52	077583	17·46	996874	·25	080710	17·72	919290	8
53	078631	17·42	996858	·25	081773	17·67	918227	7
54	079676	17·38	996843	·25	082833	17·63	917167	6
55	080719	17·33	996828	·25	083891	17·59	916109	5
56	081759	17·29	996812	·26	084947	17·55	915053	4
57	082797	17·25	996797	·26	086000	17·51	914000	3
58	083832	17·21	996782	·26	087050	17·47	912950	2
59	084864	17·17	996766	·26	088098	17·43	911902	1
60	085894	17·13	996751	·26	089144	17·38	910856	0
	Cosine.	D.	Sine.		Cotang.	D.	Tang.	M.

M.	Sine.	D.	Cosine.	D.	Tang.	D.	Cotang.	
0	9·085894	17·13	9·996751	·26	9·089144	17·88	10·910856	60
1	086922	17·09	996735	·26	090187	17·34	909813	59
2	087947	17·04	996720	·26	091228	17·30	908772	58
3	088970	17·00	996704	·26	092266	17·27	907734	57
4	089990	16·96	996688	·26	093302	17·22	906698	56
5	091008	16·92	996673	·26	094336	17·19	905664	55
6	092024	16·88	996657	·26	095367	17·15	904633	54
7	093037	16·84	996641	·26	096395	17·11	903605	53
8	094047	16·80	996625	·26	097422	17·07	902578	52
9	095056	16·76	996610	·26	098446	17·03	901554	51
10	096062	16·73	996594	·26	099468	16·99	900532	50
11	9·097065	16·68	9·996578	·27	9·100487	16·95	10·899513	49
12	098066	16·65	996562	·27	101504	16·91	898496	48
13	099065	16·61	996546	·37	102519	16·87	897481	47
14	100062	16·57	996530	·27	103532	16·84	896468	46
15	101056	16·53	996514	·27	104542	16·80	895458	45
16	102048	16·49	996498	·27	105550	16·76	894450	44
17	103037	16·45	996482	·27	106556	16·72	893444	43
18	104025	16·41	996465	·27	107559	16·69	892441	42
19	105010	16·38	996449	·27	108560	16·65	891440	41
20	105992	16·34	996433	·27	109559	16·61	890441	40
21	9·106973	16·30	9·996417	·27	9·110556	16·58	10·889444	39
22	107951	16·27	996400	·27	111551	16·54	888449	38
23	108927	16·23	996384	·27	112543	16·50	887457	37
24	109901	16·19	996368	·27	113533	16·46	886467	36
25	110873	16·16	996351	·27	114521	16·43	885479	35
26	111842	16·12	996335	·27	115507	16·39	884493	34
27	112809	16·08	996318	·27	116491	16·36	883509	33
28	113774	16·05	996302	·28	117472	16·32	882528	32
29	114737	16·01	996285	·28	118452	16·29	881548	31
30	115698	15·97	996269	·28	119429	16·25	880571	30
31	9·116656	15·94	9·996252	·28	9·120404	16·22	10·879596	29
32	117613	15·90	996235	·28	121377	16·18	878623	28
33	118567	15·87	996219	·28	122348	16·15	877652	27
34	119519	15·83	996202	·28	123317	16·11	876683	26
35	120469	15·80	996185	·28	124284	16·07	875716	25
36	121417	15·76	996168	·28	125249	16·04	874751	24
37	122362	15·73	996151	·28	126211	16·01	873789	23
38	123306	15·69	996134	·28	127172	15·97	872828	22
39	124248	15·66	996117	·28	128130	15·94	871870	21
40	125187	15·62	996100	·28	129087	15·91	870913	20
41	9·126125	15·59	9·996083	·29	9·130041	15·87	10·869959	19
42	127060	15·56	996066	·29	130994	15·84	869006	18
43	127993	15·52	996049	·29	131944	15·81	868056	17
44	128925	15·49	996032	·29	132893	15·77	867107	16
45	129854	15·45	996015	·29	133839	15·74	866161	15
46	130781	15·42	995998	·29	134784	15·71	865216	14
47	131706	15·39	995980	·29	135726	15·67	864274	13
48	132630	15·35	995963	·29	136667	15·64	863333	12
49	133551	15·32	995946	·29	137605	15·61	862395	11
50	134470	15·29	995928	·29	138542	15·58	861458	10
51	9·135387	15·25	9·995911	·29	9·139476	15·55	10·860524	9
52	136303	15·22	995894	·29	140409	15·51	859591	8
53	137216	15·19	995876	·29	141340	15·48	858660	7
54	138128	15·16	995859	·29	142269	15·45	857731	6
55	139037	15·12	995841	·29	143196	15·42	856804	5
56	139944	15·09	995823	·29	144121	15·39	855879	4
57	140850	15·06	995806	·29	145044	15·35	854956	3
58	141754	15·03	995788	·29	145966	15·32	854034	2
59	142655	15·00	995771	·29	146885	15·29	853115	1
60	143555	14·96	995753	·29	147803	15·26	852197	0
	Cosine.	D.	Sine.		Cotang.	D.	Tang.	M.

M.	Sine.	D.	Cosine.	D.	Tang.	D.	Cotang.	
0	9·143555	14·96	9·995753	·30	9·147803	15·26	10·852197	60
1	144453	14·93	995735	·30	148718	15·23	851282	59
2	145349	14·90	995717	·30	149632	15·20	850368	58
3	146243	14·87	995699	·30	150544	15·17	849456	57
4	147136	14·84	995681	·30	151454	15·14	848546	56
5	148026	14·81	995664	·30	152363	15·11	847637	55
6	148915	14·78	995646	·30	153269	15·08	846731	54
7	149802	14·75	995628	·30	154174	15·05	845826	53
8	150686	14·72	995610	·30	155077	15·02	844923	52
9	151569	14·69	995591	·30	155978	14·99	844022	51
10	152451	14·66	995573	·30	156877	14·96	843123	50
11	9·153330	14·63	9·995555	·30	9·157775	14·93	10·842225	49
12	154208	14·60	995537	·30	158671	14·90	841339	48
13	155088	14·57	995519	·30	159565	14·87	840435	47
14	155957	14·54	995501	·31	160457	14·84	839543	46
15	156830	14·51	995482	·31	161347	14·81	838653	45
16	157700	14·48	995464	·31	162236	14·79	837764	44
17	158569	14·45	995446	·31	163123	14·76	836877	43
18	159435	14·42	995427	·31	164008	14·73	835992	42
19	160301	14·39	995409	·31	164892	14·70	835108	41
20	161164	14·36	995390	·31	165774	14·67	834226	40
21	9·162025	14·33	9·995372	·31	9·166654	14·64	10·833346	39
22	162885	14·30	995353	·31	167532	14·61	832468	38
23	163743	14·27	995334	·31	168409	14·58	831591	37
24	164600	14·24	995316	·31	169284	14·55	830716	36
25	165454	14·22	995297	·31	170157	14·53	829843	35
26	166307	14·19	995278	·31	171029	14·50	828971	34
27	167159	14·16	995260	·31	171899	14·47	828101	33
28	168008	14·13	995241	·32	172767	14·44	827233	32
29	168856	14·10	995222	·32	173634	14·42	826366	31
30	169702	14·07	995203	·32	174499	14·39	825501	30
31	9·170547	14·05	9·995184	·32	9·175362	14·36	10·824638	29
32	171389	14·02	995165	·32	176224	14·33	823776	28
33	172230	13·99	995146	·32	177084	14·31	822916	27
34	173070	13·96	995127	·32	177942	14·28	822058	26
35	173908	13·94	995108	·32	178799	14·25	821201	25
36	174744	13·91	995089	·32	179655	14·23	820345	24
37	175578	13·88	995070	·32	180508	14·20	819492	23
38	176411	13·86	995051	·32	181360	14·17	818640	22
39	177242	13·83	995032	·32	182211	14·15	817789	21
40	178072	13·80	995013	·32	183059	14·12	816941	20
41	9·178900	13·77	9·994998	·32	9·183907	14·09	10·816093	19
42	179726	13·74	994974	·32	184752	14·07	815248	18
43	180551	13·72	994955	·32	185597	14·04	814403	17
44	181374	13·69	994935	·32	186439	14·02	813561	16
45	182196	13·66	994916	·33	187280	13·99	812720	15
46	183016	13·64	994896	·33	188120	13·96	811880	14
47	183834	13·61	994877	·33	188958	13·93	811042	13
48	184651	13·59	994857	·33	189794	13·91	810206	12
49	185466	13·56	994838	·33	190629	13·89	809371	11
50	186280	13·53	994818	·33	191462	13·86	808538	10
51	9·187092	13·51	9·994798	·33	9·192294	13·84	10·807706	9
52	187903	13·48	994779	·33	193124	13·81	806876	8
53	188712	13·46	994759	·33	193953	13·79	806047	7
54	189519	13·43	994739	·33	194780	13·76	805220	6
55	190325	13·41	994719	·33	195606	13·74	804394	5
56	191130	13·38	994700	·33	196430	13·71	803570	4
57	191933	13·36	994680	·33	197253	13·69	802747	3
58	192734	13·33	994660	·33	198074	13·66	801926	2
59	193534	13·30	994640	·33	198894	13·64	801106	1
60	194332	13·28	994620	·33	199713	13·61	800287	0
	Cosine.	D.	Sine.	D.	Cotang.	D.	Tang.	M.

M.	Sine.	D.	Cosine.	D.	Tang.	D.	Cotang.	
0	9·194332	13·28	9·994620	·38	9·199718	13·61	10·800287	60
1	195129	13·26	994600	·38	200529	13·59	799471	59
2	195925	13·28	994580	·38	201845	13·56	798655	58
3	196719	13·21	994560	·34	202159	13·54	797841	57
4	197511	13·18	994540	·34	202971	13·52	797029	56
5	198302	13·16	994519	·34	203782	13·49	796218	55
6	199091	13·13	994499	·34	204592	13·47	795408	54
7	199879	13·11	994479	·34	205400	13·45	794600	53
8	200666	13·08	994459	·34	206207	13·42	793793	52
9	201451	13·06	994438	·34	207013	13·40	792987	51
10	202234	13·04	994418	·34	207817	13·38	792183	50
11	9·203017	13·01	9·994397	·34	9·208619	13·35	10·791381	49
12	203797	12·99	994377	·34	209420	13·33	790580	48
13	204577	12·96	994357	·34	210220	13·31	789780	47
14	205354	12·94	994336	·34	211018	13·28	788982	46
15	206131	12·92	994316	·34	211815	13·26	788185	45
16	206906	12·89	994295	·34	212611	13·24	787389	44
17	207679	12·87	994274	·35	213405	13·21	786595	43
18	208452	12·85	994254	·35	214198	13·19	785802	42
19	209222	12·82	994233	·35	214989	13·17	785011	41
20	209992	12·80	994212	·35	215780	13·15	784220	40
21	9·210760	12·78	9·994191	·35	9·216568	13·12	10·783432	39
22	211526	12·75	994171	·35	217356	13·10	782644	38
23	212291	12·73	994150	·35	218142	13·08	781858	37
24	213055	12·71	994129	·35	218926	13·05	781074	36
25	213818	12·68	994108	·35	219710	13·03	780290	35
26	214579	12·66	994087	·35	220492	13·01	779508	34
27	215338	12·64	994066	·35	221272	12·99	778728	33
28	216097	12·61	994045	·35	222052	12·97	777948	32
29	216854	12·59	994024	·35	222830	12·94	777170	31
30	217609	12·57	994003	·35	223606	12·92	776394	30
31	9·218363	12·55	9·993981	·35	9·224382	12·90	10·775618	29
32	219116	12·53	993960	·35	225156	12·88	774844	28
33	219868	12·50	993939	·35	225929	12·86	774071	27
34	220618	12·48	993918	·35	226700	12·84	773300	26
35	221367	12·46	993896	·36	227471	12·81	772529	25
36	222115	12·44	993875	·36	228239	12·79	771761	24
37	222861	12·42	993854	·36	229007	12·77	770993	23
38	223606	12·39	993832	·36	229773	12·75	770227	22
39	224349	12·37	993811	·36	230539	12·73	769461	21
40	225092	12·35	993789	·36	231302	12·71	768698	20
41	9·225833	12·33	9·993768	·36	9·232065	12·69	10·767985	19
42	226573	12·31	993746	·36	232826	12·67	767174	18
43	227311	12·28	993725	·36	233586	12·65	766414	17
44	228048	12·26	993703	·36	234345	12·62	765655	16
45	228784	12·24	993681	·36	235103	12·60	764897	15
46	229518	12·22	993660	·36	235859	12·58	764141	14
47	230252	12·20	993638	·36	236614	12·56	763386	13
48	230984	12·18	993616	·36	237368	12·54	762632	12
49	231714	12·16	993594	·37	238120	12·52	761880	11
50	232444	12·14	993572	·37	238872	12·50	761128	10
51	9·233172	12·12	9·993550	·37	9·239622	12·48	10·760378	9
52	233899	12·09	993528	·37	240371	12·46	759629	8
53	234625	12·07	993506	·37	241118	12·44	758882	7
54	235349	12·05	993484	·37	241865	12·42	758135	6
55	236073	12·03	993462	·37	242610	12·40	757390	5
56	236795	12·01	993440	·37	243354	12·38	756646	4
57	237515	11·99	993418	·37	244097	12·36	755903	3
58	238235	11·97	993396	·37	244839	12·34	755161	2
59	238953	11·95	993374	·37	245579	12·32	754421	1
60	239670	11·93	993351	·37	246319	12·30	753681	0
	Cosine.	D.	Sine.		Cotang.	D.	Tang.	M.

M.	Sine.	D.	Cosine.	D.	Tang.	D.	Cotang.	
0	9·239670	11·93	9·993351	·37	9·246319	12·30	10·753681	60
1	240386	11·91	993329	·37	247057	12·28	752943	59
2	241101	11·89	993307	·37	247794	12·26	752206	58
3	241814	11·87	993285	·37	248530	12·24	751470	57
4	242526	11·85	993262	·37	249264	12·22	750786	56
5	243237	11·83	993240	·37	249998	12·20	750002	55
6	243947	11·81	993217	·38	250730	12·18	749270	54
7	244656	11·79	993195	·38	251461	12·17	748539	53
8	245363	11·77	993172	·38	252191	12·15	747809	52
9	246069	11·75	993149	·38	252920	12·13	747080	51
10	246775	11·73	993127	·38	253648	12·11	746352	50
11	9·247478	11·71	9·993104	·38	9·254374	12·09	10·745626	49
12	248181	11·69	993081	·38	255100	12·07	744900	48
13	248883	11·67	993059	·38	255824	12·05	744176	47
14	249583	11·65	993036	·38	256547	12·03	743453	46
15	250282	11·63	993013	·38	257269	12·01	742731	45
16	250980	11·61	992990	·38	257990	12·00	742010	44
17	251677	11·59	992967	·38	258710	11·98	741290	43
18	252373	11·58	992944	·38	259429	11·96	740571	42
19	253067	11·56	992921	·38	260146	11·94	739854	41
20	253761	11·54	992898	·38	260863	11·92	739137	40
21	9·254453	11·52	9·992875	·38	9·261578	11·90	10·738422	39
22	255144	11·50	992852	·38	262292	11·89	737708	38
23	255834	11·48	992829	·39	263005	11·87	736995	37
24	256523	11·46	992806	·39	263717	11·85	736283	36
25	257211	11·44	992783	·39	264428	11·83	735572	35
26	257898	11·42	992759	·39	265138	11·81	734862	34
27	258583	11·41	992736	·39	265847	11·79	734153	33
28	259268	11·39	992713	·39	266555	11·78	733445	32
29	259951	11·37	992690	·39	267261	11·76	732739	31
30	260633	11·35	992666	·39	267967	11·74	732033	30
31	9·261314	11·33	9·992643	·39	9·268671	11·72	10·731329	29
32	261994	11·31	992619	·39	269375	11·70	730625	28
33	262673	11·30	992596	·39	270077	11·69	729923	27
34	263351	11·28	992572	·39	270779	11·67	729221	26
35	264027	11·26	992549	·39	271479	11·65	728521	25
36	264703	11·24	992525	·39	272178	11·64	727822	24
37	265377	11·22	992501	·39	272876	11·62	727124	23
38	266051	11·20	992478	·40	273573	11·60	726427	22
39	266723	11·19	992454	·40	274269	11·58	725731	21
40	267395	11·17	992430	·40	274964	11·57	725036	20
41	9·268065	11·15	9·992406	·40	9·275658	11·55	10·724842	19
42	268734	11·13	992382	·40	276351	11·53	723649	18
43	269402	11·11	992359	·40	277043	11·51	722957	17
44	270069	11·10	992335	·40	277734	11·50	722266	16
45	270735	11·08	992311	·40	278424	11·48	721576	15
46	271400	11·06	992287	·40	279113	11·47	720887	14
47	272064	11·05	992263	·40	279801	11·45	720199	13
48	272726	11·03	992239	·40	280488	11·43	719512	12
49	273388	11·01	992214	·40	281174	11·41	718826	11
50	274049	10·99	992190	·40	281858	11·40	718142	10
51	9·274708	10·98	9·992166	·40	9·282542	11·38	10·717458	9
52	275367	10·96	992142	·40	283225	11·36	716775	8
53	276024	10·94	992117	·41	283907	11·35	716093	7
54	276681	10·92	992093	·41	284588	11·33	715412	6
55	277337	10·91	992069	·41	285268	11·31	714732	5
56	277991	10·89	992044	·41	285947	11·30	714053	4
57	278644	10·87	992020	·41	286624	11·28	713376	3
58	279297	10·86	991996	·41	287301	11·26	712699	2
59	279948	10·84	991971	·41	287977	11·25	712023	1
60	280599	10·82	991947	·41	288652	11·23	711348	0
	Cosine.	D.	Sine.		Cotang.	D.	Tang.	M.

M.	Sine.	D.	Cosine.	D.	Tang.	D.	Cotang.	
0	9·280599	10·82	9·991947	·41	9·288652	11·23	10·711848	60
1	281248	10·81	991922	·41	289326	11·22	710674	59
2	281897	10·79	991897	·41	289999	11·20	710001	58
3	282544	10·77	991873	·41	290671	11·18	709329	57
4	283190	10·76	991848	·41	291342	11·17	708658	56
5	283836	10·74	991823	·41	292013	11·15	707987	55
6	284480	10·72	991799	·41	292682	11·14	707318	54
7	285124	10·71	991774	·42	293350	11·12	706650	53
8	285766	10·69	991749	·42	294017	11·11	705983	52
9	286408	10·67	991724	·42	294684	11·09	705316	51
10	287048	10·66	991699	·42	295349	11·07	704651	50
11	9·287687	10·64	9·991674	·42	9·296013	11·06	10·703987	49
12	288326	10·63	991649	·42	296677	11·04	703323	48
13	288964	10·61	991624	·42	297339	11·03	702661	47
14	289600	10·59	991599	·42	298001	11·01	701999	46
15	290236	10·58	991574	·42	298662	11·00	701338	45
16	290870	10·56	991549	·42	299322	10·98	700678	44
17	291504	10·54	991524	·42	299980	10·96	700020	43
18	292137	10·53	991498	·42	300638	10·95	699362	42
19	292768	10·51	991473	·42	301295	10·93	698705	41
20	293399	10·50	991448	·42	301951	10·92	698049	40
21	9·294029	10·48	9·991422	·42	9·302607	10·90	10·697393	39
22	294658	10·46	991397	·42	303261	10·89	696739	38
23	295286	10·45	991372	·43	303914	10·87	696086	37
24	295913	10·43	991346	·43	304567	10·86	695433	36
25	296539	10·42	991321	·43	305218	10·84	694782	35
26	297164	10·40	991295	·43	305869	10·83	694131	34
27	297788	10·39	991270	·43	306519	10·81	693481	33
28	298412	10·37	991244	·43	307168	10·80	692832	32
29	299034	10·36	991218	·43	307815	10·78	692185	31
30	299655	10·34	991193	·43	308463	10·77	691537	30
31	9·300276	10·32	9·991167	·43	9·309109	10·75	10·690891	29
32	300895	10·31	991141	·43	309754	10·74	690246	28
33	301514	10·29	991115	·43	310398	10·73	689602	27
34	302132	10·28	991090	·43	311042	10·71	688958	26
35	302748	10·26	991064	·43	311685	10·70	688315	25
36	303364	10·25	991038	·43	312327	10·68	687673	24
37	303979	10·23	991012	·43	312967	10·67	687033	23
38	304593	10·22	990986	·43	313608	10·65	686392	22
39	305207	10·20	990960	·43	314247	10·64	685753	21
40	305819	10·19	990934	·44	314885	10·62	685115	20
41	9·306430	10·17	9·990908	·44	9·315523	10·61	10·684477	19
42	307041	10·16	990882	·44	316159	10·60	683841	18
43	307650	10·14	990855	·44	316795	10·58	683205	17
44	308259	10·13	990829	·44	317430	10·57	682570	16
45	308867	10·11	990803	·44	318064	10·55	681936	15
46	309474	10·10	990777	·44	318697	10·54	681303	14
47	310080	10·08	990750	·44	319329	10·53	680671	13
48	310685	10·07	990724	·44	319961	10·51	680039	12
49	311289	10·05	990697	·44	320592	10·50	679408	11
50	311893	10·04	990671	·44	321222	10·48	678778	10
51	9·312495	10·03	9·990644	·44	9·321851	10·47	10·678149	9
52	313097	10·01	990618	·44	322479	10·45	677521	8
53	313698	10·00	990591	·44	323106	10·44	676894	7
54	314297	9·98	990565	·44	323733	10·43	676267	6
55	314897	9·97	990538	·44	324358	10·41	675642	5
56	315495	9·96	990511	·45	324983	10·40	675017	4
57	316092	9·94	990485	·45	325607	10·39	674393	3
58	316689	9·93	990458	·45	326231	10·37	673769	2
59	317284	9·91	990431	·45	326853	10·36	673147	1
60	317879	9·90	990404	·45	327475	10·35	672525	0
	Cosine.	D.	Sine.		Cotang.	D.	Tang.	M.

(78 DEGREES.)

M.	Sine.	D.	Cosine.	D.	Tang.	D.	Cotang.	
0	9·317879	9·90	9·990404	·45	9·327474	10·35	10·672526	60
1	318473	9·88	990378	·45	328095	10·33	671905	59
2	319066	9·87	990351	·45	328715	10·32	671285	58
3	319658	9·86	990324	·45	329334	10·30	670666	57
4	320249	9·84	990297	·45	329953	10·29	670047	56
5	320840	9·83	990270	·45	330570	10·28	669430	55
6	321430	9·82	990243	·45	331187	10·26	668813	54
7	322019	9·80	990215	·45	331803	10·25	668197	53
8	322607	9·79	990188	·45	332418	10·24	667582	52
9	323194	9·77	990161	·45	333033	10·23	666967	51
10	323780	9·76	990134	·45	333646	10·21	666354	50
11	9·324366	9·75	9·990107	·46	9·334259	10·20	10·665741	49
12	324950	9·73	990079	·46	334871	10·19	665129	48
13	325534	9·72	990052	·46	335482	10·17	664518	47
14	326117	9·70	990025	·46	336093	10·16	663907	46
15	326700	9·69	989997	·46	336702	10·15	663298	45
16	327281	9·68	989970	·46	337311	10·13	662689	44
17	327862	9·66	989942	·46	337919	10·12	662081	43
18	328442	9·65	989915	·46	338527	10·11	661473	42
19	329021	9·64	989887	·46	339133	10·10	660867	41
20	329599	9·62	989860	·46	339739	10·08	660261	40
21	9·330176	9·61	9·989832	·46	9·340344	10·07	10·659656	39
22	330753	9·60	989804	·46	340948	10·06	659052	38
23	331329	9·58	989777	·46	341552	10·04	658448	37
24	331903	9·57	989749	·47	342155	10·03	657845	36
25	332478	9·56	989721	·47	342757	10·02	657243	35
26	333051	9·54	989693	·47	343358	10·00	656642	34
27	333624	9·53	989665	·47	343958	9·99	656042	33
28	334195	9·52	989637	·47	344558	9·98	655442	32
29	334766	9·50	989609	·47	345157	9·97	654843	31
30	335337	9·49	989581	·47	345755	9·96	654245	30
31	9·335906	9·48	9·989553	·47	9·346353	9·94	10·653647	29
32	336475	9·46	989525	·47	346949	9·93	653051	28
33	337043	9·45	989497	·47	347545	9·92	652455	27
34	337610	9·44	989469	·47	348141	9·91	651859	26
35	338176	9·43	989441	·47	348735	9·90	651265	25
36	338742	9·41	989413	·47	349329	9·88	650671	24
37	339306	9·40	989384	·47	349922	9·87	650078	23
38	339871	9·39	989356	·47	350514	9·86	649486	22
39	340434	9·37	989328	·47	351106	9·85	648894	21
40	340996	9·36	989300	·47	351697	9·83	648303	20
41	9·341558	9·35	9·989271	·47	9·352287	9·82	10·647713	19
42	342119	9·34	989243	·47	352876	9·81	647124	18
43	342679	9·32	989214	·47	353465	9·80	646535	17
44	343239	9·31	989186	·47	354053	9·79	645947	16
45	343797	9·30	989157	·47	354640	9·77	645360	15
46	344355	9·29	989128	·48	355227	9·76	644773	14
47	344912	9·27	989100	·48	355813	9·75	644187	13
48	345469	9·26	989071	·48	356398	9·74	643602	12
49	346024	9·25	989042	·48	356982	9·73	643018	11
50	346579	9·24	989014	·48	357566	9·71	642434	10
51	9·347134	9·22	9·988985	·48	9·358149	9·70	10·641851	9
52	347687	9·21	988956	·48	358731	9·69	641269	8
53	348240	9·20	988927	·48	359313	9·68	640687	7
54	348792	9·19	988898	·48	359893	9·67	640107	6
55	349343	9·17	988869	·48	360474	9·66	639526	5
56	349893	9·16	988840	·48	361053	9·65	638947	4
57	350443	9·15	988811	·49	361632	9·63	638368	3
58	350992	9·14	988782	·49	362210	9·62	637790	2
59	351540	9·13	988753	·49	362787	9·61	637213	1
60	352088	9·11	988724	·49	363364	9·60	636636	0
	Cosine.	D.	Sine.		Cotang.	D.	Tang.	M.

M.	Sine.	D.	Cosine.	D.	Tang.	D.	Cotang.	
0	9·352088	9·11	9·988724	·49	9·363364	9·60	10·636636	60
1	352635	9·10	988695	·49	363940	9·59	636060	59
2	353181	9·09	988666	·49	364515	9·58	635485	58
3	353726	9·08	988636	·49	365090	9·57	634910	57
4	354271	9·07	988607	·49	365664	9·55	634336	56
5	354815	9·05	988578	·49	366287	9·54	633763	55
6	355358	9·04	988548	·49	366810	9·53	633190	54
7	355901	9·03	988519	·49	367382	9·52	632618	53
8	356443	9·02	988489	·49	367953	9·51	632047	52
9	356984	9·01	988460	·49	368524	9·50	631476	51
10	357524	8·99	988430	·49	369094	9·49	630906	50
11	9·358064	8·98	9·988401	·49	9·369663	9·48	10·630337	49
12	358603	8·97	988371	·49	370232	9·46	629768	48
13	359141	8·96	988342	·49	370799	9·45	629201	47
14	359678	8·95	988312	·50	371367	9·44	628633	46
15	360215	8·93	988282	·50	371933	9·43	628067	45
16	360752	8·92	988252	·50	372499	9·42	627501	44
17	361287	8·91	988223	·50	373064	9·41	626936	43
18	361822	8·90	988193	·50	373629	9·40	626371	42
19	362356	8·89	988163	·50	374193	9·39	625807	41
20	362889	8·88	988133	·50	374756	9·38	625244	40
21	9·363422	8·87	9·988103	·50	9·375319	9·37	10·624681	39
22	363954	8·85	988073	·50	375881	9·35	624119	38
23	364485	8·84	988043	·50	376442	9·34	623558	37
24	365016	8·83	988013	·50	377003	9·33	622997	36
25	365546	8·82	987983	·50	377563	9·32	622437	35
26	366075	8·81	987953	·50	378122	9·31	621878	34
27	366604	8·80	987922	·50	378681	9·30	621319	33
28	367131	8·79	987892	·50	379239	9·29	620761	32
29	367659	8·77	987862	·50	379797	9·28	620203	31
30	368185	8·76	987832	·51	380354	9·27	619646	30
31	9·368711	8·75	9·987801	·51	9·380910	9·26	10·619090	29
32	369236	8·74	987771	·51	381466	9·25	618534	28
33	369761	8·73	987740	·51	382020	9·24	617980	27
34	370285	8·72	987710	·51	382575	9·23	617425	26
35	370808	8·71	987679	·51	383129	9·22	616871	25
36	371330	8·70	987649	·51	383682	9·21	616318	24
37	371852	8·69	987618	·51	384234	9·20	615766	23
38	372373	8·67	987588	·51	384786	9·19	615214	22
39	372894	8·66	987557	·51	385337	9·18	614663	21
40	373414	8·65	987526	·51	385888	9·17	614112	20
41	9·373933	8·64	9·987496	·51	9·386438	9·15	10·613562	19
42	374452	8·63	987465	·51	386987	9·14	613013	18
43	374970	8·62	987434	·51	387536	9·13	612464	17
44	375487	8·61	987403	·52	388084	9·12	611916	16
45	376003	8·60	987372	·52	388631	9·11	611369	15
46	376519	8·59	987341	·52	389178	9·10	610822	14
47	377035	8·58	987310	·52	389724	9·09	610276	13
48	377549	8·57	987279	·52	390270	9·08	609730	12
49	378063	8·56	987248	·52	390815	9·07	609185	11
50	378577	8·54	987217	·52	391360	9·06	608640	10
51	9·379089	8·53	9·987186	·52	9·391903	9·05	10·608097	9
52	379601	8·52	987155	·52	392447	9·04	607553	8
53	380113	8·51	987124	·52	392989	9·03	607011	7
54	380624	8·50	987092	·52	393531	9·02	606469	6
55	381134	8·49	987061	·52	394073	9·01	605927	5
56	381643	8·48	987030	·52	394614	9·00	605386	4
57	382152	8·47	986998	·52	395154	8·99	604846	3
58	382661	8·46	986967	·52	395694	8·98	604306	2
59	383168	8·45	986936	·52	396233	8·97	603767	1
60	383675	8·44	986904	·52	396771	8·96	603229	0
	Cosine.	D.	Sine.		Cotang.	D.	Tang.	M.

M.	Sine.	D.	Cosine.	D.	Tang.	D.	Cotang.	
0	9·383675	8·44	9·986904	·52	9·396771	8·96	10·603229	60
1	384182	8·43	986873	·53	397309	8·96	602691	59
2	384687	8·42	986841	·53	397846	8·95	602154	58
3	385192	8·41	986809	·53	398383	8·94	601617	57
4	385697	8·40	986778	·53	398919	8·93	601081	56
5	386201	8·39	986746	·53	399455	8·92	600545	55
6	386704	8·38	986714	·53	399990	8·91	600010	54
7	387207	8·37	986683	·53	400524	8·90	599476	53
8	387709	8·36	986651	·53	401058	8·89	598942	52
9	388210	8·35	986619	·53	401591	8·88	598409	51
10	388711	8·34	986587	·53	402124	8·87	597876	50
11	9·389211	8·33	9·986555	·53	9·402656	8·86	10·597344	49
12	389711	8·32	986523	·53	403187	8·85	596813	48
13	390210	8·31	986491	·53	403718	8·84	596282	47
14	390708	8·30	986459	·53	404249	8·83	595751	46
15	391206	8·28	986427	·53	404778	8·82	595222	45
16	391703	8·27	986395	·53	405308	8·81	594692	44
17	392199	8·26	986363	·54	405836	8·80	594164	43
18	392695	8·25	986331	·54	406364	8·79	593636	42
19	393191	8·24	986299	·54	406892	8·78	593108	41
20	393685	8·23	986266	·54	407419	8·77	592581	40
21	9·394179	8·22	9·986234	·54	9·407945	8·76	10·592055	39
22	394673	8·21	986202	·54	408471	8·75	591529	38
23	395166	8·20	986169	·54	408997	8·74	591003	37
24	395658	8·19	986137	·54	409521	8·74	590479	36
25	396150	8·18	986104	·54	410045	8·73	589955	35
26	396641	8·17	986072	·54	410569	8·72	589431	34
27	397132	8·17	986039	·54	411092	8·71	588908	33
28	397621	8·16	986007	·54	411615	8·70	588385	32
29	398111	8·15	985974	·54	412137	8·69	587863	31
30	398600	8·14	985942	·54	412658	8·68	587342	30
31	9·399088	8·13	9·985909	·55	9·413179	8·67	10·586821	29
32	399575	8·12	985876	·55	413699	8·66	586301	28
33	400062	8·11	985843	·55	414219	8·65	585781	27
34	400549	8·10	985811	·55	414738	8·64	585262	26
35	401035	8·09	985778	·55	415257	8·64	584743	25
36	401520	8·08	985745	·55	415775	8·63	584225	24
37	402005	8·07	985712	·55	416293	8·62	583707	23
38	402489	8·06	985679	·55	416810	8·61	583190	22
39	402972	8·05	985646	·55	417326	8·60	582674	21
40	403455	8·04	985613	·55	417842	8·59	582158	20
41	9·403938	8·03	9·985580	·55	9·418358	8·58	10·581642	19
42	404420	8·02	985547	·55	418873	8·57	581127	18
43	404901	8·01	985514	·55	419387	8·56	580613	17
44	405382	8·00	985480	·55	419901	8·55	580099	16
45	405862	7·99	985447	·55	420415	8·55	579585	15
46	406341	7·98	985414	·56	420927	8·54	579073	14
47	406820	7·97	985380	·56	421440	8·53	578560	13
48	407299	7·96	985347	·56	421952	8·52	578048	12
49	407777	7·95	985314	·56	422463	8·51	577537	11
50	408254	7·94	985280	·56	422974	8·50	577026	10
51	9·408731	7·94	9·985247	·56	9·423484	8·49	10·576516	9
52	409207	7·93	985213	·56	423993	8·48	576007	8
53	409682	7·92	985180	·56	424503	8·48	575497	7
54	410157	7·91	985146	·56	425011	8·47	574989	6
55	410632	7·90	985113	·56	425519	8·46	574481	5
56	411106	7·89	985079	·56	426027	8·45	573973	4
57	411579	7·88	985045	·56	426534	8·44	573466	3
58	412052	7·87	985011	·56	427041	8·43	572959	2
59	412524	7·86	984978	·56	427547	8·43	572453	1
60	412996	7·85	984944	·56	428052	8·42	571948	0
	Cosine.	D.	Sine.		Cotang.	D.	Tang.	M.

M.	Sine.	D.	Cosine.	D.	Tang.	D.	Cotang.	
0	9·412996	7·85	9·984944	·57	9·428052	8·42	10·571948	60
1	413467	7·84	984910	·57	428557	8·41	571443	59
2	413938	7·83	984876	·57	429062	8·40	570938	58
3	414408	7·83	984842	·57	429566	8·39	570434	57
4	414878	7·82	984808	·57	430070	8·38	569930	56
5	415347	7·81	984774	·57	430573	8·38	569427	55
6	415815	7·80	984740	·57	431075	8·37	568925	54
7	416283	7·79	984706	·57	431577	8·36	568423	53
8	416751	7·78	984672	·57	432079	8·35	567921	52
9	417217	7·77	984637	·57	432580	8·34	567420	51
10	417684	7·76	984603	·57	433080	8·33	566920	50
11	9·418150	7·75	9·984569	·57	9·433580	8·32	10·566420	49
12	418615	7·74	984535	·57	434080	8·32	565920	48
13	419079	7·73	984500	·57	434579	8·31	565421	47
14	419544	7·73	984466	·57	435078	8·30	564922	46
15	420007	7·72	984432	·58	435576	8·29	564424	45
16	420470	7·71	984397	·58	436073	8·28	563927	44
17	420933	7·70	984363	·58	436570	8·28	563430	43
18	421395	7·69	984328	·58	437067	8·27	562933	42
19	421857	7·68	984294	·58	437563	8·26	562437	41
20	422318	7·67	984259	·58	438059	8·25	561941	40
21	9·422778	7·67	9·984224	·58	9·438554	8·24	10·561446	39
22	423238	7·66	984190	·58	439048	8·23	560952	38
23	423697	7·65	984155	·58	439543	8·23	560457	37
24	424156	7·64	984120	·58	440036	8·22	559964	36
25	424615	7·63	984085	·58	440529	8·21	559471	35
26	425073	7·62	984050	·58	441022	8·20	558978	34
27	425530	7·61	984015	·58	441514	8·19	558486	33
28	425987	7·60	983981	·58	442006	8·19	557994	32
29	426443	7·60	983946	·58	442497	8·18	557503	31
30	426899	7·59	983911	·58	442988	8·17	557012	30
31	9·427354	7·58	9·983875	·58	9·443479	8·16	10·556521	29
32	427809	7·57	983840	·59	443968	8·16	556032	28
33	428263	7·56	983805	·59	444458	8·15	555542	27
34	428717	7·55	983770	·59	444947	8·14	555053	26
35	429170	7·54	983735	·59	445435	8·13	554565	25
36	429623	7·53	983700	·59	445923	8·12	554077	24
37	430075	7·52	983664	·59	446411	8·12	553589	23
38	430527	7·52	983629	·59	446898	8·11	553102	22
39	430978	7·51	983594	·59	447384	8·10	552616	21
40	431429	7·50	983558	·59	447870	8·09	552130	20
41	9·431879	7·49	9·983523	·59	9·448356	8·09	10·551644	19
42	432329	7·49	983487	·59	448841	8·08	551159	18
43	432778	7·48	983452	·59	449326	8·07	550674	17
44	433226	7·47	983416	·59	449810	8·06	550190	16
45	433675	7·46	983381	·59	450294	8·06	549706	15
46	434122	7·45	983345	·59	450777	8·05	549223	14
47	434569	7·44	983309	·59	451260	8·04	548740	13
48	435016	7·44	983273	·60	451743	8·03	548257	12
49	435462	7·43	983238	·60	452225	8·02	547775	11
50	435908	7·42	983202	·60	452706	8·02	547294	10
51	9·436353	7·41	9·983166	·60	9·453187	8·01	10·546813	9
52	436798	7·40	983130	·60	453668	8·00	546332	8
53	437242	7·40	983094	·60	454148	7·99	545852	7
54	437686	7·39	983058	·60	454628	7·99	545372	6
55	438129	7·38	983022	·60	455107	7·98	544893	5
56	438572	7·37	982986	·60	455586	7·97	544414	4
57	439014	7·36	982950	·60	456064	7·96	543936	3
58	439456	7·36	982914	·60	456542	7·96	543458	2
59	439897	7·35	982878	·60	457019	7·95	542981	1
60	440338	7·34	982842	·60	457496	7·94	542504	0
	Cosine.	D.	Sine.		Cotang.	D.	Tang.	M.

M.	Sine.	D.	Cosine.	D.	Tang.	D.	Cotang.	
0	9·440338	7·34	9·982842	·60	9·457496	7·94	10·542504	60
1	440778	7·33	982805	·60	457973	7·93	542027	59
2	441218	7·32	982769	·61	458449	7·93	541551	58
3	441658	7·31	982733	·61	458925	7·92	541075	57
4	442096	7·31	982696	·61	459400	7·91	540600	56
5	442535	7·30	982660	·61	459875	7·90	540125	55
6	442973	7·29	982624	·61	460349	7·90	539651	54
7	443410	7·28	982587	·61	460823	7·89	539177	53
8	443847	7·27	982551	·61	461297	7·88	538703	52
9	444284	7·27	982514	·61	461770	7·88	538230	51
10	444720	7·26	982477	·61	462242	7·87	537758	50
11	9·445155	7·25	9·982441	·61	9·462714	7·86	10·537286	49
12	445590	7·24	982404	·61	463186	7·85	536814	48
13	446025	7·23	982367	·61	463658	7·85	536342	47
14	446459	7·23	982331	·61	464129	7·84	535871	46
15	446893	7·22	982294	·61	464599	7·83	535401	45
16	447326	7·21	982257	·61	465069	7·83	534931	44
17	447759	7·20	982220	·62	465539	7·82	534461	43
18	448191	7·20	982183	·62	466008	7·81	533992	42
19	448623	7·19	982146	·62	466476	7·80	533524	41
20	449054	7·18	982109	·62	466945	7·80	533055	40
21	9·449485	7·17	9·982072	·62	9·467413	7·79	10·532587	39
22	449915	7·16	982035	·62	467880	7·78	532120	38
23	450345	7·16	981998	·62	468347	7·78	531653	37
24	450775	7·15	981961	·62	468814	7·77	531186	36
25	451204	7·14	981924	·62	469280	7·76	530720	35
26	451632	7·13	981886	·62	469746	7·75	530254	34
27	452060	7·13	981849	·62	470211	7·75	529789	33
28	452488	7·12	981812	·62	470676	7·74	529324	32
29	452915	7·11	981774	·62	471141	7·73	528859	31
30	453342	7·10	981737	·62	471605	7·73	528395	30
31	9·453768	7·10	9·981699	·63	9·472068	7·72	10·527932	29
32	454194	7·09	981662	·63	472532	7·71	527468	28
33	454619	7·08	981625	·63	472995	7·71	527005	27
34	455044	7·07	981587	·63	473457	7·70	526543	26
35	455469	7·07	981549	·63	473919	7·69	526081	25
36	455893	7·06	981512	·63	474381	7·69	525619	24
37	456316	7·05	981474	·63	474842	7·68	525158	23
38	456739	7·04	981436	·63	475303	7·67	524697	22
39	457162	7·04	981399	·63	475763	7·67	524237	21
40	457584	7·03	981361	·63	476223	7·66	523777	20
41	9·458006	7·02	9·981323	·63	9·476683	7·65	10·523317	19
42	458427	7·01	981285	·63	477142	7·65	522858	18
43	458848	7·01	981247	·63	477601	7·64	522399	17
44	459268	7·00	981209	·63	478059	7·63	521941	16
45	459688	6·99	981171	·63	478517	7·63	521483	15
46	460108	6·98	981133	·64	478975	7·62	521025	14
47	460527	6·98	981095	·64	479432	7·61	520568	13
48	460946	6·97	981057	·64	479889	7·61	520111	12
49	461364	6·96	981019	·64	480345	7·60	519655	11
50	461782	6·95	980981	·64	480801	7·59	519199	10
51	9·462199	6·95	9·980942	·64	9·481257	7·59	10·518743	9
52	462616	6·94	980904	·64	481712	7·58	518288	8
53	463032	6·93	980866	·64	482167	7·57	517833	7
54	463448	6·93	980827	·64	482621	7·57	517379	6
55	463864	6·92	980789	·64	483075	7·56	516925	5
56	464279	6·91	980750	·64	483529	7·55	516471	4
57	464694	6·90	980712	·64	483982	7·55	516018	3
58	465108	6·90	980673	·64	484435	7·54	515565	2
59	465522	6·89	980635	·64	484887	7·53	515113	1
60	465935	6·88	980596	·64	485339	7·53	514661	0
	Cosine.	D.	Sine.		Cotang.	D.	Tang.	M.

M.	Sine.	D.	Cosine.	D.	Tang.	D.	Cotang.	
0	9·465935	6·88	9·980596	·64	9·485339	7·55	10·514661	60
1	466348	6·88	980558	·64	485791	7·52	514209	59
2	406761	6·87	980519	·65	486242	7·51	513758	58
3	467178	6·86	980480	·65	486693	7·51	513307	57
4	467585	6·85	980442	·65	487143	7·50	512857	56
5	467996	6·85	980403	·65	487593	7·49	512407	55
6	468407	6·84	980364	·65	488043	7·49	511957	54
7	468817	6·83	980325	·65	488492	7·48	511508	53
8	469227	6·83	980286	·65	488941	7·47	·511059	52
9	469637	6·82	980247	·65	489390	7·47	510610	51
10	470046	6·81	980208	·65	489838	7·46	510162	50
11	9·470455	6·80	9·980169	·65	9·490286	7·46	10·509714	49
12	470863	6·80	980130	·65	490733	7·45	509267	48
13	471271	6·79	980091	·65	491180	7·44	508820	47
14	471679	6·78	980052	·65	491627	7·44	508373	46
15	472086	6·78	980012	·65	492073	7·43	507927	45
16	472492	6·77	979973	·65	492519	7·43	507481	44
17	472898	6·76	979934	·66	492965	7·42	507035	43
18	473304	6·76	979895	·66	493410	7·41	506590	42
19	473710	6·75	979855	·66	493854	7·40	506146	41
20	474115	6·74	979816	·66	494299	7·40	505701	40
21	9·474519	6·74	9·979776	·66	9·494743	7·40	10·505257	39
22	474923	6·73	979737	·66	495186	7·39	504814	38
23	475327	6·72	979697	·66	495630	7·38	504370	37
24	475730	6·72	979658	·66	496073	7·37	503927	36
25	476133	6·71	979618	·66	496515	7·37	503485	35
26	476536	6·70	979579	·66	496957	7·36	503043	34
27	476938	6·69	979539	·66	497399	7·36	502601	33
28	477340	6·69	979499	·66	497841	7·35	502159	82
29	477741	6·68	979459	·66	498282	7·34	501718	31
30	478142	6·67	979420	·66	498722	7·34	501278	30
31	9·478542	6·67	9·979380	·66	9·499163	7·33	10·500837	29
32	478942	6·66	979340	·66	499603	7·33	500397	28
33	479342	6·65	979300	·67	500042	7·32	499958	27
34	479741	6·65	979260	·67	500481	7·31	499519	26
35	480140	6·64	979220	·67	500920	7·31	499080	25
36	480539	6·63	979180	·67	501359	7·30	498641	24
37	480937	6·63	979140	·67	501797	7·30	498203	23
38	481334	6·62	979100	·67	502235	7·29	497765	22
39	481731	6·61	979059	·67	502672	7·28	497328	21
40	482128	6·61	979019	·67	503109	7·28	496891	20
41	9·482525	6·60	9·978979	·67	9·503546	7·27	10·496454	19
42	482921	6·59	978989	·67	503982	7·27	496018	18
43	483316	6·59	978898	·67	504418	7·26	495582	17
44	483712	6·58	978858	·67	504854	7·25	495146	16
45	484107	6·57	978817	·67	505289	7·25	494711	15
46	484501	6·57	978777	·67	505724	7·24	494276	14
47	484895	6·56	978736	·67	506159	7·24	493841	13
48	485289	6·55	978696	·68	506598	7·23	493407	12
49	485682	6·55	978655	·68	507027	7·22	492973	11
50	486075	6·54	978615	·68	507460	7·22	492540	10
51	9·486467	6·53	9·978574	·68	9·507893	7·21	10·492107	9
52	486860	6·53	978533	·68	508326	7·21	491674	8
53	487251	6·52	978493	·68	508759	7·20	491241	7
54	487643	6·51	978452	·68	509191	7·19	490809	6
55	488034	6·51	978411	·68	509622	7·19	490378	5
56	488424	6·50	978370	·68	510054	7·18	489946	4
57	488814	6·50	978329	·68	510485	7·18	489515	3
58	489204	6·49	978288	·68	510916	7·17	489084	2
59	489593	6·48	978247	·68	511346	7·16	488654	1
60	489982	6·48	978206	·68	511776	7·16	488224	0
	Cosine.	D.	Sine.	D.	Cotang.	D.	Tang.	M.

M.	Sine.	D.	Cosine.	D.	Tang.	D.	Cotang.	
0	9·489982	6·48	9·978206	·68	9·511776	7·16	10·488224	60
1	490371	6·48	978165	·68	512206	7·16	487794	59
2	490759	6·47	978124	·68	512635	7·15	487365	58
3	491147	6·46	978083	·69	513064	7·14	486936	57
4	491535	6·46	978042	·69	513493	7·14	486507	56
5	491922	6·45	978001	·69	513921	7·13	486079	55
6	492308	6·44	977959	·69	514349	7·13	485651	54
7	492695	6·44	977918	·69	514777	7·12	485223	53
8	493081	6·43	977877	·69	515204	7·12	484796	52
9	493466	6·42	977835	·69	515631	7·11	484369	51
10	493851	6·42	977794	·69	516057	7·10	483943	50
11	9·494236	6·41	9·977752	·69	9·516484	7·10	10·483516	49
12	494621	6·41	977711	·69	516910	7·09	483090	48
13	495005	6·40	977669	·69	517335	7·09	482665	47
14	495388	6·39	977628	·69	517761	7·08	482239	46
15	495772	6·39	977586	·69	518185	7·08	481815	45
16	496154	6·38	977544	·70	518610	7·07	481390	44
17	496537	6·37	977503	·70	519034	7·06	480966	43
18	496919	6·37	977461	·70	519458	7·06	480542	42
19	497301	6·36	977419	·70	519882	7·05	480118	41
20	497682	6·36	977377	·70	520305	7·05	479695	40
21	9·498064	6·35	9·977335	·70	9·520728	7·04	10·479272	39
22	498444	6·34	977293	·70	521151	7·03	478849	38
23	498825	6·34	977251	·70	521573	7·03	478427	37
24	499204	6·33	977209	·70	521995	7·03	478005	36
25	499584	6·32	977167	·70	522417	7·02	477583	35
26	499963	6·32	977125	·70	522838	7·02	477162	34
27	500342	6·31	977088	·70	523259	7·01	476741	33
28	500721	6·31	977041	·70	523680	7·01	476320	32
29	501099	6·30	976999	·70	524100	7·00	475900	31
30	501476	6·29	976957	·70	524520	6·99	475480	30
31	9·501854	6·29	9·976914	·70	9·524939	6·99	10·475061	29
32	502231	6·28	976872	·71	525359	6·98	474641	28
33	502607	6·28	976830	·71	525778	6·98	474222	27
34	502984	6·27	976787	·71	526197	6·97	473803	26
35	503360	6·26	976745	·71	526615	6·97	473385	25
36	503735	6·26	976702	·71	527033	6·96	472967	24
37	504110	6·25	976660	·71	527451	6·96	472549	23
38	504485	6·25	976617	·71	527868	6·95	472132	22
39	504860	6·24	976574	·71	528285	6·95	471715	21
40	505234	6·23	976532	·71	528702	6·94	471298	20
41	9·505608	6·23	9·976489	·71	9·529119	6·93	10·470881	19
42	505981	6·22	976446	·71	529535	6·93	470465	18
43	506354	6·22	976404	·71	529950	6·93	470050	17
44	506727	6·21	976361	·71	530366	6·92	469634	16
45	507099	6·20	976318	·71	530781	6·91	469219	15
46	507471	6·20	976275	·71	531196	6·91	468804	14
47	507843	6·19	976232	·72	531611	6·90	468389	13
48	508214	6·19	976189	·72	532025	6·90	467975	12
49	508585	6·18	976146	·72	532439	6·89	467561	11
50	508956	6·18	976103	·72	532853	6·89	467147	10
51	9·509326	6·17	9·976060	·72	9·533266	6·88	10·466734	9
52	509696	6·16	976017	·72	533679	6·88	466321	8
53	510065	6·16	975974	·72	534092	6·87	465908	7
54	510434	6·15	975930	·72	534504	6·87	465496	6
55	510803	6·15	975887	·72	534916	6·86	465084	5
56	511172	6·14	975844	·72	535328	6·86	464672	4
57	511540	6·13	975800	·72	535739	6·85	464261	3
58	511907	6·13	975757	·72	536150	6·85	463850	2
59	512275	6·12	975714	·72	536561	6·84	463439	1
60	512642	6·12	975670	·72	536972	6·84	463028	0
	Cosine.	D.	Sine.	D.	Cotang.	D.	Tang.	M.

M.	Sine.	D.	Cosine.	D.	Tang.	D.	Cotang.	
0	9·512642	6·12	9·975670	·73	9·536972	6·84	10·463028	60
1	513009	6·11	975627	·73	537382	6·83	462618	59
2	513375	6·11	975583	·73	537792	6·83	462208	58
3	513741	6·10	975539	·73	538202	6·82	461798	57
4	514107	6·09	975496	·73	538611	6·82	461389	56
5	514472	6·09	975452	·73	539020	6·81	460980	55
6	514837	6·08	975408	·73	539429	6·81	460571	54
7	515202	6·08	975365	·73	539837	6·80	460163	53
8	515566	6·07	975321	·73	540245	6·80	459755	52
9	515930	6·07	975277	·73	540653	6·79	459347	51
10	516294	6·06	975233	·73	541061	6·79	458939	50
11	9·516657	6·05	9·975189	·73	9·541468	6·78	10·458532	49
12	517020	6·05	975145	·73	541875	6·78	458125	48
13	517382	6·04	975101	·73	542281	6·77	457719	47
14	517745	6·04	975057	·73	542688	6·77	457312	46
15	518107	6·03	975013	·73	543094	6·76	456906	45
16	518468	6·03	974969	·74	543499	6·76	456501	44
17	518829	6·02	974925	·74	543905	6·75	456095	43
18	519190	6·01	974880	·74	544310	6·75	455690	42
19	519551	6·01	974836	·74	544715	6·74	455285	41
20	519911	6·00	974792	·74	545119	6·74	454881	40
21	9·520271	6·00	9·974748	·74	9·545524	6·73	10·454476	39
22	520631	5·99	974703	·74	545928	6·73	454072	38
23	520990	5·99	974659	·74	546331	6·72	453669	37
24	521349	5·98	974614	·74	546735	6·72	453265	36
25	521707	5·98	974570	·74	547138	6·71	452862	35
26	522066	5·97	974525	·74	547540	6·71	452460	34
27	522424	5·96	974481	·74	547943	6·70	452057	33
28	522781	5·96	974436	·74	548345	6·70	451655	32
29	523138	5·95	974391	·74	548747	6·69	451253	31
30	523495	5·95	974347	·75	549149	6·69	450851	30
31	9·523852	5·94	9·974302	·75	9·549550	6·68	10·450450	29
32	524208	5·94	974257	·75	549951	6·68	450049	28
33	524564	5·93	974212	·75	550352	6·67	449648	27
34	524920	5·93	974167	·75	550752	6·67	449248	26
35	525275	5·92	974122	·75	551152	6·66	448848	25
36	525630	5·91	974077	·75	551552	6·66	448448	24
37	525984	5·91	974032	·75	551952	6·65	448048	23
38	526339	5·90	973987	·75	552351	6·65	447649	22
39	526693	5·90	973942	·75	552750	6·65	447250	21
40	527046	5·89	973897	·75	553149	6·64	446851	20
41	9·527400	5·89	9·973852	·75	9·553548	6·64	10·446452	19
42	527753	5·88	973807	·75	553946	6·63	446054	18
43	528105	5·88	973761	·75	554344	6·63	445656	17
44	528458	5·87	973716	·76	554741	6·62	445259	16
45	528810	5·87	973671	·76	555139	6·62	444861	15
46	529161	5·86	973625	·76	555536	6·61	444464	14
47	529513	5·86	973580	·76	555933	6·61	444067	13
48	529864	5·85	973535	·76	556329	6·60	443671	12
49	530215	5·85	973489	·76	556725	6·60	443275	11
50	530565	5·84	973444	·76	557121	6·59	442879	10
51	9·530915	5·84	9·973398	·76	9·557517	6·59	10·442483	9
52	531265	5·83	973352	·76	557913	6·59	442087	8
53	531614	5·82	973307	·76	558308	6·58	441692	7
54	531963	5·82	973261	·76	558702	6·58	441298	6
55	532312	5·81	973215	·76	559097	6·57	440903	5
56	532661	5·81	973169	·76	559491	6·57	440509	4
57	533009	5·80	973124	·76	559885	6·56	440115	3
58	533357	5·80	973078	·76	560279	6·56	439721	2
59	533704	5·79	973032	·77	560673	6·55	439327	1
60	534052	5·78	972986	·77	561066	6·55	438934	0
	Cosine.	D.	Sine.	D.	Cotang.	D.	Tang.	M.

M.	Sine.	D.	Cosine.	D.	Tang.	D.	Cotang.	
0	9·534052	5·78	9·972986	·77	9·561066	6·55	10·438934	60
1	534399	5·77	972940	·77	561459	6·54	438541	59
2	534745	5·77	972894	·77	561851	6·54	438149	58
3	535092	5·77	972848	·77	562244	6·53	437756	57
4	535438	5·76	972802	·77	562636	6·53	437364	56
5	535783	5·76	972755	·77	563028	6·53	436972	55
6	536129	5·75	972709	·77	563419	6·52	436581	54
7	536474	5·74	972663	·77	563811	6·52	436189	53
8	536818	5·74	972617	·77	564202	6·51	435798	52
9	537163	5·73	972570	·77	564592	6·51	435408	51
10	537507	5·73	972524	·77	564983	6·50	435017	50
11	9·537851	5·72	9·972478	·77	9·565373	6·50	10·434627	49
12	538194	5·72	972431	·78	565763	6·49	434237	48
13	538538	5·71	972385	·78	566153	6·49	433847	47
14	538880	5·71	972338	·78	566542	6·49	433458	46
15	539223	5·70	972291	·78	566932	6·48	433068	45
16	539565	5·70	972245	·78	567320	6·48	432680	44
17	539907	5·69	972198	·78	567709	6·47	432291	43
18	540249	5·69	972151	·78	568098	6·47	431902	42
19	540590	5·68	972105	·78	568486	6·46	431514	41
20	540931	5·68	972058	·78	568873	6·46	431127	40
21	9·541272	5·67	9·972011	·78	9·569261	6·45	10·430739	39
22	541613	5·67	971964	·78	569648	6·45	430352	38
23	541953	5·66	971917	·78	570035	6·45	429965	37
24	542293	5·66	971870	·78	570422	6·44	429578	36
25	542632	5·65	971823	·78	570809	6·44	429191	35
26	542971	5·65	971776	·78	571195	6·43	428805	34
27	543310	5·64	971729	·79	571581	6·43	428419	33
28	543649	5·64	971682	·79	571967	6·42	428033	32
29	543987	5·63	971635	·79	572352	6·42	427648	31
30	544325	5·63	971588	·79	572738	6·42	427262	30
31	9·544663	5·62	9·971540	·79	9·573123	6·41	10·426877	29
32	545000	5·62	971493	·79	573507	6·41	426493	28
33	545338	5·61	971446	·79	573892	6·40	426108	27
34	545674	5·61	971398	·79	574276	6·40	425724	26
35	546011	5·60	971351	·79	574660	6·39	425340	25
36	546347	5·60	971303	·79	575044	6·39	424956	24
37	546683	5·59	971256	·79	575427	6·39	424573	23
38	547019	5·59	971208	·79	575810	6·38	424190	22
39	547354	5·58	971161	·79	576193	6·38	423807	21
40	547689	5·58	971113	·79	576576	6·37	423424	20
41	9·548024	5·57	9·971066	·80	9·576958	6·37	10·423041	19
42	548359	5·57	971018	·80	577341	6·36	422659	18
43	548693	5·56	970970	·80	577723	6·36	422277	17
44	549027	5·56	970922	·80	578104	6·36	421896	16
45	549360	5·55	970874	·80	578486	6·35	421514	15
46	549693	5·55	970827	·80	578867	6·35	421133	14
47	550026	5·54	970779	·80	579248	6·34	420752	13
48	550359	5·54	970731	·80	579629	6·34	420371	12
49	550692	5·53	970683	·80	580009	6·34	419991	11
50	551024	5·53	970635	·80	580389	6·33	419611	10
51	9·551356	5·52	9·970586	·80	9·580769	6·33	10·419231	9
52	551687	5·52	970538	·80	581149	6·32	418851	8
53	552018	5·52	970490	·80	581528	6·32	418472	7
54	552349	5·51	970442	·80	581907	6·32	418093	6
55	552680	5·51	970394	·80	582286	6·31	417714	5
56	553010	5·50	970345	·81	582665	6·31	417335	4
57	553341	5·50	970297	·81	583043	6·30	416957	3
58	553670	5·49	970249	·81	583422	6·30	416578	2
59	554000	5·49	970200	·81	583800	6·29	416200	1
60	554329	5·48	970152	·81	584177	6·29	415823	0
	Cosine.	D.	Sine.	D.	Cotang.	D.	Tang.	M.

M.	Sine.	D.	Cosine.	D.	Tang.	D.	Cotang.	
0	9·554329	5·48	9·970152	·81	9·584177	6·29	10·415823	60
1	554658	5·48	970108	·81	584555	6·29	415445	59
2	554987	5·47	970055	·81	584932	6·28	415068	58
3	555315	5·47	970006	·81	585309	6·28	414691	57
4	555643	5·46	969957	·81	585686	6·27	414314	56
5	555971	5·46	969909	·81	586062	6·27	413938	55
6	556299	5·45	969860	·81	586439	6·27	413561	54
7	556626	5·45	969811	·81	586815	6·26	413185	53
8	556953	5·44	969762	·81	587190	6·26	412810	52
9	557280	5·44	969714	·81	587566	6·25	412434	51
10	557606	5·43	969665	·81	587941	6·25	412059	50
11	9·557932	5·43	9·969616	·82	9·588316	6·25	10·411684	49
12	558258	5·43	969567	·82	588691	6·24	411309	48
13	558583	5·42	969518	·82	589066	6·24	410934	47
14	558909	5·42	969469	·82	589440	6·23	410560	46
15	559234	5·41	969420	·82	589814	6·23	410186	45
16	559558	5·41	969370	·82	590188	6·23	409812	44
17	559883	5·40	969321	·82	590562	6·22	409438	43
18	560207	5·40	969272	·82	590935	6·22	409065	42
19	560531	5·39	969223	·82	591308	6·22	408692	41
20	560855	5·39	969173	·82	591681	6·21	408319	40
21	9·561178	5·38	9·969124	·82	9·592054	6·21	10·407946	39
22	561501	5·38	969075	·82	592426	6·20	407574	38
23	561824	5·37	969025	·82	592798	6·20	407202	37
24	562146	5·37	968976	·82	593170	6·19	406829	36
25	562468	5·36	968926	·83	593542	6·19	406458	35
26	562790	5·36	968877	·83	593914	6·18	406086	34
27	563112	5·36	968827	·83	594285	6·18	405715	33
28	563433	5·35	968777	·83	594656	6·18	405344	32
29	563755	5·35	968728	·83	595027	6·17	404973	31
30	564075	5·34	968678	·83	595398	6·17	404602	30
31	9·564396	5·34	9·968628	·83	9·595768	6·17	10·404232	29
32	564716	5·33	968578	·83	596138	6·16	403862	28
33	565036	5·33	968528	·83	596508	6·16	403492	27
34	565356	5·32	968479	·83	596878	6·16	403122	26
35	565676	5·32	968429	·83	597247	6·15	402753	25
36	565995	5·31	968379	·83	597616	6·15	402384	24
37	566314	5·31	968329	·83	597985	6·15	402015	23
38	566632	5·31	968278	·83	598354	6·14	401646	22
39	566951	5·30	968228	·84	598722	6·14	401278	21
40	567269	5·30	968178	·84	599091	6·13	400909	20
41	9·567587	5·29	9·968128	·84	9·599459	6·13	10·400541	19
42	567904	5·29	968078	·84	599827	6·13	400173	18
43	568222	5·28	968027	·84	600194	6·12	399806	17
44	568539	5·28	967977	·84	600562	6·12	399438	16
45	568856	5·28	967927	·84	600929	6·11	399071	15
46	569172	5·27	967876	·84	601296	6·11	398704	14
47	569488	5·27	967826	·84	601662	6·11	398338	13
48	569804	5·26	967775	·84	602029	6·10	397971	12
49	570120	5·26	967725	·84	602395	6·10	397605	11
50	570435	5·25	967674	·84	602761	6·10	397239	10
51	9·570751	5·25	9·967624	·84	9·603127	6·09	10·396873	9
52	571066	5·24	967573	·84	603493	6·09	396507	8
53	571380	5·24	967522	·85	603858	6·09	396142	7
54	571695	5·23	967471	·85	604223	6·08	395777	6
55	572009	5·23	967421	·85	604588	6·08	395412	5
56	572323	5·23	967370	·85	604953	6·07	395047	4
57	572636	5·22	967319	·85	605317	6·07	394683	3
58	572950	5·22	967268	·85	605682	6·07	394318	2
59	573263	5·21	967217	·85	606046	6·06	393954	1
60	573575	5·21	967166	·85	606410	6·06	393590	0
	Cosine.	D.	Sine.	D.	Cotang.	D.	Tang.	M.

M.	Sine.	D.	Cosine.	D.	Tang.	D.	Cotang.	
0	9·573575	5·21	9·967166	·85	9·606410	6·06	10·393590	60
1	578888	5·20	967115	·85	606773	6·06	393227	59
2	574200	5·20	967064	·85	607137	6·05	392863	58
3	574512	5·19	967013	·85	607500	6·05	392500	57
4	574824	5·19	966961	·85	607863	6·04	392137	56
5	575136	5·19	966910	·85	608225	6·04	391775	55
6	575447	5·18	966859	·85	608588	6·04	391412	54
7	575758	5·18	966808	·85	608950	6·03	391050	53
8	576069	5·17	966756	·86	609312	6·03	390688	52
9	576379	5·17	966705	·86	609674	6·03	390326	51
10	576689	5·16	966653	·86	610036	6·02	389964	50
11	9·576999	5·16	9·966602	·86	9·610397	6·02	10·389603	49
12	577309	5·16	966550	·86	610759	6·02	389241	48
13	577618	5·15	966499	·86	611120	6·01	388880	47
14	577927	5·15	966447	·86	611480	6·01	388520	46
15	578236	5·14	966395	·86	611841	6·01	388159	45
16	578545	5·14	966344	·86	612201	6·00	387799	44
17	578853	5·13	966292	·86	612561	6·00	387439	43
18	579162	5·13	966240	·86	612921	6·00	387079	42
19	579470	5·13	966188	·86	613281	5·99	386719	41
20	579777	5·12	966136	·86	613641	5·99	386359	40
21	9·580085	5·12	9·966085	·87	9·614000	5·98	10·386000	39
22	580392	5·11	966033	·87	614359	5·98	385641	38
23	580699	5·11	965981	·87	614718	5·98	385282	37
24	581005	5·11	965928	·87	615077	5·97	384923	36
25	581312	5·10	965876	·87	615435	5·97	384565	35
26	581618	5·10	965824	·87	615793	5·97	384207	34
27	581924	5·09	965772	·87	616151	5·96	383849	33
28	582229	5·09	965720	·87	616509	5·96	383491	32
29	582535	5·09	965668	·87	616867	5·96	383133	31
30	582840	5·08	965615	·87	617224	5·95	382776	30
31	9·583145	5·08	9·965563	·87	9·617582	5·95	10·382418	29
32	583449	5·07	965511	·87	617989	5·95	382061	28
33	583754	5·07	965458	·87	618295	5·94	381705	27
34	584058	5·06	965406	·87	618652	5·94	381348	26
35	584361	5·06	965353	·88	619008	5·94	380992	25
36	584665	5·06	965301	·88	619364	5·93	380636	24
37	584968	5·05	965248	·88	619721	5·93	380279	23
38	585272	5·05	965195	·88	620076	5·93	379924	22
39	585574	5·04	965143	·88	620432	5·92	379568	21
40	585877	5·04	965090	·88	620787	5·92	379213	20
41	9·586179	5·03	9·965037	·88	9·621142	5·92	10·378858	19
42	586482	5·03	964984	·88	621497	5·91	378503	18
43	586783	5·03	964931	·88	621852	5·91	378148	17
44	587085	5·02	964879	·88	622207	5·90	377793	16
45	587386	5·02	964826	·88	622561	5·90	377439	15
46	587688	5·01	964773	·88	622915	5·90	377085	14
47	587989	5·01	964719	·88	623269	5·89	376731	13
48	588289	5·01	964666	·89	623623	5·89	376377	12
49	588590	5·00	964613	·89	623976	5·89	376024	11
50	588890	5·00	964560	·89	624330	5·88	375670	10
51	9·589190	4·99	9·964507	·89	9·624683	5·88	10·375317	9
52	589489	4·99	964454	·89	625036	5·88	374964	8
53	589789	4·99	964400	·89	625388	5·87	374612	7
54	590088	4·98	964347	·89	625741	5·87	374259	6
55	590387	4·98	964294	·89	626093	5·87	373907	5
56	590686	4·97	964240	·89	626445	5·86	373555	4
57	590984	4·97	964187	·89	626797	5·86	373203	3
58	591282	4·97	964133	·89	627149	5·86	372851	2
59	591580	4·96	964080	·89	627501	5·85	372499	1
60	591878	4·96	964026	·89	627852	5·85	372148	0
	Cosine.	D.	Sine.	D.	Cotang.	D.	Tang.	M.

M.	Sine.	D.	Cosine.	D.	Tang.	D.	Cotang.	
0	9·591878	4·96	9·964026	·89	9·627852	5·85	10·872148	60
1	592176	4·95	963972	·89	628203	5·85	871797	59
2	592478	4·95	963919	·89	628554	5·85	871446	58
3	592770	4·95	963865	·90	628905	5·84	371095	57
4	593067	4·94	963811	·90	629255	5·84	370745	56
5	593363	4·94	963757	·90	629606	5·83	370394	55
6	593659	4·93	963704	·90	629956	5·83	370044	54
7	593955	4·93	963650	·90	630306	5·83	369694	53
8	594251	4·93	963596	·90	630656	5·83	369344	52
9	594547	4·92	963542	·90	631005	5·82	368995	51
10	594842	4·92	963488	·90	631355	5·82	368645	50
11	9·595137	4·91	9·963434	·90	9·631704	5·82	10·368296	49
12	595432	4·91	963379	·90	632053	5·81	367947	48
13	595727	4·91	963325	·90	632401	5·81	367599	47
14	596021	4·90	963271	·90	632750	5·81	367250	46
15	596315	4·90	963217	·90	633098	5·80	366902	45
16	596609	4·89	963163	·90	633447	5·80	366553	44
17	596903	4·89	963108	·91	633795	5·80	366205	43
18	597196	4·89	963054	·91	634143	5·79	365857	42
19	597490	4·88	962999	·91	634490	5·79	365510	41
20	597783	4·88	962945	·91	634838	5·79	365162	40
21	9·598075	4·87	9·962890	·91	9·635185	5·78	10·364815	39
22	598368	4·87	962836	·91	635532	5·78	364468	38
23	598660	4·87	962781	·91	635879	5·78	364121	37
24	598952	4·86	962727	·91	636226	5·77	363774	36
25	599244	4·86	962672	·91	636572	5·77	363428	35
26	599536	4·85	962617	·91	636919	5·77	363081	34
27	599827	4·85	962562	·91	637265	5·77	362735	33
28	600118	4·85	962508	·91	637611	5·76	362389	32
29	600409	4·84	962453	·91	637956	5·76	362044	31
30	600700	4·84	962398	·92	638302	5·76	361698	30
31	9·600990	4·84	9·962343	·92	9·638647	5·75	10·361353	29
32	601280	4·83	962288	·92	638992	5·75	361008	28
33	601570	4·83	962233	·92	639337	5·75	360663	27
34	601860	4·82	962178	·92	639682	5·74	360318	26
35	602150	4·82	962123	·92	640027	5·74	359973	25
36	602439	4·82	962067	·92	640371	5·74	359629	24
37	602728	4·81	962012	·92	640716	5·73	359284	23
38	603017	4·81	961957	·92	641060	5·73	358940	22
39	603305	4·81	961902	·92	641404	5·73	358596	21
40	603594	4·80	961846	·92	641747	5·72	358253	20
41	9·603882	4·80	9·961791	·92	9·642091	5·72	10·357909	19
42	604170	4·79	961735	·92	642434	5·72	357566	18
43	604457	4·79	961680	·92	642777	5·72	357223	17
44	604745	4·79	961624	·93	643120	5·71	356880	16
45	605032	4·78	961569	·93	643463	5·71	356537	15
46	605319	4·78	961513	·93	643806	5·71	356194	14
47	605606	4·78	961458	·93	644148	5·70	355852	13
48	605892	4·77	961402	·93	644490	5·70	355510	12
49	606179	4·77	961346	·93	644832	5·70	355168	11
50	606465	4·76	961290	·93	645174	5·69	354826	10
51	9·606751	4·76	9·961235	·93	9·645516	5·69	10·354484	9
52	607036	4·76	961179	·93	645857	5·69	354143	8
53	607322	4·75	961123	·93	646199	5·69	353801	7
54	607607	4·75	961067	·93	646540	5·68	353460	6
55	607892	4·74	961011	·93	646881	5·68	353119	5
56	608177	4·74	960955	·93	647222	5·68	352778	4
57	608461	4·74	960899	·93	647562	5·67	352438	3
58	608745	4·73	960843	·94	647903	5·67	352097	2
59	609029	4·73	960786	·94	648243	5·67	351757	1
60	609313	4·73	960730	·94	648583	5·66	351417	0
	Cosine.	D.	Sine.	D.	Cotang.	D.	Tang.	M.

M.	Sine.	D.	Cosine.	D.	Tang.	D.	Cotang.	
0	9·609813	4·78	9·960730	·94	9·648583	5·66	10·351417	60
1	609597	4·72	960674	·94	648923	5·66	351077	59
2	609880	4·72	960618	·94	649263	5·66	350737	58
3	610164	4·72	960561	·94	649602	5·66	350398	57
4	610447	4·71	960505	·94	649942	5·65	350058	56
5	610729	4·71	960448	·94	650281	5·65	349719	55
6	611012	4·70	960392	·94	650620	5·65	349380	54
7	611294	4·70	960335	·94	650959	5·64	349041	53
8	611576	4·70	960279	·94	651297	5·64	348703	52
9	611858	4·69	960222	·94	651636	5·64	348364	51
10	612140	4·69	960165	·94	651974	5·63	348026	50
11	9·612421	4·69	9·960109	·95	9·652312	5·63	10·347688	49
12	612702	4·68	960052	·95	652650	5·63	347350	48
13	612983	4·68	959995	·95	652988	5·63	347012	47
14	613264	4·67	959938	·95	653326	5·62	346674	46
15	613545	4·67	959882	·95	653663	5·62	346337	45
16	613825	4·67	959825	·95	654000	5·62	346000	44
17	614105	4·66	959768	·95	654337	5·61	345663	43
18	614385	4·66	959711	·95	654674	5·61	345326	42
19	614665	4·66	959654	·95	655011	5·61	344989	41
20	614944	4·65	959596	·95	655348	5·61	344652	40
21	9·615223	4·65	9·959539	·95	9·655684	5·60	10·344316	39
22	615502	4·65	959482	·95	656020	5·60	343980	38
23	615781	4·64	959425	·95	656356	5·60	343644	37
24	616060	4·64	959368	·95	656692	5·59	343308	36
25	616338	4·64	959310	·96	657028	5·59	342972	35
26	616616	4·63	959253	·96	657364	5·59	342636	34
27	616894	4·63	959195	·96	657699	5·59	342301	33
28	617172	4·62	959138	·96	658034	5·58	341966	32
29	617450	4·62	959081	·96	658369	5·58	341631	31
30	617727	4·62	959023	·96	658704	5·58	341296	30
31	9·618004	4·61	9·958965	·96	9·659089	5·58	10·340961	29
32	618281	4·61	958908	·96	659373	5·57	340627	28
33	618558	4·61	958850	·96	659708	5·57	340292	27
34	618834	4·60	958792	·96	660042	5·57	339958	26
35	619110	4·60	958734	·96	660376	5·57	339624	25
36	619386	4·60	958677	·96	660710	5·56	339290	24
37	619662	4·59	958619	·96	661043	5·56	338957	23
38	619938	4·59	958561	·96	661377	5·56	338623	22
39	620213	4·59	958503	·97	661710	5·55	338290	21
40	620488	4·58	958445	·97	662043	5·55	337957	20
41	9·620763	4·58	9·958387	·97	9·662376	5·55	10·337624	19
42	621088	4·57	958329	·97	662709	5·54	337291	18
43	621313	4·57	958271	·97	663042	5·54	336958	17
44	621587	4·57	958213	·97	663375	5·54	336625	16
45	621861	4·56	958154	·97	663707	5·54	336293	15
46	622135	4·56	958096	·97	664039	5·53	335961	14
47	622409	4·56	958038	·97	664371	5·53	335629	13
48	622682	4·55	957979	·97	664703	5·53	335297	12
49	622956	4·55	957921	·97	665035	5·53	334965	11
50	623229	4·55	957863	·97	665366	5·52	334634	10
51	9·623502	4·54	9·957804	·97	9·665697	5·52	10·334303	9
52	623774	4·54	957746	·98	666029	5·52	333971	8
53	624047	4·54	957687	·98	666360	5·51	333640	7
54	624319	4·53	957628	·98	666691	5·51	333309	6
55	624591	4·53	957570	·98	667021	5·51	332979	5
56	624863	4·53	957511	·98	667352	5·51	332648	4
57	625135	4·52	957452	·98	667682	5·50	332318	3
58	625406	4·52	957393	·98	668013	5·50	331987	2
59	625677	4·52	957335	·98	668343	5·50	331657	1
60	625948	4·51	957276	·98	668672	5·50	331328	0
	Cosine.	D.	Sine.	D.	Cotang.	D.	Tang.	M.

M.	Sine.	D.	Cosine.	D.	Tang.	D.	Cotang.	
0	9·625948	4·51	9·957276	·98	9·668673	5·50	10·331327	60
1	626219	4·51	957217	·98	669002	5·49	330998	59
2	626490	4·51	957158	·98	669332	5·49	330668	58
3	626760	4·50	957099	·98	669661	5·49	330339	57
4	627030	4·50	957040	·98	669991	5·48	330009	56
5	627300	4·50	956981	·98	670320	5·48	329680	55
6	627570	4·49	956921	·99	670649	5·48	329351	54
7	627840	4·49	956862	·99	670977	5·48	329023	53
8	628109	4·49	956803	·99	671306	5·47	328694	52
9	628378	4·48	956744	·99	671634	5·47	328366	51
10	628647	4·48	956684	·99	671963	5·47	328037	50
11	9·628916	4·47	9·956625	·99	9·672291	5·47	10·327709	49
12	629185	4·47	956566	·99	672619	5·46	327381	48
13	629453	4·47	956506	·99	672947	5·46	327053	47
14	629721	4·46	956447	·99	673274	5·46	326726	46
15	629989	4·46	956387	·99	673602	5·46	326398	45
16	630257	4·46	956327	·99	673929	5·45	326071	44
17	630524	4·46	956268	·99	674257	5·45	325743	43
18	630792	4·45	956208	·1·00	674584	5·45	325416	42
19	631059	4·45	956148	1·00	674910	5·44	325090	41
20	631326	4·45	956089	1·00	675237	5·44	324763	40
21	0·631593	4·44	9·956029	1·00	9·675564	5·44	10·324436	39
22	631859	4·44	955969	1·00	675890	5·44	324110	38
23	632125	4·44	955909	1·00	676216	5·43	323784	37
24	632392	4·43	955849	1·00	676543	5·43	323457	36
25	632658	4·43	955789	1·00	676869	5·43	323131	35
26	632923	4·43	955729	1·00	677194	5·43	322806	34
27	633189	4·42	955669	1·00	677520	5·42	322480	33
28	633454	4·42	955609	1·00	677846	5·42	322154	32
29	633719	4·42	955548	1·00	678171	5·42	321829	31
30	633984	4·41	955488	1·00	678496	5·42	321504	30
31	9·634249	4·41	9·955428	1·01	9·678821	5·41	10·321179	29
32	634514	4·40	955368	1·01	679146	5·41	320854	28
33	634778	4·40	955307	1·01	679471	5·41	320529	27
34	635042	4·40	955247	1·01	679795	5·41	320205	26
35	635306	4·39	955186	1·01	680120	5·40	319880	25
36	635570	4·39	955126	1·01	680444	5·40	319556	24
37	635834	4·39	955065	1·01	680768	5·40	319232	23
38	636097	4·38	955005	1·01	681092	5·40	318908	22
39	636360	4·38	954944	1·01	681416	5·39	318584	21
40	636623	4·38	954883	1·01	681740	5·39	318260	20
41	9·636886	4·87	9·954823	1·01	9·682063	5·39	10·317937	19
42	637148	4·87	954762	1·01	682387	5·39	317613	18
43	637411	4·87	−954701	1·01	682710	5·38	317290	17
44	637673	4·87	954640	1·01	683033	5·38	316967	16
45	637935	4·86	954579	1·01	683356	5·38	316644	15
46	638197	4·86	954518	1·02	683679	5·38	316321	14
47	638458	4·86	954457	1·02	684001	5·37	315999	13
48	638720	4·85	954396	1·02	684324	5·37	315676	12
49	638981	4·85	954335	1·02	684646	5·37	315354	11
50	639242	4·85	954274	1·02	684968	5·37	315032	10
51	9·639503	4·84	9·954213	1·02	9·685290	5·36	10·314710	9
52	639764	4·84	954152	1·02	685612	5·36	314388	8
53	640024	4·84	954090	1·02	685934	5·36	314066	7
54	640284	4·83	954029	1·02	686255	5·36	313745	6
55	640544	4·83	953968	1·02	686577	5·35	313423	5
56	640804	4·83	953906	1·02	686898	5·35	313102	4
57	641064	4·82	953845	1·02	687219	5·35	312781	3
58	641324	4·82	953783	1·02	687540	5·35	312460	2
59	641584	4·82	953722	1·03	687861	5·34	312139	1
60	641842	4·81	953660	1·03	688182	5·34	311818	0
	Cosine.	D.	Sine.	D.	Cotang.	D.	Tang.	M.

M.	Sine.	D.	Cosine.	D.	Tang.	D. .	Cotang.	
0	9·641842	4·31	9·953660	1·03	9·688182	5·84	10·811818	60
1	642101	4·31	953599	1·03	688502	5·34	311498	59
2	642360	4·31	953537	1·03	688823	5·84	311177	58
3	642618	4·30	953475	1·03	689148	5·33	310857	57
4	642877	4·30	953413	1·03	689463	5·33	310537	56
5	643185	4·30	953352	1·03	689783	5·33	810217	55
6	643393	4·30	953290	1·03	690103	5·33	809897	54
7	643650	4·29	953228	1·03	690423	5·33	809577	53
8	643908	4·29	953166	1·03	690742	5·32	809258	52
9	644165	4·29	953104	1·03	691062	5·32	808938	51
10	644423	4·28	953042	1·03	691381	5·32	308619	50
11	9·644680	4·28	9·952980	1·04	9·691700	5·81	10·808300	49
12	644936	4·28	952918	1·04	692019	5·81	307981	48
13	645193	4·27	952855	1·04	692338	5·81	307662	47
14	645450	4·27	952793	1·04	692656	5·31	307344	46
15	645706	4·27	952731	1·04	692975	5·81	307025	45
16	645962	4·26	952669	1·04	693293	5·30	306707	44
17	646218	4·26	952606	1·04	693612	5·30	306388	43
18	646474	4·26	952544	1·04	693930	5·80	306070	42
19	646729	4·25	952481	1·04	694248	5·30	805752	41
20	646984	4·25	952419	1·04	694566	5·29	305434	40
21	9·647240	4·25	9·952356	1·04	9·694883	5·29	10·805117	39
22	647494	4·24	952294	1·04	695201	5·29	304799	38
23	647749	4·24	952231	1·04	695518	5·29	804482	37
24	648004	4·24	952168	1·05	695836	5·29	804164	36
25	648258	4·24	952106	1·05	696153	5·28	803847	35
26	648512	4·23	952043	1·05	696470	5·28	308530	34
27	648766	4·23	951980	1·05	696787	5·28	803213	83
28	649020	4·23	951917	1·05	697103	5·28	802897	32
29	649274	4·22	951854	1·05	697420	5·27	302580	31
30	649527	4·22	951791	1·05	697736	5·27	302264	30
31	9·649781	4·22	9·951728	1·05	9·698053	5·27	10·301947	29
32	650034	4·22	951665	1·05	698369	5·27	301631	28
33	650287	4·21	951602	1·05	698685	5·26	801315	27
34	650539	4·21	951539	1·05	699001	5·26	300999	26
35	650792	4·21	951476	1·05	699316	5·26	800684	25
36	651044	4·20	951412	1·05	699632	5·26	300368	24
37	651297	4·20	951349	1·06	699947	5·26	800053	23
38	651549	4·20	951286	1·06	700263	5·25	299737	22
39	651800	4·19	951222	1·06	700578	5·25	299422	21
40	652052	4·19	951159	1·06	700893	5·25	299107	20
41	9·652304	4·19	9·951096	1·06	9·701208	5·24	10·298792	19
42	652555	4·18	951032	1·06	701523	5·24	298477	18
43	652806	4·18	950968	1·06	701837	5·24	298163	17
44	653057	4·18	950905	1·06	702152	5·24	297848	16
45	653308	4·18	950841	1·06	702466	5·24	297534	15
46	653558	4·17	950778	1·06	702780	5·23	297220	14
47	653808	4·17	950714	1·06	703095	5·23	296905	13
48	654059	4·17	950650	1·06	703409	5·23	296591	12
49	654309	4·16	950586	1·06	703723	5·23	296277	11
50	654558	4·16	950522	1·07	704036	5·22	295964	10
51	9·654808	4·16	9·950458	1·07	9·704350	5·22	10·295650	9
52	655058	4·16	950394	1·07	704663	5·22	295337	8
53	655307	4·15	950330	1·07	704977	5·22	295023	7
54	655556	4·15	950266	1·07	705290	5·22	294710	6
55	655805	4·15	950202	1·07	705603	5·21	294397	5
56	656054	4·14	950138	1·07	705916	5·21	294084	4
57	656302	4·14	950074	1·07	706228	5·21	293772	3
58	656551	4·14	950010	1·07	706541	5·21	293459	2
59	656799	4·13	949945	1·07	706854	5·21	293146	1
60	657047	4·13	949881	1·07	707166	5·20	292834	0
	Cosine.	D.	Sine.	D.	Cotang.	D.	Tang.	M.

M.	Sine.	D.	Cosine.	D.	Tang.	D.	Cotang.	
0	9·657047	4·18	9·949881	1·07	9·707166	5·20	10·292834	60
1	657295	4·18	949816	1·07	707478	5·20	292622	59
2	657542	4·12	949752	1·07	707790	5·20	292210	58
3	657790	4·12	949688	1·08	708102	5·20	291898	57
4	658037	4·12	949623	1·08	708414	5·19	291586	56
5	658284	4·12	949558	1·08	708726	5·19	291274	55
6	658531	4·11	949494	1·08	709087	5·19	290963	54
7	658778	4·11	949429	1·08	709349	5·19	290651	53
8	659025	4·11	949364	1·08	709660	5·19	290340	52
9	659271	4·10	949300	1·08	709971	5·18	290029	51
10	659517	4·10	949235	1·08	710282	5·18	289718	50
11	9·659763	4·10	9·949170	1·08	9·710593	5·18	10·289407	49
12	660009	4·09	949105	1·08	710904	5·18	289096	48
13	660255	4·09	949040	1·08	711215	5·18	288785	47
14	660501	4·09	948975	1·08	711525	5·17	288475	46
15	660746	4·09	948910	1·08	711836	5·17	288164	45
16	660991	4·08	948845	1·08	712146	5·17	287854	44
17	661236	4·08	948780	1·09	712456	5·17	287544	43
18	661481	4·08	948715	1·09	712766	5·16	287234	42
19	661726	4·07	948650	1·09	713076	5·16	286924	41
20	661970	4·07	948584	1·09	713386	5·16	286614	40
21	9·662214	4·07	9·948519	1·09	9·713696	5·16	10·286304	39
22	662459	4·07	948454	1·09	714005	5·16	285995	38
23	662703	4·06	948388	1·09	714314	5·15	285686	37
24	662946	4·06	948323	1·09	714624	5·15	285376	36
25	663190	4·06	948257	1·09	714933	5·15	285067	35
26	663433	4·05	948192	1·09	715242	5·15	284758	34
27	663677	4·05	948126	1·09	715551	5·14	284449	33
28	663920	4·05	948060	1·09	715860	5·14	284140	32
29	664163	4·05	947995	1·10	716168	5·14	283832	31
30	664406	4·04	947929	1·10	716477	5·14	283523	30
31	9·664648	4·04	9·947863	1·10	9·716785	5·14	10·283215	29
32	664891	4·04	947797	1·10	717093	5·13	282907	28
33	665133	4·03	947731	1·10	717401	5·13	282599	27
34	665375	4·03	947665	1·10	717709	5·13	282291	26
35	665617	4·03	947600	1·10	718017	5·13	281983	25
36	665859	4·02	947533	1·10	718325	5·12	281670	24
37	666100	4·02	947467	1·10	718633	5·12	281367	23
38	666342	4·02	947401	1·10	718940	5·12	281060	22
39	666583	4·02	947335	1·10	719248	5·12	280752	21
40	666824	4·01	947269	1·10	719555	5·12	280445	20
41	9·667065	4·01	9·947203	1·10	9·719862	5·12	10·280138	19
42	667305	4·01	947136	1·11	720169	5·11	279831	18
43	667546	4·01	947070	1·11	720476	5·11	279524	17
44	667786	4·00	947004	1·11	720783	5·11	279217	16
45	668027	4·00	946937	1·11	721089	5·11	278911	15
46	668267	4·00	946871	1·11	721396	5·11	278604	14
47	668506	3·99	946804	1·11	721702	5·10	278298	13
48	668746	3·99	946738	1·11	722009	5·10	277991	12
49	668986	3·99	946671	1·11	722315	5·10	277685	11
50	669225	3·99	946604	1·11	722621	5·10	277379	10
51	9·669464	3·98	9·946538	1·11	9·722927	5·10	10·277073	9
52	669703	3·98	946471	1·11	723232	5·09	276768	8
53	669942	3·98	946404	1·11	723538	5·09	276462	7
54	670181	3·97	946337	1·11	723844	5·09	276156	6
55	670419	3·97	946270	1·12	724149	5·09	275851	5
56	670658	3·97	946203	1·12	724454	5·09	275546	4
57	670896	3·97	946136	1·12	724759	5·08	275241	3
58	671134	3·96	946069	1·12	725065	5·08	274935	2
59	671372	3·96	946002	1·12	725369	5·08	274631	1
60	671609	3·96	945935	1·12	725674	5·08	274326	0
	Cosine.	D.	Sine.	D.	Cotang.	D.	Tang.	M.

M.	Sine.	D.	Cosine.	D.	Tang.	D.	Cotang.	
0	9·671609	3·96	9·945935	1·12	9·725674	5·08	10·274326	60
1	671847	3·95	945868	1·12	725979	5·08	274021	59
2	672084	3·95	945800	1·12	726284	5·07	273716	58
3	672321	3·95	945733	1·12	726588	5·07	273412	57
4	672558	3·95	945666	1·12	726892	5·07	273108	56
5	672795	3·94	945598	1·12	727197	5·07	272803	55
6	673032	3·94	945531	1·12	727501	5·07	272499	54
7	673268	3·94	945464	1·13	727805	5·06	272195	53
8	673505	3·94	945396	1·13	728109	5·06	271891	52
9	673741	3·93	945328	1·13	728412	5·06	271588	51
10	673977	3·93	945261	1·13	728716	5·06	271284	50
11	9·674213	3·93	9·945193	1·13	9·729020	5·06	10·270980	49
12	674448	3·92	945125	1·13	729323	5·05	270677	48
13	674684	3·92	945058	1·13	729626	5·05	270374	47
14	674919	3·92	944990	1·13	729929	5·05	270071	46
15	675155	3·92	944922	1·13	730233	5·05	269767	45
16	675390	3·91	944854	1·13	730535	5·05	269465	44
17	675624	3·91	944786	1·13	730838	5·04	269162	43
18	675859	3·91	944718	1·13	731141	5·04	268859	42
19	676094	3·91	944650	1·13	731444	5·04	268556	41
20	676328	3·90	944582	1·14	731746	5·04	268254	40
21	0·676562	3·90	9·944514	1·14	9·732048	5·04	10·267952	39
22	676796	3·90	944446	1·14	732351	5·03	267649	38
23	677080	3·90	944377	1·14	732653	5·03	267347	37
24	677264	3·89	944309	1·14	732955	5·03	267045	36
25	677498	3·89	944241	1·14	733257	5·03	266743	35
26	677731	3·89	944172	1·14	733558	5·03	266442	34
27	677964	3·88	944104	1·14	733860	5·02	266140	33
28	678197	3·88	944036	1·14	734162	5·02	265838	32
29	678430	3·88	943967	1·14	734463	5·02	265537	31
30	678663	3·88	943899	1·14	734764	5·02	265236	30
31	9·678895	3·87	9·943830	1·14	9·735066	5·02	10·264934	29
32	679128	3·87	943761	1·14	735367	5·02	264633	28
33	679360	3·87	943693	1·15	735668	5·01	264332	27
34	679592	3·87	943624	1·15	735969	5·01	264031	26
35	679824	3·86	943555	1·15	736269	5·01	263731	25
36	680056	3·86	943486	1·15	736570	5·01	263430	24
37	680288	3·86	943417	1·15	736871	5·01	263129	23
38	680519	3·85	943348	1·15	737171	5·00	262829	22
39	680750	3·85	943279	1·15	737471	5·00	262529	21
40	680982	3·85	943210	1·15	737771	5·00	262229	20
41	9·681213	3·85	9·943141	1·15	9·738071	5·00	10·261929	19
42	681443	3·84	943072	1·15	738371	5·00	261629	18
43	681674	3·84	943003	1·15	738671	4·99	261329	17
44	681905	3·84	942934	1·15	738971	4·99	261029	16
45	682135	3·84	942864	1·15	739271	4·99	260729	15
46	682365	3·83	942795	1·16	739570	4·99	260430	14
47	682595	3·83	942726	1·16	739870	4·99	260130	13
48	682825	3·83	942656	1·16	740169	4·99	259831	12
49	683055	3·83	942587	1·16	740468	4·98	259532	11
50	683284	3·82	942517	1·16	740767	4·98	259233	10
51	9·683514	3·82	9·942448	1·16	9·741066	4·98	10·258934	9
52	683743	3·82	942378	1·16	741365	4·98	258635	8
53	683972	3·82	942308	1·16	741664	4·98	258336	7
54	684201	3·81	942239	1·16	741962	4·97	258038	6
55	684430	3·81	942169	1·16	742261	4·97	257739	5
56	684658	3·81	942099	1·16	742559	4·97	257441	4
57	684887	3·80	942029	1·16	742858	4·97	257142	3
58	685115	3·80	941959	1·16	743156	4·97	256844	2
59	685343	3·80	941889	1·17	743454	4·97	256546	1
60	685571	3·80	941819	1·17	743752	4·96	256248	0
	Cosine.	D.	Sine.	D.	Cotang.	D.	Tang.	M.

M.	Sine.	D.	Cosine.	D.	Tang.	D.	Cotang.	
0	9·685571	3·80	9·941819	1·17	9·743752	4·96	10·256248	60
1	685799	3·79	941749	1·17	744050	4·96	255950	59
2	686027	3·79	941679	1·17	744348	4·96	255652	58
3	686254	3·79	941609	1·17	744645	4·96	255355	57
4	686482	3·79	941539	1·17	744943	4·96	255057	56
5	686709	3·78	941469	1·17	745240	4·96	254760	55
6	686936	3·78	941398	1·17	745538	4·95	254462	54
7	687163	3·78	941328	1·17	745835	4·95	254165	53
8	687389	3·78	941258	1·17	746132	4·95	253868	52
9	687616	3·77	941187	1·17	746429	4·95	253571	51
10	687843	3·77	941117	1·17	746726	4·95	253274	50
11	9·688069	3·77	9·941046	1·18	9·747023	4·94	10·252977	49
12	688295	3·77	940975	1·18	747319	4·94	252681	48
13	688521	3·76	940905	1·18	747616	4·94	252384	47
14	688747	3·76	940834	1·18	747913	4·94	252087	46
15	688972	3·76	940763	1·18	748209	4·94	251791	45
16	689198	3·76	940693	1·18	748505	4·93	251495	44
17	689423	3·75	940622	1·18	748801	4·93	251199	43
18	689648	3·75	940551	1·18	749097	4·93	250903	42
19	689873	3·75	940480	1·18	749393	4·93	250607	41
20	690098	3·75	940409	1·18	749689	4·93	250311	40
21	9·690323	3·74	9·940338	1·18	9·749985	4·93	10·250015	39
22	690548	3·74	940267	1·18	750281	4·92	249719	38
23	690772	3·74	940196	1·18	750576	4·92	249424	37
24	690996	3·74	940125	1·19	750872	4·92	249128	36
25	691220	3·73	940054	1·19	751167	4·92	248833	35
26	691444	3·73	939982	1·19	751462	4·92	248538	34
27	691668	3·73	939911	1·19	751757	4·92	248243	33
28	691892	3·73	939840	1·19	752052	4·91	247948	32
29	692115	3·72	939768	1·19	752347	4·91	247653	31
30	692339	3·72	939697	1·19	752642	4·91	247358	30
31	9·692562	3·72	9·939625	1·19	9·752937	4·91	10·247063	29
32	692785	3·71	939554	1·19	753231	4·91	246769	28
33	693008	3·71	939482	1·19	753526	4·91	246474	27
34	693231	3·71	939410	1·19	753820	4·90	246180	26
35	693453	3·71	939339	1·19	754115	4·90	245885	25
36	693676	3·70	939267	1·20	754409	4·90	245591	24
37	693898	3·70	939195	1·20	754703	4·90	245297	23
38	694120	3·70	939123	1·20	754997	4·90	245003	22
39	694342	3·70	939052	1·20	755291	4·90	244709	21
40	694564	3·69	938980	1·20	755585	4·89	244415	20
41	9·694786	3·69	9·938908	1·20	9·755878	4·89	10·244122	19
42	695007	3·69	938836	1·20	756172	4·89	243828	18
43	695229	3·69	938763	1·20	756465	4·89	243535	17
44	695450	3·68	938691	1·20	756759	4·89	243241	16
45	695671	3·68	938619	1·20	757052	4·89	242948	15
46	695892	3·68	938547	1·20	757345	4·88	242655	14
47	696113	3·68	938475	1·20	757638	4·88	242362	13
48	696334	3·67	938402	1·21	757931	4·88	242069	12
49	696554	3·67	938330	1·21	758224	4·88	241776	11
50	696775	3·67	938258	1·21	758517	4·88	241483	10
51	9·696995	3·67	9·938185	1·21	9·758810	4·88	10·241190	9
52	697215	3·66	938113	1·21	759102	4·87	240898	8
53	697435	3·66	938040	1·21	759395	4·87	240605	7
54	697654	3·66	937967	1·21	759687	4·87	240313	6
55	697874	3·66	937895	1·21	759979	4·87	240021	5
56	698094	3·65	937822	1·21	760272	4·87	239728	4
57	698313	3·65	937749	1·21	760564	4·87	239436	3
58	698532	3·65	937676	1·21	760856	4·86	239144	2
59	698751	3·65	937604	1·21	761148	4·86	238852	1
60	698970	3·64	937531	1·21	761439	4·86	238561	0
	Cosine.	D.	Sine.	D.	Cotang.	D.	Tang.	M.

M.	Sine.	D.	Cosine.	D.	Tang.	D.	Cotang.	
0	9·698970	3·64	9·937531	1·21	9·761439	4·86	10·238561	60
1	699189	3·64	937458	1·22	761731	4·86	238269	59
2	699407	3·64	937385	1·22	762023	4·86	237977	58
3	699626	3·64	937312	1·22	762314	4·86	237686	57
4	699844	3·63	937238	1·22	762606	4·85	237394	56
5	700062	3·63	937165	1·22	762897	4·85	237103	55
6	700280	3·63	937092	1·22	763188	4·85	236812	54
7	700498	3·63	937019	1·22	763479	4·85	236521	53
8	700716	3·63	936946	1·22	763770	4·85	236230	52
9	700933	3·62	936872	1·22	764061	4·85	235939	51
10	701151	3·62	936799	1·22	764352	4·84	235648	50
11	9·701368	3·62	9·936725	1·22	9·764643	4·84	10·235357	49
12	701585	3·62	936652	1·23	764933	4·84	235067	48
13	701802	3·61	936578	1·23	765224	4·84	234776	47
14	702019	3·61	936505	1·23	765514	4·84	234486	46
15	702236	3·61	936431	1·23	765805	4·84	234195	45
16	702452	3·61	936357	1·23	766095	4·84	233905	44
17	702669	3·60	936284	1·23	766385	4·83	233615	43
18	702885	3·60	936210	1·23	766675	4·83	233325	42
19	703101	3·60	936136	1·23	766965	4·83	233035	41
20	703317	3·60	936062	1·23	767255	4·83	232745	40
21	9·703533	3·59	9·935988	1·23	9·767545	4·83	10·232455	39
22	703749	3·59	·· 935914	1·23	767834	4·83	232166	38
23	703964	3·59	935840	1·23	768124	4·82	231876	37
24	704179	3·59	935766	1·24	768413	4·82	231587	36
25	704395	3·59	935692	1·24	768703	4·82	231297	35
26	704610	3·58	935618	1·24	768992	4·82	231008	34
27	704825	3·58	935543	1·24	769281	4·82	230719	33
28	705040	3·58	935469	1·24	769570	4·82	230430	32
29	705254	3·58	935395	1·24	769860	4·81	230140	31
30	705469	3·57	935320	1·24	770148	4·81	229852	30
31	9·705683	3·57	9·935246	1·24	9·770437	4·81	10·229563	29
32	705898	3·57	935171	1·24	770726	4·81	229274	28
33	706112	3·57	935097	1·24	771015	4·81	228985	27
34	706326	3·56	935022	1·24	771303	4·81	228697	26
35	706539	3·56	934948	1·24	771592	4·81	228408	25
36	706753	3·56	934873	1·24	771880	4·80	228120	24
37	706967	3·56	934798	1·25	772168	4·80	227832	23
38	707180	3·55	934723	1·25	772457	4·80	227543	22
39	707393	3·55	934649	1·25	772745	4·80	227255	21
40	707606	3·55	934574	1·25	773033	4·80	226967	20
41	9·707819	3·55	9·934499	1·25	9·773321	4·80	10·226679	19
42	708032	3·54	934424	1·25	773608	4·79	226392	18
43	708245	3·54	934349	1·25	773896	4·79	226104	17
44	708458	3·54	934274	1·25	774184	4·79	225816	16
45	708670	3·54	934199	1·25	774471	4·79	225529	15
46	708882	3·53	934123	1·25	774759	4·79	225241	14
47	709094	3·53	934048	1·25	775046	4·79	224954	13
48	709306	3·53	933973	1·25	775333	4·79	224667	12
49	709518	3·53	933898	1·26	775621	4·78	224379	11
50	709730	3·53	933822	1·26	775908	4·78	224092	10
51	9·709941	3·52	9·933747	1·26	9·776195	4·78	10·223805	9
52	710153	3·52	933671	1·26	776482	4·78	223518	8
53	710364	3·52	933596	1·26	776769	4·78	223231	7
54	710575	3·52	933520	1·26	777055	4·78	222945	6
55	710786	3·51	933445	1·26	777342	4·78	222658	5
56	710997	3·51	933369	1·26	777628	4·77	222372	4
57	711208	3·51	933293	1·26	777915	4·77	222085	3
58	711419	3·51	933217	1·26	778201	4·77	221799	2
59	711629	3·50	933141	1·26	778487	4·77	221512	1
60	711839	3·50	933066	1·26	778774	4·77	221226	0
	Cosine.	D.	Sine.	D.	Cotang.	D.	Tang.	M.

M.	Sine.	D.	Cosine.	D.	Tang.	D.	Cotang.	
0	9·711839	3·50	9·933066	1·26	9·778774	4·77	10·221226	60
1	712050	3·50	932990	1·27	779060	4·77	220940	59
2	712260	3·50	932914	1·27	779346	4·76	220654	58
3	712469	3·49	932838	1·27	779632	4·76	220368	57
4	712679	3·49	932762	1·27	779918	4·76	220082	56
5	712889	3·49	932685	1·27	780203	4·76	219797	55
6	713098	3·49	932609	1·27	780489	4·76	219511	54
7	713308	3·49	932533	1·27	780775	4·76	219225	53
8	713517	3·48	932457	1·27	781060	4·76	218940	52
9	713726	3·48	932380	1·27	781346	4·75	218654	51
10	713935	3·48	932304	1·27	781631	4·75	218369	50
11	9·714144	3·48	9·932228	1·27	9·781916	4·75	10·218084	49
12	714352	3·47	932151	1·27	782201	4·75	217799	48
13	714561	3·47	932075	1·28	782486	4·75	217514	47
14	714769	3·47	931998	1·28	782771	4·75	217229	46
15	714978	3·47	931921	1·28	783056	4·75	216944	45
16	715186	3·47	931845	1·28	783341	4·75	216659	44
17	715394	3·46	931768	1·28	783626	4·74	216374	43
18	715602	3·46	931691	1·28	783910	4·74	216090	42
19	715809	3·46	931614	1·28	784195	4·74	215805	41
20	716017	3·46	931537	1·28	784479	4·74	215521	40
21	9·716224	3·45	9·931460	1·28	9·784764	4·74	10·215236	39
22	716432	3·45	931383	1·28	785048	4·74	214952	38
23	716639	3·45	931306	1·28	785332	4·73	214668	37
24	716846	3·45	931229	1·29	785616	4·73	214384	36
25	717053	3·45	931152	1·29	785900	4·73	214100	35
26	717259	3·44	931075	1·29	786184	4·73	213816	34
27	717466	3·44	930998	1·29	786468	4·73	213532	33
28	717673	3·44	930921	1·29	786752	4·73	213248	32
29	717879	3·44	930843	1·29	787036	4·73	212964	31
30	718085	3·43	930766	1·29	787319	4·72	212681	30
31	9·718291	3·43	9·930688	1·29	9·787603	4·72	10·212397	29
32	718497	3·43	930611	1·29	787886	4·72	212114	28
33	718703	3·43	930533	1·29	788170	4·72	211830	27
34	718909	3·43	930456	1·29	788453	4·72	211547	26
35	719114	3·42	930378	1·29	788736	4·72	211264	25
36	719320	3·42	930300	1·30	789019	4·72	210981	24
37	719525	3·42	930223	1·30	789302	4·71	210698	23
38	719730	3·42	930145	1·30	789585	4·71	210415	22
39	719935	3·41	930067	1·30	789868	4·71	210132	21
40	720140	3·41	929989	1·30	790151	4·71	209849	20
41	9·720345	3·41	9·929911	1·30	9·790433	4·71	10·209567	19
42	720549	3·41	929833	1·30	790716	4·71	209284	18
43	720754	3·40	929755	1·30	790999	4·71	209001	17
44	720958	3·40	929677	1·30	791281	4·71	208719	16
45	721162	3·40	929599	1·30	791563	4·70	208437	15
46	721366	3·40	929521	1·30	791846	4·70	208154	14
47	721570	3·40	929442	1·30	792128	4·70	207872	13
48	721774	3·39	929364	1·31	792410	4·70	207590	12
49	721978	3·39	929286	1·31	792692	4·70	207308	11
50	722181	3·39	929207	1·31	792974	4·70	207026	10
51	9·722385	3·39	9·929129	1·31	9·793256	4·70	10·206744	9
52	722588	3·39	929050	1·31	793538	4·69	206462	8
53	722791	3·38	928972	1·31	793819	4·69	206181	7
54	722994	3·38	928893	1·31	794101	4·69	205899	6
55	723197	3·38	928815	1·31	794383	4·69	205617	5
56	723400	3·38	928736	1·31	794664	4·69	205336	4
57	723603	3·37	928657	1·31	794945	4·69	205055	3
58	723805	3·37	928578	1·31	795227	4·69	204773	2
59	724007	3·37	928499	1·31	795508	4·68	204492	1
60	724210	3·37	928420	1·31	795789	4·68	204211	0
	Cosine.	D.	Sine.	D.	Cotang.	D.	Tang.	M.

M.	Sine.	D.	Cosine.	D.	Tang.	D.	Cotang.	
0	9·724210	3·37	9·928420	1·32	9·795789	4·68	10·204211	60
1	724412	3·37	928342	1·32	796070	4·68	203930	59
2	724614	3·36	928263	1·32	796351	4·68	203649	58
3	724816	8·36	928183	1·32	796632	4·68	203368	57
4	725017	3·36	928104	1·32	796913	4·68	203087	56
5	725219	8·36	928025	1·32	797194	4·68	202806	55
6	725420	3·35	927946	1·32	797475	4·68	202525	54
7	725622	3·35	927867	1·32	797755	4·68	202245	53
8	725823	3·35	927787	1·32	798036	4·67	201964	52
9	726024	3·35	927708	1·32	798316	4·67	201684	51
10	726225	3·35	927629	1·32	798596	4·67	201404	50
11	9·726426	3·34	9·927549	1·32	9·798877	4·67	10·201123	49
12	726626	3·34	927470	1·33	799157	4·67	200843	48
13	726827	3·34	927390	1·33	799437	4·67	200563	47
14	727027	3·34	927310	1·33	799717	4·67	200283	46
15	727228	3·34	927231	1·33	799997	4·66	200003	45
16	727428	3·33	927151	1·33	800277	4·66	199723	44
17	727628	3·33	927071	1·33	800557	4·66	199443	43
18	727828	3·33	926991	1·33	800836	4·66	199164	42
19	728027	3·33	926911	1·33	801116	4·66	198884	41
20	728227	3·33	926831	1·33	801396	4·66	198604	40
21	9·728427	3·32	9·926751	1·33	9·801675	4·66	10·198325	39
22	728626	3·32	926671	1·33	801955	4·66	198045	38
23	728825	3·32	926591	1·33	802234	4·65	197766	37
24	729024	3·32	926511	1·34	802513	4·65	197487	36
25	729223	3·31	926431	1·34	802792	4·65	197208	35
26	729422	3·31	926351	1·34	803072	4·65	196928	34
27	729621	3·31	926270	1·34	803351	4·65	196649	33
28	729820	3·31	926190	1·34	803630	4·65	196370	32
29	730018	3·30	926110	1·34	803908	4·65	196092	31
30	730216	3·30	926029	1·34	804187	4·65	195813	30
31	9·730415	3·30	9·925949	1·34	9·804466	4·64	10·195534	29
32	730613	3·30	925868	1·34	804745	4·64	195255	28
33	730811	3·30	925788	1·34	805023	4·64	194977	27
34	731009	8·29	925707	1·34	805302	4·64	194698	26
35	731206	3·29	925626	1·34	805580	4·64	194420	25
36	731404	3·29	925545	1·35	805859	4·64	194141	24
37	731602	3·29	925465	1·35	806137	4·64	193863	23
38	731799	3·29	925384	1·35	806415	4·63	193585	22
39	731996	3·28	925303	1·35	806693	4·63	193307	21
40	732193	3·28	925222	1·35	806971	4·63	193029	20
41	9·732390	3·28	9·925141	1·35	9·807249	4·63	10·192751	19
42	732587	3·28	925060	1·35	807527	4·63	192473	18
43	732784	3·28	924979	1·35	807805	4·63	192195	17
44	732980	3·27	924897	1·35	808083	4·63	191917	16
45	733177	3·27	924816	1·35	808361	4·63	191639	15
46	733373	3·27	924735	1·36	808638	4·62	191362	14
47	733569	3·27	924654	1·36	808916	4·62	191084	13
48	733765	3·27	924572	1·36	809193	4·62	190807	12
49	733961	3·26	924491	1·36	809471	4·62	190529	11
50	734157	3·26	924409	1·36	809748	4·62	190252	10
51	9·734353	3·26	9·924328	1·36	9·810025	4·62	10·189975	9
52	734549	3·26	924246	1·36	810302	4·62	189698	8
53	734744	3·25	924164	1·36	810580	4·62	189420	7
54	734939	3·25	924083	1·36	810857	4·62	189143	6
55	735135	3·25	924001	1·36	811134	4·61	188866	5
56	735330	3·25	923919	1·36	811410	4·61	188590	4
57	735525	3·25	923837	1·36	811687	4·61	188313	3
58	735719	3·24	923755	1·37	811964	4·61	188036	2
59	735914	3·24	923673	1·37	812241	4·61	187759	1
60	736109	8·24	923591	1·37	812517	4·61	187483	0
	Cosine.	D.	Sine.	D.	Cotang.	D.	Tang.	M.

M.	Sine.	D.	Cosine.	D.	Tang.	D.	Cotang.	
0	9·786109	3·24	9·923591	1·37	9·812517	4·61	10·187482	60
1	736303	3·24	923509	1·37	812794	4·61	187206	59
2	736498	3·24	923427	1·37	813070	4·61	186930	58
3	736692	3·23	923345	1·37	813347	4·60	186653	57
4	736886	3·23	923263	1·37	813623	4·60	186377	56
5	737080	3·23	923181	1·37	813899	4·60	186101	55
6	737274	3·23	923098	1·37	814175	4·60	185825	54
7	737467	3·23	923016	1·37	814452	4·60	185548	53
8	737661	3·22	922933	1·37	814728	4·60	185272	52
9	737855	3·22	922851	1·37	815004	4·60	184996	51
10	738048	3·22	922768	1·38	815279	4·60	184721	50
11	9·738241	3·22	9·922686	1·38	9·815555	4·59	10·184445	49
12	738434	3·22	922603	1·38	815831	4·59	184169	48
13	738627	3·21	922520	1·38	816107	4·59	183893	47
14	738820	3·21	922438	1·38	816382	4·59	183618	46
15	739013	3·21	922355	1·38	816658	4·59	183342	45
16	739206	3·21	922272	1·38	816933	4·59	183067	44
17	739398	3·21	922189	1·38	817209	4·59	182791	43
18	739590	3·20	922106	1·38	817484	4·59	182516	42
19	739783	3·20	922023	1·38	817759	4·59	182241	41
20	739975	3·20	921940	1·38	818035	4·58	181965	40
21	9·740167	3·20	9·921857	1·39	9·818310	4·58	10·181690	39
22	740359	3·20	921774	1·39	818585	4·58	181415	38
23	740550	3·19	921691	1·39	818860	4·58	181140	37
24	740742	3·19	921607	1·39	819135	4·58	180865	36
25	740934	3·19	921524	1·39	819410	4·58	180590	35
26	741125	3·19	921441	1·39	819684	4·58	180316	34
27	741316	3·19	921357	1·39	819959	4·58	180041	33
28	741508	3·18	921274	1·39	820234	4·58	179766	32
29	741699	3·18	921190	1·39	820508	4·57	179492	31
30	741889	3·18	921107	1·39	820783	4·57	179217	30
31	9·742080	3·18	9·921023	1·39	9·821057	4·57	10·178943	29
32	742271	3·18	920939	1·40	821332	4·57	178668	28
33	742462	3·17	920856	1·40	821606	4·57	178394	27
34	742652	3·17	920772	1·40	821880	4·57	178120	26
35	742842	3·17	920688	1·40	822154	4·57	177846	25
36	743033	3·17	920604	1·40	822429	4·57	177571	24
37	743223	3·17	920520	1·40	822703	4·57	177297	23
38	743413	3·16	920436	1·40	822977	4·56	177023	22
39	743602	3·16	920352	1·40	823250	4·56	176750	21
40	743792	3·16	920268	1·40	823524	4·56	176476	20
41	9·743982	3·16	9·920184	1·40	9·823798	4·56	10·176202	19
42	744171	3·16	920099	1·40	824072	4·56	175928	18
43	744361	3·15	920015	1·40	824345	4·56	175655	17
44	744550	3·15	919931	1·41	824619	4·56	175381	16
45	744739	3·15	919846	1·41	824893	4·56	175107	15
46	744928	3·15	919762	1·41	825166	4·56	174834	14
47	745117	3·15	919677	1·41	825439	4·55	174561	13
48	745306	3·14	919593	1·41	825713	4·55	174287	12
49	745494	3·14	919508	1·41	825986	4·55	174014	11
50	745683	3·14	919424	1·41	826259	4·55	173741	10
51	9·745871	3·14	9·919339	1·41	9·826532	4·55	10·173468	9
52	746059	3·14	919254	1·41	826805	4·55	173195	8
53	746248	3·13	919169	1·41	827078	4·55	172922	7
54	746436	3·13	919085	1·41	827351	4·55	172649	6
55	746624	3·13	919000	1·41	827624	4·55	172376	5
56	746812	3·13	918915	1·42	827897	4·54	172103	4
57	746999	3·13	918830	1·42	828170	4·54	171880	3
58	747187	3·12	918745	1·42	828442	4·54	171558	2
59	747374	3·12	918659	1·42	828715	4·54	171285	1
60	747562	3·12	918574	1·42	828987	4·54	171013	0
	Cosine.	D.	Sine.	D.	Cotang.	D.	Tang.	M.

(56 DEGREES.)

M.	Sine.	D.	Cosine.	D.	Tang.	D.	Cotang.	
0	9·747562	3·12	9·918574	1·42	9·828987	4·54	10·171013	60
1	747749	3·12	918489	1·42	829260	4·54	170740	59
2	747936	3·12	918404	1·42	829532	4·54	170468	58
3	748123	3·11	918318	1·42	829805	4·54	170195	57
4	748310	3·11	918233	1·42	830077	4·54	169923	56
5	748497	3·11	918147	1·42	830349	4·53	169651	55
6	748683	3·11	918062	1·42	830621	4·53	169379	54
7	748870	3·11	917976	1·43	830893	4·53	169107	53
8	749056	3·10	917891	1·43	831165	4·53	168835	52
9	749243	3·10	917805	1·43	831437	4·53	168563	51
10	749429	3·10	917719	1·43	831709	4·53	168291	50
11	9·749615	3·10	9·917634	1·43	9·831981	4·53	10·168019	49
12	749801	3·10	917548	1·43	832253	4·53	167747	48
13	749987	3·09	917462	1·43	832525	4·53	167475	47
14	750172	3·09	917376	1·43	832796	4·53	167204	46
15	750358	3·09	917290	1·43	833068	4·52	166932	45
16	750543	3·09	917204	1·43	833339	4·52	166661	44
17	750729	3·09	917118	1·44	833611	4·52	166389	43
18	750914	3·08	917032	1·44	833882	4·52	166118	42
19	751099	3·08	916946	1·44	834154	4·52	165846	41
20	751284	3·08	916859	1·44	834425	4·52	165575	40
21	9·751469	3·08	9·916773	1·44	9·834696	4·52	10·165304	39
22	751654	3·08	916687	1·44	834967	4·52	165033	38
23	751839	3·08	916600	1·44	835238	4·52	164762	37
24	752023	3·07	916514	1·44	835509	4·52	164491	36
25	752208	3·07	916427	1·44	835780	4·51	164220	35
26	752392	3·07	916341	1·44	836051	4·51	163949	34
27	752576	3·07	916254	1·44	836322	4·51	163678	33
28	752760	3·07	916167	1·45	836593	4·51	163407	32
29	752944	3·06	916081	1·45	836864	4·51	163136	31
30	753128	3·06	915994	1·45	837134	4·51	162866	30
31	9·753312	3·06	9·915907	1·45	9·837405	4·51	10·162595	29
32	753495	3·06	915820	1·45	837675	4·51	162325	28
33	753679	3·06	915733	1·45	837946	4·51	162054	27
34	753862	3·05	915646	1·45	838216	4·51	161784	26
35	754046	3·05	915559	1·45	838487	4·50	161513	25
36	754229	3·05	915472	1·45	838757	4·50	161243	24
37	754412	3·05	915385	1·45	839027	4·50	160973	23
38	754595	3·05	915297	1·45	839297	4·50	160703	22
39	754778	3·04	915210	1·45	839568	4·50	160432	21
40	754960	3·04	915123	1·46	839838	4·50	160162	20
41	9·755143	3·04	9·915035	1·46	9·840108	4·50	10·159892	19
42	755326	3·04	914948	1·46	840378	4·50	159622	18
43	755508	3·04	914860	1·46	840647	4·50	159353	17
44	755690	3·04	914773	1·46	840917	4·49	159083	16
45	755872	3·03	914685	1·46	841187	4·49	158813	15
46	756054	3·03	914598	1·46	841457	4·49	158543	14
47	756236	3·03	914510	1·46	841726	4·49	158274	13
48	756418	3·03	914422	1·46	841996	4·49	158004	12
49	756600	3·03	914334	1·46	842266	4·49	157734	11
50	756782	3·02	914246	1·47	842535	4·49	157465	10
51	9·756963	3·02	9·914158	1·47	9·842805	4·49	10·157195	9
52	757144	3·02	914070	1·47	843074	4·49	156926	8
53	757326	3·02	913982	1·47	843343	4·49	156657	7
54	757507	3·02	913894	1·47	843612	4·49	156388	6
55	757688	3·01	913806	1·47	843882	4·48	156118	5
56	757869	3·01	913718	1·47	844151	4·48	155849	4
57	758050	3·01	913630	1·47	844420	4·48	155580	3
58	758230	3·01	913541	1·47	844689	4·48	155311	2
59	758411	3·01	913453	1·47	844958	4·48	155042	1
60	758591	3·01	913365	1·47	845227	4·48	154773	0
	Cosine.	D.	Sine.	D.	Cotang.	D.	Tang.	M.

M.	Sine.	D.	Cosine.	D.	Tang.	D.	Cotang.	
0	9·758591	3·01	9·918365	1·47	9·845227	4·48	10·154773	60
1	758772	3·00	918276	1·47	845496	4·48	154504	59
2	758952	3·00	918187	1·48	845764	4·48	154236	58
3	759132	3·00	918099	1·48	846033	4·48	153967	57
4	759312	3·00	918010	1·48	846302	4·48	153698	56
5	759492	3·00	912922	1·48	846570	4·47	153430	55
6	759672	2·99	912833	1·48	846839	4·47	153161	54
7	759852	2·99	912744	1·48	847107	4·47	152893	53
8	760031	2·99	912655	1·48	847376	4·47	152624	52
9	760211	2·99	912566	1·48	847644	4·47	152356	51
10	760390	2·99	912477	1·48	847913	4·47	152087	50
11	9·760569	2·98	9·912388	1·48	9·848181	4·47	10·151819	49
12	760748	2·98	912299	1·49	848449	4·47	151551	48
13	760927	2·98	912210	1·49	848717	4·47	151283	47
14	761106	2·98	912121	1·49	848986	4·47	151014	46
15	761285	2·98	912031	1·49	849254	4·47	150746	45
16	761464	2·98	911942	1·49	849522	4·47	150478	44
17	761642	2·97	911853	1·49	849790	4·46	150210	43
18	761821	2·97	911763	1·49	850058	4·46	149942	42
19	761999	2·97	911674	1·49	850325	4·46	149675	41
20	762177	2·97	911584	1·49	850593	4·46	149407	40
21	9·762356	2·97	9·911495	1·49	9·850861	4·46	10·149139	39
22	762534	2·96	911405	1·49	851129	4·46	148871	38
23	762712	2·96	911315	1·50	851396	4·46	148604	37
24	762889	2·96	911226	1·50	851664	4·46	148336	36
25	763067	2·96	911136	1·50	851931	4·46	148069	35
26	763245	2·96	911046	1·50	852199	4·46	147801	34
27	763422	2·96	910956	1·50	852466	4·46	147534	33
28	763600	2·95	910866	1·50	852733	4·45	147267	32
29	763777	2·95	910776	1·50	853001	4·45	146999	31
30	763954	2·95	910686	1·50	853268	4·45	146732	30
31	9·764131	2·95	9·910596	1·50	9·853535	4·45	10·146465	29
32	764308	2·95	910506	1·50	853802	4·45	146198	28
33	764485	2·94	910415	1·50	854069	4·45	145931	27
34	764662	2·94	910325	1·51	854336	4·45	145664	26
35	764838	2·94	910235	1·51	854603	4·45	145397	25
36	765015	2·94	910144	1·51	854870	4·45	145130	24
37	765191	2·94	910054	1·51	855137	4·45	144863	23
38	765367	2·94	909963	1·51	855404	4·45	144596	22
39	765544	2·93	909873	1·51	855671	4·44	144329	21
40	765720	2·93	909782	1·51	855938	4·44	144062	20
41	9·765896	2·93	9·909691	1·51	9·856204	4·44	10·143796	19
42	766072	2·93	909601	1·51	856471	4·44	143529	18
43	766247	2·93	909510	1·51	856737	4·44	143263	17
44	766423	2·93	909419	1·51	857004	4·44	142996	16
45	766598	2·92	909328	1·52	857270	4·44	142730	15
46	766774	2·92	909237	1·52	857537	4·44	142463	14
47	766949	2·92	909146	1·52	857803	4·44	142197	13
48	767124	2·92	909055	1·52	858069	4·44	141931	12
49	767300	2·92	908964	1·52	858336	4·44	141664	11
50	767475	2·91	908873	1·52	858602	4·43	141398	10
51	9·767649	2·91	9·908781	1·52	9·858868	4·43	10·141132	9
52	767824	2·91	908690	1·52	859184	4·43	140866	8
53	767999	2·91	908599	1·52	859400	4·43	140600	7
54	768173	2·91	908507	1·52	859666	4·43	140334	6
55	768348	2·90	908416	1·53	859932	4·43	140068	5
56	768522	2·90	908324	1·53	860198	4·43	139802	4
57	768697	2·90	908233	1·53	860464	4·43	139536	3
58	768871	2·90	908141	1·53	860730	4·43	139270	2
59	769045	2·90	908049	1·53	860995	4·43	139005	1
60	769219	2·90	907958	1·53	861261	4·43	138739	0
	Cosine.	D.	Sine.	D.	Cotang.	D.	Tang.	M.

M.	Sine.	D.	Cosine.	D.	Tang.	D.	Cotang.	
0	9·769219	2·90	9·907958	1·53	9·861261	4·48	10·188739	60
1	769393	2·89	907866	1·53	861527	4·43	188473	59
2	769566	2·89	907774	1·53	861792	4·42	138208	58
3	769740	2·89	907682	1·53	862058	4·42	187942	57
4	769913	2·89	907590	1·53	862323	4·42	187677	56
5	770087	2·89	907498	1·53	862589	4·42	137411	55
6	770260	2·88	907406	1·53	862854	4·42	187146	54
7	770433	2·88	907314	1·54	863119	4·42	186881	53
8	770606	2·88	007222	1·54	863385	4·42	186615	52
9	770779	2·88	907129	1·54	863650	4·42	136350	51
10	770952	2·88	907037	1·54	863915	4·42	136085	50
11	9·771125	2·88	9·906945	1·54	9·864180	4·42	10·135820	49
12	771298	2·87	906852	1·54	864445	4·42	185555	48
13	771470	2·87	906760	1·54	864710	4·42	135290	47
14	771643	2·87	906667	1·54	864975	4·41	135025	46
15	771815	2·87	906575	1·54	865240	4·41	184760	45
16	771987	2·87	906482	1·54	865505	4·41	184495	44
17	772159	2·87	906389	1·55	865770	4·41	134230	43
18	772331	2·86	906296	1·55	866035	4·41	133965	42
19	772503	2·86	906204	1·55	866300	4·41	133700	41
20	772675	2·86	906111	1·55	866564	4·41	133436	40
21	9·772847	2·86	9·906018	1·55	9·866829	4·41	10·133171	39
22	773018	2·86	905925	1·55	867094	4·41	132906	38
23	773190	2·86	905832	1·55	867358	4·41	132642	37
24	773361	2·85	905739	1·55	867623	4·41	182377	36
25	773533	2·85	905645	1·55	867887	4·41	132113	35
26	773704	2·85	905552	1·55	868152	4·40	131848	34
27	773875	2·85	905459	1·55	868416	4·40	131584	33
28	774046	2·85	905366	1·56	868680	4·40	131320	32
29	774217	2·85	905272	1·56	868945	4·40	181055	31
30	774388	2·84	905179	1·56	869209	4·40	180794	30
31	9·774558	2·84	9·905085	1·56	9·869473	4·40	10·130527	29
32	774729	2·84	904992	1·56	869737	4·40	130263	28
33	774899	2·84	904898	1·56	870001	4·40	129999	27
34	775070	2·84	904804	1·56	870265	4·40	129735	26
35	775240	2·84	904711	1·56	870529	4·40	129471	25
36	775410	2·83	904617	1·56	870793	4·40	129207	24
37	775580	2·83	904523	1·56	871057	4·40	128943	23
38	775750	2·83	904429	1·57	871321	4·40	128679	22
39	775920	2·83	904335	1·57	871585	4·40	128415	21
40	776090	2·83	904241	1·57	871849	4·39	128151	20
41	9·776259	2·83	9·904147	1·57	9·872112	4·39	10·127888	19
42	776429	2·82	904053	1·57	872376	4·39	127624	18
43	776598	2·82	903959	1·57	872640	4·39	127360	17
44	776768	2·82	903864	1·57	872903	4·39	127097	16
45	776937	2·82	903770	1·57	873167	4·39	126833	15
46	777106	2·82	903676	1·57	873430	4·39	126570	14
47	777275	2·81	903581	1·57	873694	4·39	126306	13
48	777444	2·81	903487	1·57	873957	4·39	126043	12
49	777613	2·81	903392	1·58	874220	4·39	125780	11
50	777781	2·81	903298	1·58	874484	4·39	125516	10
51	9·777950	2·81	0·903203	1·58	9·874747	4·39	10·125253	9
52	778119	2·81	903108	1·58	875010	4·39	124990	8
53	778287	2·80	903014	1·58	875273	4·38	124727	7
54	778455	2·80	902919	1·58	875536	4·38	124464	6
55	778624	2·80	902824	1·58	875800	4·38	124200	5
56	778792	2·80	902729	1·58	876063	4·38	123937	4
57	778960	2·80	902634	1·58	876326	4·38	123674	3
58	779128	2·80	902539	1·59	876589	4·38	123411	2
59	779295	2·79	902444	1·59	876851	4·38	123149	1
60	779463	2·79	902349	1·59	877114	4·38	122886	0
	Cosine.	D.	Sine.	D.	Cotang.	D.	Tang.	M.

M.	Sine.	D.	Cosine.	D.	Tang.	D.	Cotang.	
0	9·779468	2·79	9·902349	1·59	9·877114	4·38	10·122886	60
1	779631	2·79	902253	1·59	877377	4·38	122623	59
2	779798	2·79	902158	1·59	877640	4·38	122360	58
3	779966	2·79	902063	1·59	877903	4·38	122097	57
4	780133	2·79	901967	1·59	878165	4·38	121835	56
5	780300	2·78	901872	1·59	878428	4·38	121572	55
6	780467	2·78	901776	1·59	878691	4·38	121309	54
7	780634	2·78	901681	1·59	878953	4·37	121047	53
8	780801	2·78	901585	1·59	879216	4·37	120784	52
9	780968	2·78	901490	1·59	879478	4·37	120522	51
10	781134	2·78	901394	1·60	879741	4·37	120259	50
11	9·781301	2·77	9·901298	1·60	9·880003	4·37	10·119997	49
12	781468	2·77	901202	1·60	880265	4·37	119785	48
13	781634	2·77	901106	1·60	880528	4·37	119472	47
14	781800	2·77	901010	1·60	880790	4·37	119210	46
15	781966	2·77	900914	1·60	881052	4·37	118948	45
16	782132	2·77	900818	1·60	881314	4·37	118686	44
17	782298	2·76	900722	1·60	881576	4·37	118424	43
18	782464	2·76	900626	1·60	881839	4·37	118161	42
19	782630	2·76	900529	1·60	882101	4·37	117899	41
20	782796	2·76	900433	1·61	882363	4·36	117637	40
21	9·782961	2·76	9·900337	1·61	9·882625	4·36	10·117375	39
22	783127	2·76	900240	1·61	882887	4·36	117113	38
23	783292	2·75	900144	1·61	883148	4·36	116852	37
24	783458	2·75	900047	1·61	883410	4·36	116590	36
25	783623	2·75	899951	1·61	883672	4·36	116328	35
26	783788	2·75	899854	1·61	883934	4·36	116066	34
27	783953	2·75	899757	1·61	884196	4·36	115804	33
28	784118	2·75	899660	1·61	884457	4·36	115543	32
29	784282	2·74	899564	1·61	884719	4·36	115281	31
30	784447	2·74	899467	1·62	884980	4·36	115020	30
31	9·784612	2·74	9·899370	1·62	9·885242	4·36	10·114758	29
32	784776	2·74	899273	1·62	885503	4·36	114497	28
33	784941	2·74	899176	1·62	885765	4·36	114235	27
34	785105	2·74	899078	1·62	886026	4·36	113974	26
35	785269	2·73	898981	1·62	886288	4·36	113712	25
36	785433	2·73	898884	1·62	886549	4·35	113451	24
37	785597	2·73	898787	1·62	886810	4·35	113190	23
38	785761	2·73	898689	1·62	887072	4·35	112928	22
39	785925	2·73	898592	1·62	887333	4·35	112667	21
40	786089	2·73	898494	1·63	887594	4·35	112406	20
41	9·786252	2·72	9·898397	1·63	9·887855	4·35	10·112145	19
42	786416	2·72	898299	1·63	888116	4·35	111884	18
43	786579	2·72	898202	1·63	888377	4·35	111623	17
44	786742	2·72	898104	1·63	888639	4·35	111361	16
45	786906	2·72	898006	1·63	888900	4·35	111100	15
46	787069	2·72	897908	1·63	889160	4·35	110840	14
47	787232	2·71	897810	1·63	889421	4·35	110579	13
48	787395	2·71	897712	1·63	889682	4·35	110318	12
49	787557	2·71	897614	1·63	889943	4·35	110057	11
50	787720	2·71	897516	1·63	890204	4·34	109796	10
51	9·787883	2·71	9·897418	1·64	9·890465	4·34	10·109535	9
52	788045	2·71	897320	1·64	890725	4·34	109275	8
53	788208	2·71	897222	1·64	890986	4·34	109014	7
54	788370	2·70	897123	1·64	891247	4·34	108753	6
55	788532	2·70	897025	1·64	891507	4·34	108493	5
56	788694	2·70	896926	1·64	891768	4·34	108232	4
57	788856	2·70	896828	1·64	892028	4·34	107972	3
58	789018	2·70	896729	1·64	892289	4·34	107711	2
59	789180	2·70	896631	1·64	892549	4·34	107451	1
60	789342	2·69	896532	1·64	892810	4·34	107190	0
	Cosine.	D.	Sine.	D.	Cotang.	D.	Tang.	M.

M.	Sine.	D.	Cosine.	D.	Tang.	D.	Cotang.	
0	9·789342	2·69	9·896532	1·64	9·892810	4·34	10·107190	60
1	789504	2·69	896433	1·65	893070	4·34	106930	59
2	789665	2·69	896335	1·65	893331	4·34	106669	58
3	789827	2·69	896236	1·65	893591	4·34	106409	57
4	789988	2·69	896137	1·65	893851	4·34	106149	56
5	790149	2·69	896038	1·65	894111	4·34	105889	55
6	790310	2·68	895939	1·65	894371	4·34	105629	54
7	790471	2·68	895840	1·65	894632	4·33	105368	53
8	790632	2·68	895741	1·65	894892	4·33	105108	52
9	790793	2·68	895641	1·65	895152	4·33	104848	51
10	790954	2·68	895542	1·65	895412	4·33	104588	50
11	9·791115	2·68	9·895443	1·66	9·895672	4·33	10·104328	49
12	791275	2·67	895343	1·66	895932	4·33	104068	48
13	791436	2·67	895244	1·66	896192	4·33	103808	47
14	791596	2·67	895145	1·66	896452	4·33	103548	46
15	791757	2·67	895045	1·66	896712	4·33	103288	45
16	791917	2·67	894945	1·66	896971	4·33	103029	44
17	792077	2·67	894846	1·66	897231	4·33	102769	43
18	792237	2·66	894746	1·66	897491	4·33	102509	42
19	792397	2·66	894646	1·66	897751	4·33	102249	41
20	792557	2·66	894546	1·66	898010	4·33	101990	40
21	9·792716	2·66	9·894446	1·67	9·898270	4·23	10·101730	39
22	792876	2·66	894346	1·67	898530	4·33	101470	38
23	793035	2·66	894246	1·67	898789	4·33	101211	37
24	793195	2·65	894146	1·67	899049	4·32	100951	36
25	793354	2·65	894046	1·67	899308	4·32	100692	35
26	793514	2·65	893946	1·67	899568	4·32	100432	34
27	793673	2·65	893846	1·67	899827	4·32	100173	33
28	793832	2·65	893745	1·67	900086	4·32	099914	32
29	793991	2·65	893645	1·67	900346	4·32	099654	31
30	794150	2·64	893544	1·67	900605	4·32	099395	30
31	9·794308	2·64	9·893444	1·68	9·900864	4·32	10·099136	29
32	794467	2·64	893343	1·68	901124	4·32	098876	28
33	794626	2·64	893243	1·68	901383	4·32	098617	27
34	794784	2·64	893142	1·68	901642	4·32	098358	26
35	794942	2·64	893041	1·68	901901	4·32	098099	25
36	795101	2·64	892940	1·68	902160	4·32	097840	24
37	795259	2·63	892839	1·68	902419	4·32	097581	23
38	795417	2·63	892739	1·68	902679	4·32	097321	22
39	795575	2·63	892638	1·68	902938	4·32	097062	21
40	795733	2·63	892536	1·68	903197	4·31	096803	20
41	795891	2·63	9·892435	1·69	9·903455	4·31	10·096545	19
42	796049	2·63	892334	1·69	903714	4·31	096286	18
43	796206	2·63	892233	1·69	903973	4·31	096027	17
44	796364	2·62	892132	1·69	904232	4·31	095768	16
45	796521	2·62	892030	1·69	904491	4·31	095509	15
46	796679	2·62	891929	1·69	904750	4·31	095250	14
47	796836	2·62	891827	1·69	905008	4·31	094992	13
48	796993	2·62	891726	1·69	905267	4·31	094733	12
49	797150	2·61	891624	1·69	905526	4·31	094474	11
50	797307	2·61	891523	1·70	905784	4·31	094216	10
51	9·797464	2·61	9·891421	1·70	9·906043	4·31	10·093957	9
52	797621	2·61	891319	1·70	906302	4·31	093698	8
53	797777	2·61	891217	1·70	906560	4·31	093440	7
54	797934	2·61	891115	1·70	906819	4·31	093181	6
55	798091	2·61	891013	1·70	907077	4·31	092923	5
56	798247	2·61	890911	1·70	907336	4·31	092664	4
57	798403	2·60	890809	1·70	907594	4·31	092406	3
58	798560	2·60	890707	1·70	907852	4·31	092148	2
59	798716	2·60	890605	1·70	908111	4·30	091889	1
60	798872	2·60	890503	1·70	908369	4·30	091631	0
	Cosine.	D.	Sine.	D.	Cotang.	D.	Tang.	M.

M.	Sine.	D.	Cosine.	D.	Tang.	D.	Cotang.	
0	9·798872	2·60	9·890503	1·70	9·908369	4·30	10·091631	60
1	799028	2·60	890400	1·71	908628	4·30	091372	59
2	799184	2·60	890298	1·71	908886	4·30	091114	58
3	799339	2·59	890195	1·71	909144	4·30	090856	57
4	799495	2·59	890093	1·71	909402	4·30	090598	56
5	799651	2·59	889990	1·71	909660	4·30	090340	55
6	799806	2·59	889888	1·71	909918	4·30	090082	54
7	799962	2·59	889785	1·71	910177	4·30	089823	53
8	800117	2·59	889682	1·71	910435	4·30	089565	52
9	800272	2·58	889579	1·71	910693	4·30	089307	51
10	800427	2·58	889477	1·71	910951	4·30	089049	50
11	9·800582	2·58	9·889374	1·72	9·911209	4·30	10·088791	49
12	800737	2·58	889271	1·72	911467	4·30	088533	48
13	800892	2·58	889168	1·72	911724	4·30	088276	47
14	801047	2·58	889064	1·72	911982	4·30	088018	46
15	801201	2·58	888961	1·72	912240	4·30	087760	45
16	801356	2·57	888858	1·72	912498	4·30	087502	44
17	801511	2·57	888755	1·72	912756	4·30	087244	43
18	801665	2·57	888651	1·72	913014	4·29	086986	42
19	801819	2·57	888548	1·72	913271	4·29	086729	41
20	801973	2·57	888444	1·73	913529	4·29	086471	40
21	9·802128	2·57	9·888341	1·73	9·913787	4·29	10·086213	39
22	802282	2·56	888237	1·73	914044	4·29	085956	38
23	802436	2·56	888134	1·73	914302	4·29	085698	37
24	802589	2·56	888030	1·73	914560	4·29	085440	36
25	802743	2·56	887926	1·73	914817	4·29	085183	35
26	802897	2·56	887822	1·73	915075	4·29	084925	34
27	803050	2·56	887718	1·73	915332	4·29	084668	33
28	803204	2·56	887614	1·73	915590	4·29	084410	32
29	803357	2·55	887510	1·73	915847	4·29	084153	31
30	803511	2·55	887406	1·74	916104	4·29	083896	30
31	9·803664	2·55	9·887302	1·74	9·916362	4·29	10·083638	29
32	803817	2·55	887198	1·74	916619	4·29	083381	28
33	803970	2·55	887093	1·74	916877	4·29	083123	27
34	804123	2·55	886989	1·74	917134	4·29	082866	26
35	804276	2·54	886885	1·74	917391	4·29	082609	25
36	804428	2·54	886780	1·74	917648	4·29	082352	24
37	804581	2·54	886676	1·74	917905	4·29	082095	23
38	804734	2·54	886571	1·74	918163	4·28	081837	22
39	804886	2·54	886466	1·74	918420	4·28	081580	21
40	805039	2·54	886362	1·75	918677	4·28	081323	20
41	9·805191	2·54	9·886257	1·75	9·918934	4·28	10·081066	19
42	805343	2·53	886152	1·75	919191	4·28	080809	18
43	805495	2·53	886047	1·75	919448	4·28	080552	17
44	805647	2·53	885942	1·75	919705	4·28	080295	16
45	805799	2·53	885837	1·75	919962	4·28	080038	15
46	805951	2·53	885732	1·75	920219	4·28	079781	14
47	806103	2·53	885627	1·75	920476	4·28	079524	13
48	806254	2·53	885522	1·75	920733	4·28	079267	12
49	806406	2·52	885416	1·75	920990	4·28	079010	11
50	806557	2·52	885311	1·76	921247	4·28	078753	10
51	9·806709	2·52	9·885205	1·76	9·921503	4·28	10·078497	9
52	806860	2·52	885100	1·76	921760	4·28	078240	8
53	807011	2·52	884994	1·76	922017	4·28	077983	7
54	807163	2·52	884889	1·76	922274	4·28	077726	6
55	807314	2·52	884783	1·76	922530	4·28	077470	5
56	807465	2·51	884677	1·76	922787	4·28	077213	4
57	807615	2·51	884572	1·76	923044	4·28	076956	3
58	807766	2·51	884466	1·76	923300	4·28	076700	2
59	807917	2·51	884360	1·76	923557	4·27	076443	1
60	808067	2·51	884254	1·77	923813	4·27	076187	0
	Cosine.	D.	Sine.	D.	Cotang.	D.	Tang.	M.

M.	Sine.	D.	Cosine.	D.	Tang.	D.	Cotang.	
0	9·808067	2·51	9·884254	1·77	9·923813	4·27	10·076187	60
1	808218	2·51	884148	1·77	924070	4·27	075930	59
2	808368	2·51	884042	1·77	924327	4·27	075673	58
3	808519	2·50	883936	1·77	924583	4·27	075417	57
4	808669	2·50	883829	1·77	924840	4·27	075160	56
5	808819	2·50	883723	1·77	925096	4·27	074904	55
6	808969	2·50	883617	1·77	925352	4·27	074648	54
7	809119	2·50	883510	1·77	925609	4·27	074391	53
8	809269	2·50	883404	1·77	925865	4·27	074135	52
9	809419	2·49	883297	1·78	926122	4·27	073878	51
10	809569	2·49	883191	1·78	926378	4·27	073622	50
11	9·809718	2·49	9·883084	1·78	9·926634	4·27	10·073366	49
12	809868	2·49	882977	1·78	926890	4·27	073110	48
13	810017	2·49	882871	1·78	927147	4·27	072853	47
14	810167	2·49	882764	1·78	927403	4·27	072597	46
15	810316	2·48	882657	1·78	927659	4·27	072341	45
16	810465	2·48	882550	1·78	927915	4·27	072085	44
17	810614	2·48	882443	1·78	928171	4·27	071829	43
18	810763	2·48	882336	1·79	928427	4·27	071573	42
19	810912	2·48	882229	1·79	928683	4·27	071817	41
20	811061	2·48	882121	1·79	928940	4·27	071060	40
21	9·811210	2·48	9·882014	1·79	9·929196	4·27	10·070804	39
22	811358	2·47	881907	1·79	929452	4·27	070548	38
23	811507	2·47	881799	1·79	929708	4·27	070292	37
24	811655	2·47	881692	1·79	929964	4·26	070036	36
25	811804	2·47	881584	1·79	930220	4·26	069780	35
26	811952	2·47	881477	1·79	930475	4·26	069525	34
27	812100	2·47	881369	1·79	930731	4·26	069269	33
28	812248	2·47	881261	1·80	930987	4·26	069013	32
29	812396	2·46	881153	1·80	931243	4·26	068757	31
30	812544	2·46	881046	1·80	931499	4·26	068501	30
31	9·812692	2·46	9·880938	1·80	9·931755	4·26	10·068245	29
32	812840	2·46	880830	1·80	932010	4·26	067990	28
33	812988	2·46	880722	1·80	932266	4·26	067734	27
34	813135	2·46	880613	1·80	932522	4·26	067478	26
35	813283	2·46	880505	1·80	932778	4·26	067222	25
36	813430	2·45	880397	1·80	933033	4·26	066967	24
37	813578	2·45	880289	1·81	933289	4·26	066711	23
38	813725	2·45	880180	1·81	933545	4·26	066455	22
39	813872	2·45	880072	1·81	933800	4·26	066200	21
40	814019	2·45	879963	1·81	934056	4·26	065944	20
41	9·814166	2·45	9·879855	1·81	9·934311	4·26	10·065689	19
42	814313	2·45	879746	1·81	934567	4·26	065433	18
43	814460	2·44	879637	1·81	934823	4·26	065177	17
44	814607	2·44	879529	1·81	935078	4·26	064922	16
45	814753	2·44	879420	1·81	935333	4·26	064667	15
46	814900	2·44	879311	1·81	935589	4·26	064411	14
47	815046	2·44	879202	1·82	935844	4·26	064156	13
48	815193	2·44	879093	1·82	936100	4·26	063900	12
49	815339	2·44	878984	1·82	936355	4·26	063645	11
50	815485	2·43	878875	1·82	936610	4·26	063390	10
51	9·815631	2·43	9·878766	1·82	9·936866	4·25	10·063134	9
52	815778	2·43	878656	1·82	937121	4·25	062879	8
53	815924	2·43	878547	1·82	937376	4·25	062624	7
54	816069	2·43	878438	1·82	937632	4·25	062368	6
55	816215	2·43	878328	1·82	937887	4·25	062113	5
56	816361	2·43	878219	1·83	938142	4·25	061858	4
57	816507	2·42	878109	1·83	938398	4·25	061602	3
58	816652	2·42	877999	1·83	938653	4·25	061347	2
59	816798	2·42	877890	1·83	938908	4·25	061092	1
60	816943	2·42	877780	1·83	939163	4·25	060837	0
	Cosine.	D.	Sine.	D.	Cotang.	D.	Tang.	M.

M.	Sine.	D.	Cosine.	D.	Tang.	D.	Cotang.	
0	9·816943	2·42	9·877780	1·88	9·939163	4·25	10·060837	60
1	817088	2·42	877670	1·88	939418	4·25	060582	59
2	817238	2·42	877560	1·88	939673	4·25	060327	58
3	817379	2·42	877450	1·88	939928	4·25	060072	57
4	817524	2·41	877340	1·83	940183	4·25	059817	56
5	817668	2·41	877230	1·84	940438	4·25	059562	55
6	817813	2·41	877120	1·84	940694	4·25	059306	54
7	817958	2·41	877010	1·84	940949	4·25	059051	53
8	818103	2·41	876899	1·84	941204	4·25	058796	52
9	818247	2·41	876789	1·84	941458	4·25	058542	51
10	818392	2·41	876678	1·84	941714	4·25	058286	50
11	9·818536	2·40	9·876568	1·84	9·941968	4·25	10·058032	49
12	818681	2·40	876457	1·84	942223	4·25	057777	48
13	818825	2·40	876347	1·84	942478	4·25	057522	47
14	818969	2·40	876236	1·85	942733	4·25	057267	46
15	819113	2·40	876125	1·85	942988	4·25	057012	45
16	819257	2·40	876014	1·85	943243	4·25	056757	44
17	819401	2·40	875904	1·85	943498	4·25	056502	43
18	819545	2·39	875793	1·85	943752	4·25	056248	42
19	819689	2·39	875682	1·85	944007	4·25	055993	41
20	819832	2·39	875571	1·85	944262	4·25	055738	40
21	9·819976	2·39	9·875459	1·85	9·944517	4·25	10·055483	39
22	820120	2·39	875348	1·85	944771	4·24	055229	38
23	820263	2·39	875237	1·85	945026	4·24	054974	37
24	820406	2·39	875126	1·86	945281	4·24	054719	36
25	820550	2·38	875014	1·86	945535	4·24	054465	35
26	820693	2·38	874903	1·86	945790	4·24	054210	34
27	820836	2·38	874791	1·86	946045	4·24	053955	33
28	820979	2·38	874680	1·86	946299	4·24	053701	32
29	821122	2·38	874568	1·86	946554	4·24	053446	31
30	821266	2·38	874456	1·86	946808	4·24	053192	30
31	9·821407	2·38	9·874344	1·86	9·947063	4·24	10·052937	29
32	821550	2·38	874232	1·87	947318	4·24	052682	28
33	821693	2·37	874121	1·87	947572	4·24	052428	27
34	821835	2·37	874009	1·87	947826	4·24	052174	26
35	821977	2·37	873896	1·87	948081	4·24	051919	25
36	822120	2·37	873784	1·87	948336	4·24	051664	24
37	822262	2·37	873672	1·87	948590	4·24	051410	23
38	822404	2·37	873560	1·87	948844	4·24	051156	22
39	822546	2·37	873448	1·87	949099	4·24	050901	21
40	822688	2·36	873335	1·87	949353	4·24	050647	20
41	9·822830	2·36	9·873223	1·87	9·949607	4·24	10·050393	19
42	822972	2·36	873110	1·88	949862	4·24	050138	18
43	823114	2·36	872998	1·88	950116	4·24	049884	17
44	823255	2·36	872885	1·88	950370	4·24	049630	16
45	823397	2·36	872772	1·88	950625	4·24	049375	15
46	823539	2·36	872659	1·88	950879	4·24	049121	14
47	823680	2·35	872547	1·88	951133	4·24	048867	13
48	823821	2·35	872434	1·88	951388	4·24	048612	12
49	823963	2·35	872321	1·88	951642	4·24	048358	11
50	824104	2·35	872208	1·88	951896	4·24	048104	10
51	9·824245	2·35	9·872095	1·89	9·952150	4·24	10·047850	9
52	824386	2·35	871981	1·89	952405	4·24	047595	8
53	824527	2·35	871868	1·89	952659	4·24	047341	7
54	824668	2·34	871755	1·89	952913	4·24	047087	6
55	824808	2·34	871641	1·89	953167	4·23	046833	5
56	824949	2·34	871528	1·89	953421	4·23	046579	4
57	825090	2·34	871414	1·89	953675	4·23	046325	3
58	825230	2·34	871301	1·89	953929	4·23	046071	2
59	825371	2·34	871187	1·89	954183	4·23	045817	1
60	825511	2·34	871073	1·90	954437	4·23	045563	0
	Cosine.	D.	Sine.	D.	Cotang.	D.	Tang.	M.

(48 DEGREES.)

M.	Sine.	D.	Cosine.	D.	Tang.	D.	Cotang.	
0	9·825511	2·34	9·871073	1·90	9·954437	4·23	10·045563	60
1	825651	2·33	870960	1·90	954691	4·23	045309	59
2	825791	2·33	870846	1·90	954945	4·23	045055	58
3	825931	2·33	870732	1·90	955200	4·23	044800	57
4	826071	2·33	870618	1·90	955454	4·23	044546	56
5	826211	2·33	870504	1·90	955707	4·23	044293	55
6	826351	2·33	870390	1·90	955961	4·23	044039	54
7	826491	2·33	870276	1·90	956215	4·23	043785	53
8	826631	2·33	870161	1·90	956469	4·23	043531	52
9	826770	2·32	870047	1·91	956723	4·23	043277	51
10	826910	2·32	869933	1·91	956977	4·23	043023	50
11	9·827049	2·32	9·869818	1·91	9·957281	4·23	10·042769	49
12	827189	2·32	869704	1·91	957485	4·23	042515	48
13	827328	2·32	869589	1·91	957739	4·23	042261	47
14	827467	2·32	869474	1·91	957993	4·23	042007	46
15	827606	2·32	869360	1·91	958246	4·23	041754	45
16	827745	2·32	869245	1·91	958500	4·23	041500	44
17	827884	2·31	869130	1·91	958754	4·23	041246	43
18	828023	2·31	869015	1·92	959008	4·23	040992	42
19	828162	2·31	868900	1·92	959262	4·23	040738	41
20	828301	2·31	868785	1·92	959516	4·23	040484	40
21	9·828439	2·31	9·868670	1·92	9·959769	4·23	10·040231	39
22	828578	2·31	868555	1·92	960023	4·23	039977	38
23	828716	2·31	868440	1·92	960277	4·23	039723	37
24	828855	2·30	868324	1·92	960531	4·23	039469	36
25	828993	2·30	868209	1·92	960784	4·23	039216	35
26	829131	2·30	868093	1·92	961038	4·23	038962	34
27	829269	2·30	867978	1·93	961291	4·23	038709	33
28	829407	2·30	867862	1·93	961545	4·23	038455	32
29	829545	2·30	867747	1·93	961799	4·23	038201	31
30	829683	2·30	867631	1·93	962052	4·23	037948	30
31	9·829821	2·29	9·867515	1·93	9·962306	4·23	10·037694	29
32	829959	2·29	867399	1·93	962560	4·23	037440	28
33	830097	2·29	867283	1·93	962813	4·23	037187	27
34	830234	2·29	867167	1·93	963067	4·23	036933	26
35	830372	2·29	867051	1·93	963320	4·23	036680	25
36	830509	2·29	866935	1·94	963574	4·23	036426	24
37	830646	2·29	866819	1·94	963827	4·23	036173	23
38	830784	2·29	866703	1·94	964081	4·23	035919	22
39	830921	2·28	866586	1·94	964335	4·23	035665	21
40	831058	2·28	866470	1·94	964588	4·22	035412	20
41	9·831195	2·28	9·866353	1·94	9·964842	4·22	10·035158	19
42	831332	2·28	866237	1·94	965095	4·22	034905	18
43	831469	2·28	866120	1·94	965349	4·22	034651	17
44	831606	2·28	866004	1·95	965602	4·22	034398	16
45	831742	2·28	865887	1·95	965855	4·22	034145	15
46	831879	2·28	865770	1·95	966105	4·22	033891	14
47	832015	2·27	865653	1·95	966362	4·22	033638	13
48	832152	2·27	865536	1·95	966616	4·22	033384	12
49	832288	2·27	865419	1·95	966869	4·22	033131	11
50	832425	2·27	865302	1·95	967123	4·22	032877	10
51	9·832561	2·27	9·865185	1·95	9·967376	4·22	10·032624	9
52	832697	2·27	865068	1·95	967629	4·22	032371	8
53	832833	2·27	864950	1·95	967883	4·22	032117	7
54	832969	2·26	864833	1·96	968136	4·22	031864	6
55	833105	2·26	864716	1·96	968389	4·22	031611	5
56	833241	2·26	864598	1·96	968643	4·22	031357	4
57	833377	2·26	864481	1·96	968896	4·22	031104	3
58	833512	2·26	864363	1·96	969149	4·22	030851	2
59	833648	2·26	864245	1·96	969403	4·22	030597	1
60	833783	2·26	864127	1·96	969656	4·22	030344	0
	Cosine.	D.	Sine.	D.	Cotang.	D.	Tang.	M.

M.	Sine.	D.	Cosine.	D.	Tang.	D.	Cotang.	
0	9·833783	2·26	9·864127	1·96	9·969656	4·22	10·080344	60
1	833919	2·25	864010	1·96	969909	4·22	030091	59
2	834054	2·25	863892	1·97	970162	4·22	029838	58
3	834189	2·25	863774	1·97	970416	4·22	029584	57
4	834325	2·25	863656	1·97	970669	4·22	029331	56
5	834460	2·25	863538	1·97	970922	4·22	029078	55
6	834595	2·25	863419	1·97	971175	4·22	028825	54
7	834730	2·25	863301	1·97	971429	4·22	028571	53
8	834865	2·25	863183	1·97	971682	4·22	028318	52
9	834999	2·24	863064	1·97	971935	4·22	028065	51
10	835134	2·24	862946	1·98	972188	4·22	027812	50
11	9·835269	2·24	9·862827	1·98	9·972441	4·22	10·027559	49
12	835403	2·24	862709	1·98	972694	4·22	027306	48
13	835538	2·24	862590	1·98	972948	4·22	027052	47
14	835672	2·24	862471	1·98	973201	4·22	026799	46
15	835807	2·24	862353	1·98	973454	4·22	026546	45
16	835941	2·24	862234	1·98	973707	4·22	026293	44
17	836075	2·23	862115	1·98	973960	4·22	026040	43
18	836209	2·23	861996	1·98	974213	4·22	025787	42
19	836343	2·23	861877	1·98	974466	4·22	025534	41
20	836477	2·23	861758	1·99	974719	4·22	025281	40
21	9·836611	2·23	9·861638	1·99	9·974973	4·22	10·025027	39
22	836745	2·23	861519	1·99	975226	4·22	024774	38
23	836878	2·23	861400	1·99	975479	4·22	024521	37
24	837012	2·22	861280	1·99	975732	4·22	024268	36
25	837146	2·22	861161	1·99	975985	4·22	024015	35
26	837279	2·22	861041	1·99	976238	4·22	023762	34
27	837412	2·22	860922	1·99	976491	4·22	023509	33
28	837546	2·22	860802	1·99	976744	4·22	023256	32
29	837679	2·22	860682	2·00	976997	4·22	023003	31
30	837812	2·22	860562	2·00	977250	4·22	022750	30
31	9·837945	2·22	9·860442	2·00	9·977503	4·22	10·022497	29
32	838078	2·21	860322	2·00	977756	4·22	022244	28
33	838211	2·21	860202	2·00	978009	4·22	021991	27
34	838344	2·21	860082	2·00	978262	4·22	021738	26
35	838477	2·21	859962	2·00	978515	4·22	021485	25
36	838610	2·21	859842	2·00	978768	4·22	021232	24
37	838742	2·21	859721	2·01	979021	4·22	020979	23
38	838875	2·21	859601	2·01	979274	4·22	020726	22
39	839007	2·21	859480	2·01	979527	4·22	020473	21
40	839140	2·20	859360	2·01	979780	4·22	020220	20
41	9·839272	2·20	9·859239	2·01	9·980033	4·22	10·019967	19
42	839404	2·20	859119	2·01	980286	4·22	019714	18
43	839536	2·20	858998	2·01	980538	4·22	019462	17
44	839668	2·20	858877	2·01	980791	4·21	019209	16
45	839800	2·20	858756	2·02	981044	4·21	018956	15
46	839932	2·20	858635	2·02	981297	4·21	018703	14
47	840064	2·19	858514	2·02	981550	4·21	018450	13
48	840196	2·19	858393	2·02	981803	4·21	018197	12
49	840328	2·19	858272	2·02	982056	4·21	017944	11
50	840459	2·19	858151	2·02	982309	4·21	017691	10
51	9·840591	2·19	9·858029	2·02	9·982562	4·21	10·017438	9
52	840722	2·19	857908	2·02	982814	4·21	017186	8
53	840854	2·19	857786	2·02	983067	4·21	016933	7
54	840985	2·19	857665	2·03	983320	4·21	016680	6
55	841116	2·18	857543	2·03	983573	4·21	016427	5
56	841247	2·18	857422	2·03	983826	4·21	016174	4
57	841378	2·18	857300	2·03	984079	4·21	015921	3
58	841509	2·18	857178	2·03	984331	4·21	015669	2
59	841640	2·18	857056	2·03	984584	4·21	015416	1
60	841771	2·18	856934	2·03	984837	4·21	015163	0
	Cosine.	D.	Sine.	D.	Cotang.	D.	Tang.	M.

(46 DEGREES.)

M.	Sine.	D.	Cosine.	D.	Tang.	D.	Cotang.	
0	9·841771	2·18	9·856934	2·03	9·984837	4·21	10·015168	60
1	841902	2·18	856812	2·03	985090	4·21	014910	59
2	842033	2·18	856690	2·04	985343	4·21	014657	58
3	842163	2·17	856568	2·04	985596	4·21	014404	57
4	842294	2·17	856446	2·04	985848	4·21	014152	56
5	842424	2·17	856323	2·04	986101	4·21	013899	55
6	842555	2·17	856201	2·04	986354	4·21	013646	54
7	842685	2·17	856078	2·04	986607	4·21	013393	53
8	842815	2·17	855956	2·04	986860	4·21	013140	52
9	842946	2·17	855833	2·04	987112	4·21	012888	51
10	843076	2·17	855711	2·05	987365	4·21	012635	50
11	9·843206	2·16	9·855588	2·05	9·987618	4·21	10·012382	49
12	843336	2·16	855465	2·05	987871	4·21	012129	48
13	843466	2·16	855342	2·05	988123	4·21	011877	47
14	843595	2·16	855219	2·05	988376	4·21	011624	46
15	843725	2·16	855096	2·05	988629	4·21	011371	45
16	843855	2·16	854973	2·05	988882	4·21	011118	44
17	843984	2·16	854850	2·05	989134	4·21	010866	43
18	844114	2·15	854727	2·06	989387	4·21	010613	42
19	844243	2·15	854603	2·06	989640	4·21	010360	41
20	844372	2·15	854480	2·06	989893	4·21	010107	40
21	9·844502	2·15	9·854356	2·06	9·990145	4·21	10·009855	39
22	844631	2·15	854233	2·06	990398	4·21	009602	38
23	844760	2·15	854109	2·06	990651	4·21	009349	37
24	844889	2·15	853986	2·06	990903	4·21	009097	36
25	845018	2·15	853862	2·06	991156	4·21	008844	35
26	845147	2·15	853738	2·06	991409	4·21	008591	34
27	845276	2·14	853614	2·07	991662	4·21	008338	33
28	845405	2·14	853490	2·07	991914	4·21	008086	32
29	845533	2·14	853366	2·07	992167	4·21	007833	31
30	845662	2·14	853242	2·07	992420	4·21	007580	30
31	9·845790	2·14	9·853118	2·07	9·992672	4·21	10·007328	29
32	845919	2·14	852994	2·07	992925	4·21	007075	28
33	846047	2·14	852869	2·07	993178	4·21	006822	27
34	846175	2·14	852745	2·07	993430	4·21	006570	26
35	846304	2·14	852620	2·07	993683	4·21	006317	25
36	846432	2·13	852496	2·08	993936	4·21	006064	24
37	846560	2·13	852371	2·08	994189	4·21	005811	23
38	846688	2·13	852247	2·08	994441	4·21	005559	22
39	846816	2·13	852122	2·08	994694	4·21	005306	21
40	846944	2·13	851997	2·08	994947	4·21	005053	20
41	9·847071	2·13	9·851872	2·08	9·995199	4·21	10·004801	19
42	847199	2·13	851747	2·08	995452	4·21	004548	18
43	847327	2·13	851622	2·08	995705	4·21	004295	17
44	847454	2·12	851497	2·09	995957	4·21	004043	16
45	847582	2·12	851372	2·09	996210	4·21	003790	15
46	847709	2·12	851246	2·09	996463	4·21	003537	14
47	847836	2·12	851121	2·09	996715	4·21	003285	13
48	847964	2·12	850996	2·09	996968	4·21	003032	12
49	848091	2·12	850870	2·09	997221	4·21	002779	11
50	848218	2·12	850745	2·09	997473	4·21	002527	10
51	9·848345	2·12	9·850619	2·09	9·997726	4·21	10·002274	9
52	848472	2·11	850493	2·10	997979	4·21	002021	8
53	848599	2·11	850368	2·10	998231	4·21	001769	7
54	848726	2·11	850242	2·10	998484	4·21	001516	6
55	848852	2·11	850116	2·10	998737	4·21	001263	5
56	848979	2·11	849990	2·10	998989	4·21	001011	4
57	849106	2·11	849864	2·10	999242	4·21	000758	3
58	849232	2·11	849738	2·10	999495	4·21	000505	2
59	849359	2·11	849611	2·10	999748	4·21	000253	1
60	849485	2·11	849485	2·10	10·000000	4·21	10·000000	0
	Cosine.	D.	Sine.	D.	Cotang.	D.	Tang.	M.

A TABLE OF NATURAL SINES.

M	0 Deg. S.	0 Deg. C.S.	1 Deg. S.	1 Deg. C.S.	2 Deg. S.	2 Deg. C.S.	3 Deg. S.	3 Deg. C.S.	4 Deg. S.	4 Deg. C.S.	M
0	00000	Unit.	01745	99985	03490	99939	05234	99863	06976	99756	60
1	00029	1·0000	01774	99984	03519	99938	05263	99861	07005	99754	59
2	00058	1·0000	01803	99984	03548	99937	05292	99860	07034	99752	58
3	00087	1·0000	01832	99983	03577	99936	05321	99858	07063	99750	57
4	00116	1·0000	01862	99983	03606	99935	05350	99857	07092	99748	56
5	00145	1·0000	01891	99982	03635	99934	05379	99855	07121	99746	55
6	00175	1·0000	01920	99982	03664	99933	05408	99854	07150	99744	54
7	00204	1·0000	01949	99981	03693	99932	05437	99852	07179	99742	53
8	00233	1·0000	01978	99980	03723	99931	05466	99851	07208	99740	52
9	00262	1·0000	02007	99980	03752	99930	05495	99849	07237	99738	51
10	00291	1·0000	02036	99979	03781	99929	05524	99847	07266	99736	50
11	00320	99999	02065	99979	03810	99927	05553	99846	07295	99734	49
12	00349	99999	02094	99978	03839	99926	05582	99844	07324	99731	48
13	00378	99999	02123	99977	03868	99925	05611	99842	07353	99729	47
14	00407	99999	02152	99977	03897	99924	05640	99841	07382	99727	46
15	00436	99999	02181	99976	03926	99923	05669	99839	07411	99725	45
16	00465	99999	02211	99976	03955	99922	05698	99838	07440	99723	44
17	00495	99999	02240	99975	03984	99921	05727	99836	07469	99721	43
18	00524	99999	02269	99974	04013	99919	05756	99834	07498	99719	42
19	00553	99998	02298	99974	04042	99918	05785	99833	07527	99716	41
20	00582	99998	02327	99973	04071	99917	05814	99831	07556	99714	40
21	00611	99998	02356	99972	04100	99916	05844	99829	07585	99712	39
22	00640	99998	02385	99972	04129	99915	05873	99827	07614	99710	38
23	00669	99998	02414	99971	04159	99913	05902	99826	07643	99708	37
24	00698	99998	02443	99970	04188	99912	05931	99824	07672	99705	36
25	00727	99997	02472	99969	04217	99911	05960	99822	07701	99703	35
26	00756	99997	02501	99969	04246	99910	05989	99821	07730	99701	34
27	00785	99997	02530	99968	04275	99909	06018	99819	07759	99699	33
28	00814	99997	02560	99967	04304	99907	06047	99817	07788	99696	32
29	00844	99996	02589	99966	04333	99906	06076	99815	07817	99694	31
30	00873	99996	02618	99966	04362	99905	06105	99813	07846	99692	30
31	00902	99996	02647	99965	04391	99904	06134	99812	07875	99689	29
32	00931	99996	02676	99964	04420	99902	06163	99810	07904	99687	28
33	00960	99995	02705	99963	04449	99901	06192	99808	07933	99685	27
34	00989	99995	02734	99963	04478	99900	06221	99806	07962	99683	26
35	01018	99995	02763	99962	04507	99898	06250	99804	07991	99680	25
36	01047	99995	02792	99961	04536	99897	06279	99803	08020	99678	24
37	01076	99994	02821	99960	04565	99896	06308	99801	08049	99676	23
38	01105	99994	02850	99959	04594	99894	06337	99799	08078	99673	22
39	01134	99994	02879	99959	04623	99893	06366	99797	08107	99671	21
40	01164	99993	02908	99958	04653	99892	06395	99795	08136	99668	20
41	01193	99993	02938	99957	04682	99890	06424	99793	08165	99666	19
42	01222	99993	02967	99956	04711	99888	06453	99792	08194	99664	18
43	01251	99992	02996	99955	04740	99888	06482	99790	08223	99661	17
44	01280	99992	03025	99954	04769	99886	06511	99788	08252	99659	16
45	01309	99991	03054	99953	04798	99885	06540	99786	08281	99657	15
46	01338	99991	03083	99952	04827	99883	06569	99784	08310	99654	14
47	01367	99991	03112	99952	04856	99882	06598	99782	08339	99652	13
48	01396	99990	03141	99951	04885	99881	06627	99780	08368	99649	12
49	01425	99990	03170	99950	04914	99879	06656	99778	08397	99647	11
50	01454	99989	03199	99949	04943	99878	06685	99776	08426	99644	10
51	01483	99989	03228	99948	04972	99876	06714	99774	08455	99642	9
52	01513	99989	03257	99947	05001	99875	06743	99772	08484	99639	8
53	01542	99988	03286	99946	05030	99873	06773	99770	08513	99637	7
54	01571	99988	03316	99945	05059	99872	06802	99768	08542	99635	6
55	01600	99987	03345	99944	05088	99870	06831	99766	08571	99632	5
56	01629	99987	03374	99943	05117	99869	06860	99764	08600	99630	4
57	01658	99986	03403	99942	05146	99867	06889	99762	08629	99627	3
58	01687	99986	03432	99941	05175	99866	06918	99760	08658	99625	2
59	01716	99985	03461	99940	05205	99864	06947	99758	08687	99622	1
M	C.S.	S.	C.S.	S.	C.S.	S.	C.S.	S.	C.S.	S	M
	89 Deg.		88 Deg.		87 Deg.		86 Deg.		85 Deg.		

M	5 Deg. S.	5 Deg. C.S.	6 Deg. S.	6 Deg. C.S.	7 Deg. S.	7 Deg. C.S.	8 Deg. S.	8 Deg. C.S.	9 Deg. S.	9 Deg. C.S.	M
0	08716	99619	10453	99452	12187	99255	13917	99027	15643	98769	60
1	08745	99617	10482	99449	12216	99251	13946	99023	15672	98764	59
2	08774	99614	10511	99446	12245	99248	13975	99019	15701	98760	58
3	08803	99612	10540	99443	12274	99244	14004	99015	15730	98755	57
4	08831	99609	10569	99440	12302	99240	14033	99011	15758	98751	56
5	08860	99607	10597	99437	12331	99237	14061	99006	15787	98746	55
6	08889	99604	10626	99434	12360	99233	14090	99002	15816	98741	54
7	08918	99602	10655	99431	12389	99230	14119	98998	15845	98737	53
8	08947	99599	10684	99428	12418	99226	14148	98994	15873	98732	52
9	08976	99596	10713	99424	12447	99222	14177	98990	15902	98728	51
10	09005	99594	10742	99421	12476	99219	14205	98986	15931	98723	50
11	09034	99591	10771	99418	12504	99215	14234	98982	15959	98718	49
12	09063	99588	10800	99415	12533	99211	14263	98978	15988	98714	48
13	09092	99586	10829	99412	12562	99208	14292	98973	16017	98709	47
14	09121	99583	10858	99409	12591	99204	14320	98969	16046	98704	46
15	09150	99580	10887	99406	12620	99200	14349	98965	16074	98700	45
16	09179	99578	10916	99402	12649	99197	14378	98961	16103	98695	44
17	09208	99575	10945	99399	12678	99193	14407	98957	16132	98690	43
18	09237	99572	10973	99396	12706	99189	14436	98953	16160	98686	42
19	09266	99570	11002	99393	12735	99186	14464	98948	16189	98681	41
20	09295	99567	11031	99390	12764	99182	14493	98944	16218	98676	40
21	09324	99564	11060	99386	12793	99178	14522	98940	16246	98671	39
22	09353	99562	11089	99383	12822	99175	14551	98936	16275	98667	38
23	09382	99559	11118	99380	12851	99171	14580	98931	16304	98662	37
24	09411	99556	11147	99377	12880	99167	14608	98927	16333	98657	36
25	09440	99553	11176	99374	12908	99163	14637	98923	16361	98652	35
26	09469	99551	11205	99370	12937	99160	14666	98919	16390	98648	34
27	09498	99548	11234	99367	12966	99156	14695	98914	16419	98643	33
28	09527	99545	11263	99364	12995	99152	14723	98910	16447	98638	32
29	09556	99542	11291	99360	13024	99148	14752	98906	16476	98633	31
30	09585	99540	11320	99357	13053	99144	14781	98902	16505	98629	30
31	09614	99537	11349	99354	13081	99141	14810	98897	16533	98624	29
32	09642	99534	11378	99351	13110	99137	14838	98893	16562	98619	28
33	09671	99531	11407	99347	13139	99133	14867	98889	16591	98614	27
34	09700	99528	11436	99344	13168	99129	14896	98884	16620	98609	26
35	09729	99526	11465	99341	13197	99125	14925	98880	16648	98604	25
36	09758	99523	11494	99337	13226	99122	14954	98876	16677	98600	24
37	09787	99520	11523	99334	13254	99118	14982	98871	16706	98595	23
38	09816	99517	11552	99331	13283	99114	15011	98867	16734	98590	22
39	09845	99514	11580	99327	13312	99110	15040	98863	16763	98585	21
40	09874	99511	11609	99324	13341	99106	15069	98858	16792	98580	20
41	09903	99508	11638	99320	13370	99102	15097	98854	16820	98575	19
42	09932	99506	11667	99317	13399	99098	15126	98849	16849	98570	18
43	09961	99503	11696	99314	13427	99094	15155	98845	16878	98565	17
44	09990	99500	11725	99310	13456	99091	15184	98841	16906	98561	16
45	10019	99497	11754	99307	13485	99087	15212	98836	16935	98556	15
46	10048	99494	11783	99303	13514	99083	15241	98832	16964	98551	14
47	10077	99491	11812	99300	13543	99079	15270	98827	16992	98546	13
48	10106	99488	11840	99297	13572	99075	15292	98823	17021	98541	12
49	10135	99485	11869	99293	13600	99071	15327	98818	17050	98536	11
50	10164	99482	11898	99290	13629	99067	15356	98814	17078	98531	10
51	10192	99479	11927	99286	13658	99063	15385	98809	17107	98526	9
52	10221	99476	11956	99283	13687	99059	15414	98805	17136	98521	8
53	10250	99473	11985	99279	13716	99055	15442	98800	17164	98516	7
54	10279	99470	12014	99276	13744	99051	15471	98796	17193	98511	6
55	10308	99467	12043	99272	13773	99047	15500	98791	17222	98506	5
56	10337	99464	12071	99269	13902	99043	15529	98787	17250	98501	4
57	10366	99461	12100	99265	13331	99039	15557	98782	17279	98496	3
58	10395	99458	12129	99262	13860	99035	15586	98778	17308	98491	2
59	10424	99455	12158	99258	13889	99031	15615	98773	17336	98486	1
M	C.S.	S.	C.S.	S.	C.S.	S.	C.S.	S.	C.S.	S.	M
	84 Deg.		83 Deg.		82 Deg.		81 Deg.		80 Deg.		

M	10 Deg. S.	10 Deg. C.S.	11 Deg. S.	11 Deg. C.S.	12 Deg. S.	12 Deg. C.S.	13 Deg. S.	13 Deg. C.S.	14 Deg. S.	14 Deg. C.S.	M
0	17365	98481	19081	98163	20791	97815	22495	97437	24192	97030	60
1	17393	98476	19109	98157	20820	97809	22523	97430	24220	97023	59
2	17422	98471	19138	98152	20848	97803	22552	97424	24249	97015	58
3	17451	98466	19167	98146	20877	97797	22580	97417	24277	97008	57
4	17479	98461	19195	98140	20905	97791	22608	97411	24305	97001	56
5	17508	98455	19224	98135	20933	97784	22637	97404	24333	96994	55
6	17537	98450	19252	98129	20962	97778	22665	97398	24362	96987	54
7	17565	98445	19281	98124	20990	97772	22693	97391	24390	96980	53
8	17594	98440	19309	98118	21019	97766	22722	97384	24418	96973	52
9	17623	98435	19338	98112	21047	97760	22750	97378	24446	96966	51
10	17651	98430	19366	98107	21076	97754	22778	97371	24474	96959	50
11	17680	98425	19395	98101	21104	97748	22807	97365	24503	96952	49
12	17708	98420	19423	98096	21132	97742	22835	97358	24531	96945	48
13	17737	98414	19452	98090	21161	97735	22863	97351	24559	96937	47
14	17766	98409	19481	98084	21189	97729	22892	97345	24587	96930	46
15	17794	98404	19509	98079	21218	97723	22920	97318	24615	96923	45
16	17823	98399	19538	98073	21246	97717	22948	973?1	24644	96916	44
17	17852	98394	19566	98067	21275	97711	22977	973?5	24672	96909	43
18	17880	98389	19595	98061	21303	97705	23005	97318	24700	96902	42
19	17909	98383	19623	98056	21331	97698	23033	97311	24728	96894	41
20	17937	98378	19652	98050	21360	97692	23062	97304	24756	96887	40
21	17966	98373	19680	98044	21388	97686	23090	972?1	24784	96880	39
22	17995	98368	19709	98039	21417	97680	23118	97291	24813	96873	38
23	18023	98362	19737	98033	21445	97673	23146	97284	24841	96866	37
24	18052	98357	19766	98027	21474	97667	23175	97278	24869	96858	36
25	18081	98352	19794	98021	21502	97661	23203	97271	24897	96851	35
26	18109	98347	19823	98016	21530	97655	23231	97264	24925	96844	34
27	18138	98341	19851	98010	21559	97648	23260	97257	24953	96837	33
28	18166	98336	19880	98004	21587	97642	23288	97251	24982	96829	32
29	18195	98331	19908	97998	21616	97636	23316	97244	25010	96822	31
30	18224	98325	19937	97992	21644	97630	23345	97237	25038	96815	30
31	18252	98320	19965	97987	21672	97623	23373	97230	25066	96807	29
32	18281	98315	19994	97981	21701	97617	23401	97223	25094	96800	28
33	18309	98310	20022	97975	21729	97611	23429	97217	25122	96793	27
34	18338	98304	20051	97969	21758	97604	23458	97210	25151	96786	26
35	18367	98299	20079	97963	21786	97598	23486	97203	25179	96778	25
36	18395	98294	20108	97958	21814	97592	23514	97196	25207	96771	24
37	18424	98288	20136	97952	21843	97585	23542	97189	25235	96764	23
38	18452	98283	20165	97946	21871	97579	23571	97182	25263	96756	22
39	18481	98277	20193	97940	21899	97573	23599	97176	25291	96749	21
40	18509	98272	20222	97934	21928	97566	23627	97169	25320	96742	20
41	18538	98267	20250	97928	21956	97560	23656	97162	25348	96734	19
42	18567	98261	20279	97922	21985	97553	23684	97155	25376	96727	18
43	18595	98256	20307	97916	22013	97547	23712	97148	25404	96719	17
44	18624	98250	20336	97910	22041	97541	23740	97141	25432	96712	16
45	18652	98245	20364	97905	22070	97534	23769	97134	25460	96705	15
46	18681	98240	20393	97899	22098	97528	23797	97127	25488	96697	14
47	18710	98234	20421	97893	22126	97521	23825	97120	25516	96690	13
48	18738	98229	20450	97887	22155	97515	23853	97113	25545	96682	12
49	18767	98223	20478	97881	22183	97508	23882	97106	25573	96675	11
50	18795	98218	20507	97875	22212	97502	23910	97100	25601	96667	10
51	18824	98212	20535	97869	22240	97496	23938	97093	25629	96660	9
52	18852	98207	20563	97863	22268	97489	23966	97086	25657	96653	8
53	18881	98201	20592	97857	22297	97483	23995	97079	25685	96645	7
54	18910	98196	20620	97851	22325	97476	24023	97072	25713	96638	6
55	18938	98190	20649	97845	22353	97470	24051	97065	25741	96630	5
56	18967	98185	20677	97839	22382	97463	24079	97058	25769	96623	4
57	18995	98179	20706	97833	22410	97457	24108	97051	25798	96615	3
58	19024	98174	20734	97827	22438	97450	24136	97044	25826	96608	2
59	19052	98168	20763	97821	22467	97444	24164	97037	25854	96600	1
M	C.S.	S.	C.S.	S.	C.S.	S.	C.S.	S.	C.S.	S.	M
	79 Deg.		78 Deg.		77 Deg.		76 Deg.		75 Deg.		

M	15 Deg. S.	C.S.	16 Deg. S.	C.S.	17 Deg. S.	C.S.	18 Deg. S.	C.S.	19 Deg. S.	S.C.	M
0	25882	96593	27564	96126	29237	95630	30902	95106	32557	94552	60
1	25910	96585	27592	96118	29265	95622	30929	95097	32584	94542	59
2	25938	96578	27620	96110	29293	95613	30957	95088	32612	94533	58
3	25966	96570	27648	96102	29321	95605	30985	95079	32639	94523	57
4	25994	96562	27676	96094	29348	95596	31012	95070	32667	94514	56
5	26022	96555	27704	96086	29376	95588	31040	95061	32694	94504	55
6	26050	96547	27731	96078	29404	95579	31068	95052	32722	94495	54
7	26079	96540	27759	96070	29432	95571	31095	95043	32749	94485	53
8	26107	96532	27787	96062	29460	95562	31123	95033	32777	94476	52
9	26135	96524	27815	96054	29487	95554	31151	95024	32804	94466	51
10	26163	96517	27843	96046	29515	95545	31178	95015	32832	94457	50
11	26191	96509	27871	96037	29543	95536	31206	95006	32859	94447	49
12	26219	96502	27899	96029	29571	95528	31233	94997	32887	94438	48
13	26247	96494	27927	96021	29599	95519	31261	94988	32914	94428	47
14	26275	96486	27955	96013	29626	95511	31289	94979	32942	94418	46
15	26303	96479	27983	96005	29654	95502	31316	94970	32969	94409	45
16	26331	96471	28011	95997	29682	95493	31344	94961	32997	94399	44
17	26359	96463	28039	95989	29710	95485	31372	94952	33024	94390	43
18	26387	96456	28067	95981	29737	95476	31399	94943	33051	94380	42
19	26415	96448	28095	95972	29765	95467	31427	94933	33079	94370	41
20	26443	96440	28123	95964	29793	95459	31454	94924	33106	94361	40
21	26471	96433	28150	95956	29821	95450	31482	94915	33134	94351	39
22	26500	96425	28178	95948	29849	95441	31510	94906	33161	94342	38
23	26528	96417	28206	95940	29876	95433	31537	94897	33189	94332	37
24	26556	96410	28234	95931	29904	95424	31565	94888	33216	94322	36
25	26584	96402	28262	95923	29932	95415	31593	94878	33244	94313	35
26	26612	96394	28290	95915	29960	95407	31620	94869	33271	94303	34
27	26640	96386	28318	95907	29987	95398	31648	94860	33298	94293	33
28	26668	96379	28346	95898	30015	95389	31675	94851	33326	94284	32
29	26696	96371	28374	95890	30043	95380	31703	94842	33353	94274	31
30	26724	96363	28402	95882	30071	95372	31730	94832	33381	94264	30
31	26752	96355	28429	95874	30098	95363	31758	94823	33408	94254	29
32	26780	96347	28457	95865	30126	95354	31786	94814	33436	94245	28
33	26808	96340	28485	95857	30154	95345	31813	94805	33463	94235	27
34	26836	96332	28513	95849	30182	95337	31841	94795	33490	94225	26
35	26864	96324	28541	95841	30209	95328	31868	94786	33518	94215	25
36	26892	96316	28569	95832	30237	95319	31896	94777	33545	94206	24
37	26920	96308	28597	95824	30265	95310	31923	94768	33573	94196	23
38	26948	96301	28625	95816	30292	95301	31951	94758	33600	94186	22
39	26976	96293	28652	95807	30320	95293	31979	94749	33627	94176	21
40	27004	96285	28680	95799	30348	95284	32006	94740	33655	94167	20
41	27032	96277	28708	95791	30376	95275	32034	94730	33682	94157	19
42	27060	96269	28736	95782	30403	95266	32061	94721	33710	94147	18
43	27088	96261	28764	95774	30431	95257	32089	94712	33737	94137	17
44	27116	96253	28792	95766	30459	95248	32116	94702	33764	94127	16
45	27144	96246	28820	95757	30486	95240	32144	94693	33792	94118	15
46	27172	96238	28847	95749	30514	95231	32171	94684	33819	94108	14
47	27200	96230	28875	95740	30542	95222	32199	94674	33846	94098	13
48	27228	96222	28903	95732	30570	95213	32227	94665	33874	94088	12
49	27256	96214	28931	95724	30597	95204	32254	94656	33901	94078	11
50	27284	96206	28959	95715	30625	95195	32282	94646	33929	94068	10
51	27312	96198	28987	95707	30653	95186	32309	94637	33956	94058	9
52	27340	96190	29015	95698	30680	95177	32337	94627	33983	94049	8
53	27368	96182	29042	95690	30708	95168	32364	94618	34011	94039	7
54	27396	96174	29070	95681	30736	95159	32392	94609	34038	94029	6
55	27424	96166	29098	95673	30763	95150	32419	94599	34065	94019	5
56	27452	96158	29126	95664	30791	95142	32447	94590	34093	94009	4
57	27480	96150	29154	95656	30819	95133	32474	94580	34120	93999	3
58	27508	96142	29182	95647	30846	95124	32502	94571	34147	93989	2
59	27536	96134	29209	95639	30874	95115	32529	94561	34175	93979	1
M	C.S.	S.	C.S.	S.	C.S.	S.	C.S.	S.	C.S.	S.	M
	74 Deg.		78 Deg.		72 Deg.		71 Deg.		70 Deg.		

M	20 Deg. S.	20 Deg. C.S.	21 Deg. S.	21 Deg. C.S.	22 Deg. S.	22 Deg. C.S.	23 Deg. S.	23 Deg. C.S.	24 Deg. S.	24 Deg. C.S.	M
0	34202	93969	35837	93358	37461	92718	39073	92050	40674	91355	60
1	34229	93959	35864	93348	37488	92707	39100	92039	40700	91343	59
2	34257	93949	35891	93337	37515	92697	39127	92028	40727	91331	58
3	34284	93939	35918	93327	37542	92686	39153	92016	40753	91319	57
4	34311	93929	35945	93316	37569	9:675	39180	92005	40780	91307	56
5	34339	93919	35973	93306	37595	92 64	39207	91994	40806	91295	55
6	34366	93909	36000	93295	37622	92053	39234	91982	40833	91283	54
7	34393	93899	36027	93285	37649	92642	39260	91971	40860	91272	53
8	34421	93889	36054	93274	37676	92631	39287	91959	40886	91260	52
9	34448	93879	36081	93264	37703	92620	39314	91948	40913	91248	51
10	34475	93869	36108	93253	37730	92609	39341	91936	40939	91236	50
11	34503	93859	36135	93243	37757	92598	39367	91925	40966	01224	49
12	34530	93849	36162	93232	37784	92587	39394	91914	40992	91212	48
13	34557	93839	36190	93222	37811	92576	39421	91902	41019	91200	47
14	34584	93829	36217	93211	37838	92565	39448	91891	41045	91188	46
15	34612	93819	36244	93201	37865	92554	39474	91879	41072	91176	45
16	34639	93809	36271	93190	37892	92543	39501	91868	41098	91164	44
17	34666	93799	36298	93180	37919	92532	39528	91856	41125	91152	43
18	34694	93789	36325	93169	37946	92521	39555	91845	41151	91140	42
19	34721	93779	36352	93159	37973	92510	39581	91833	41178	91128	41
20	34748	93769	36379	93148	37999	92499	39608	91822	41204	91116	40
21	34775	93759	36406	93137	38026	92488	39635	91810	41231	91104	39
22	34803	93748	36434	93127	38053	92477	39661	91799	41257	91092	38
23	34830	93738	36461	93116	38080	92466	39688	91787	41284	91080	37
24	34857	93728	36488	93106	38107	92455	39715	91775	41310	91068	36
25	34884	93718	36515	93095	38134	92444	39741	91764	41337	91056	35
26	34912	93708	36542	93084	38161	92432	39768	91752	41363	91044	34
27	34939	93698	36569	93074	38188	92421	39795	91741	41390	91032	33
28	34966	93688	36596	93063	38215	92410	39822	91729	41416	91020	32
29	34993	93677	36623	93052	38241	92399	39848	91718	41443	91008	31
30	35021	93667	36650	93042	38268	92388	39875	91706	41469	90996	30
31	35048	93657	36677	93031	38295	92377	39902	91694	41496	90984	29
32	35075	93647	36704	93020	38322	92366	39928	91683	41522	90972	28
33	35102	93637	36731	93010	38349	92355	39955	91671	41549	90960	27
34	35130	93626	36758	92999	38376	92343	39982	91660	41575	90948	26
35	35157	93616	36785	92988	38403	92332	40008	91648	41602	90936	25
36	35183	93606	36812	92978	38430	92321	40035	91636	41628	90924	24
37	35211	93596	36839	92967	38456	92310	40062	91625	41655	90911	23
38	35239	93585	36867	92956	38483	92299	40088	91613	41681	90899	22
39	35266	93575	36894	92945	38510	92287	40115	91601	41707	90887	21
40	35293	93565	36921	92935	38537	92276	40141	91590	41734	90875	20
41	35320	93555	36948	92924	38564	92265	40168	91578	41760	90863	19
42	35347	93544	36975	92913	38591	92254	40195	91566	41787	90851	18
43	35375	93534	37002	92902	38617	92243	40221	91555	41813	90839	17
44	35402	93524	37029	92892	38644	92231	40248	91543	41840	90826	16
45	35429	93514	37056	92881	38671	92220	40275	91531	41866	90814	15
46	35456	93503	37083	92870	38698	92209	40301	91519	41892	90802	14
47	35484	93493	37110	92859	38725	92198	40328	91508	41919	90790	13
48	35511	93483	37137	92849	38752	92186	40355	91496	41945	90778	12
49	35538	93472	37164	92838	38778	92175	40381	91484	41972	90766	11
50	35565	93462	37191	92827	38805	92164	40408	91472	41998	90753	10
51	35592	93452	37218	92816	38832	92152	40434	91461	42024	90741	9
52	35619	93441	37245	92805	38859	92141	40461	91449	42051	90729	8
53	35647	93431	37272	92794	38886	92130	40488	91437	42077	90717	7
54	35674	93420	37299	92784	38912	92119	40514	91425	42104	90704	6
55	35701	93410	37326	92773	38939	92107	40541	91414	42130	90692	5
56	35728	93400	37353	92762	38966	92096	40567	91402	42156	90680	4
57	35755	93389	37380	92751	38993	92085	40594	91390	42183	90668	3
58	35782	93379	37407	92740	39020	92073	40621	91378	42209	90655	2
59	35810	93368	37434	92729	39046	92062	40647	91366	42235	90643	1
M	C.S.	S.	C.S.	S.	C.S.	S.	C.S.	S.	C.S.	S.	M
	69 Deg.		68 Deg.		67 Deg.		66 Deg.		65 Deg.		

19

M	25 Deg. S	25 Deg. C.S.	26 Deg. S	26 Deg. C.S.	27 Deg. S	27 Deg. C.S.	28 Deg. S	28 Deg. C.S.	29 Deg. S	29 Deg. C.S.	M
0	42252	90631	43837	89879	45399	89101	46947	88295	48481	87462	60
1	42288	90618	43863	89867	45425	89087	46973	88281	48506	87448	59
2	42315	90606	43889	89854	45451	89074	46999	88267	48532	87434	58
3	42341	90594	43916	89841	45477	89061	47024	88254	48557	87420	57
4	42367	90582	43942	89828	45503	89048	47050	88240	48583	87406	56
5	42394	90569	43968	89816	45529	89035	47076	88226	48608	87391	55
6	42420	90557	43994	89803	45554	89021	47101	88213	48634	87377	54
7	42446	90545	44020	89790	45580	89008	47127	88199	48659	87363	53
8	42473	90532	44046	89777	45606	88995	47153	88185	48684	87349	52
9	42499	90520	44072	89764	45632	88981	47178	88172	48710	87335	51
10	42525	90507	44098	89752	45658	88968	47204	88158	48735	87321	50
11	42552	90495	44124	89739	45684	88955	47229	88144	48761	87306	49
12	42578	90483	44151	89726	45710	88942	47255	88130	48786	87292	48
13	42604	90470	44177	89713	45736	88928	47281	88117	48811	87278	47
14	42631	90458	44203	89700	45762	88915	47306	88103	48837	87264	46
15	42657	90446	44229	89687	45787	88902	47332	88089	48862	87250	45
16	42683	90433	44255	89674	45813	88888	47358	88075	48888	87235	44
17	42709	90421	44281	89662	45839	88875	47383	88062	48913	87221	43
18	42736	90408	44307	89649	45865	88862	47409	88048	48938	87207	42
19	42762	90396	44333	89636	45891	88848	47434	88034	48964	87193	41
20	42788	90383	44359	89623	45917	88835	47460	88020	48989	87178	40
21	42815	90371	44385	89610	45942	88822	47486	88006	49014	87164	39
22	42841	90358	44411	89597	45968	88808	47511	87993	49040	87150	38
23	42867	90346	44437	89584	45994	88795	47537	87979	49065	87136	37
24	42894	90334	44464	89571	46020	88782	47562	87965	49090	87121	36
25	42920	90321	44490	89558	46046	88768	47588	87951	49116	87107	35
26	42946	90309	44516	89545	46072	88755	47614	87937	49141	87093	34
27	42972	90296	44542	89532	46097	88741	47639	87923	49166	87079	33
28	42999	90284	44568	89519	46123	88728	47665	87909	49192	87064	32
29	43025	90271	44594	89506	46149	88715	47690	87896	49217	87050	31
30	43051	90259	44620	89493	46175	88701	47716	87882	49242	87036	30
31	43077	90246	44646	89480	46201	88688	47741	87868	49268	87021	29
32	43104	90233	44672	89467	46226	88674	47767	87854	49293	87007	28
33	43130	90221	44698	89454	46252	88661	47793	87840	49318	86993	27
34	43156	90208	44724	89441	46278	88647	47818	87826	49344	86978	26
35	43182	90196	44750	89428	46304	88634	47844	87812	49369	86964	25
36	43209	90183	44776	89415	46330	88620	47869	87798	49394	86949	24
37	43235	90171	44802	89402	46355	88607	47895	87784	49419	86935	23
38	43261	90158	44828	89389	46381	88593	47920	87770	49445	86921	22
39	43287	90146	44854	89376	46407	88580	47946	87756	49470	86906	21
40	43313	90133	44880	89363	46433	88566	47971	87743	49495	86892	20
41	43340	90120	44906	89350	46458	88553	47997	87729	49521	86878	19
42	43366	90108	44932	89337	46484	88539	48022	87715	49546	86863	18
43	43392	90095	44958	89324	46510	88526	48048	87701	49571	86849	17
44	43418	90082	44984	89311	46536	88512	48073	87687	49596	86834	16
45	43445	90070	45010	89298	46561	88499	48099	87673	49622	86820	15
46	43471	90057	45036	89285	46587	88485	48124	87659	49647	86805	14
47	43497	90045	45062	89272	46613	88472	48150	87645	49672	86791	13
48	43523	90032	45088	89259	46639	88458	48175	87631	49697	86777	12
49	43549	90019	45114	89245	46664	88445	48201	87617	49723	86762	11
50	43575	90007	45140	89232	46690	88431	48226	87603	49748	86748	10
51	43602	89994	45166	89219	46716	88417	48252	87589	49773	86733	9
52	43628	89981	45192	89206	46742	88404	48277	87575	49798	86719	8
53	43654	89968	45218	89193	46767	88390	48303	87561	49824	86704	7
54	43680	89956	45243	89180	46793	88377	48328	87546	49849	86690	6
55	43706	89943	45269	89167	46819	88363	48354	87532	49874	86675	5
56	43733	89930	45295	89153	46844	88349	48379	87518	49899	86661	4
57	43759	89918	45321	89140	46870	88336	48405	87504	49924	86646	3
58	43785	89905	45347	89127	46896	88322	48430	87490	49950	86632	2
59	43811	89892	45373	89114	46921	88308	48456	87476	49975	86617	1
M	C.S.	S.	C.S.	S.	C.S.	S.	C.S.	S.	C.S.	S.	M
	64 Deg.		63 Deg.		62 Deg.		61 Deg.		60 Deg.		

M	80 Deg. S.	80 Deg. C.S.	81 Deg. S.	81 Deg. C.S.	82 Deg. S.	82 Deg. C.S.	83 Deg. S.	83 Deg. C.S	84 Deg. S.	84 Deg. C.S.	M
0	50000	86603	51504	85717	52992	84805	54464	83867	55919	82904	50
1	50025	86588	51529	85702	53017	84789	54488	83851	55943	82887	59
2	50050	86573	51554	85687	53041	84774	54513	83835	55968	92871	58
3	50076	86559	51579	85672	53066	84759	54537	83819	55992	92855	57
4	50101	86544	51604	85657	53091	84743	54561	83804	56016	82839	56
5	50126	86530	51628	85642	53115	84728	54586	83788	56040	82822	55
6	50151	86515	51653	85627	53140	84712	54610	83772	56064	82806	54
7	50176	86501	51678	85612	53164	84697	54635	83756	56088	82790	53
8	50201	86486	51703	85597	53189	84681	54659	83740	56112	82773	52
9	50227	86471	51728	85582	53214	84666	54683	83724	56136	82757	51
10	50252	86457	51753	85567	53238	84650	54708	83708	56160	82741	50
11	50277	86442	51778	85551	53263	84635	54732	83692	56184	82724	49
12	50302	86427	51803	85536	53288	84619	54756	83676	56208	82708	48
13	50327	86413	51828	85521	53312	84604	54781	83660	56232	82692	47
14	50352	86398	51852	85506	53337	84588	54805	83645	56256	82675	46
15	50377	86384	51877	85491	53361	84573	54829	83629	56280	82659	45
16	50403	86369	51902	85476	53386	84557	54854	83613	56305	82643	44
17	50428	86354	51927	85461	53411	84542	54878	83597	56329	82626	43
18	50453	86340	51952	85446	53435	84526	54902	83581	56353	82610	42
19	50478	86325	51977	85431	53460	84511	54927	83565	56377	82593	41
20	50503	86310	52002	85416	53484	84495	54951	83549	56401	82577	40
21	50528	86295	52026	85401	53509	84480	54975	83533	56425	82561	39
22	50553	86281	52051	85385	53534	84464	54999	83517	56449	82544	38
23	50578	86266	52076	85370	53558	84448	55024	83501	56473	82528	37
24	50603	86251	52101	85355	53583	84433	55048	83485	56497	82511	36
25	50628	86237	52126	85340	53607	84417	55072	83469	56521	82495	35
26	50654	86222	52151	85325	53632	84402	55097	83453	56545	82478	34
27	50679	86207	52175	85310	53656	84386	55121	83437	56569	82462	33
28	50704	86192	52200	85294	53681	84370	55145	83421	56593	82446	32
29	50729	86178	52225	85279	53705	84355	55169	83405	56617	82429	31
30	50754	86163	52250	85264	53730	84339	55194	83389	56641	82413	30
31	50779	86148	52275	85249	53754	84324	55218	83373	56665	82396	29
32	50804	86133	52299	85234	53779	84308	55242	83356	56689	82380	28
33	50829	86119	52324	85218	53804	84292	55266	83340	56713	82363	27
34	50854	86104	52349	85203	53828	84277	55291	83324	56736	82347	26
35	50879	86089	52374	85188	53853	84261	55315	83308	56760	82330	25
36	50904	86074	52399	85173	53877	84245	55339	83292	56784	82314	24
37	50929	86059	52423	85157	53902	84230	55363	83276	56808	82297	23
38	50954	86045	52448	85142	53926	84214	55388	83260	56832	82281	22
39	50979	86030	52473	85127	53951	84198	55412	83244	56856	82264	21
40	51004	86015	52498	85112	53975	84182	55436	83228	56880	82248	20
41	51029	86000	52522	85096	54000	84167	55460	83212	56904	82231	19
42	51054	85985	52547	85081	54024	84151	55484	83195	56928	82214	18
43	51079	85970	52572	85066	54049	84135	55509	83179	56952	82198	17
44	51104	85956	52597	85051	54073	84120	55533	83163	56976	82181	16
45	51129	85941	52621	85035	54097	84104	55557	83147	57000	82165	15
46	51154	85926	52646	85020	54122	84088	55581	83131	57024	82148	14
47	51179	85911	52671	85005	54146	84072	55605	83115	57047	82132	13
48	51204	85896	52696	84989	54171	84057	55630	83098	57071	82115	12
49	51229	85881	52720	84974	54195	84041	55654	83082	57095	82098	11
50	51254	85866	52745	84959	54220	84025	55678	83066	57119	82082	10
51	51279	85851	52770	84943	54244	84009	55702	83050	57143	82065	9
52	51304	85836	52794	84928	54269	83994	55726	83034	57167	82048	8
53	51329	85821	52819	84913	54293	83978	55750	83017	57191	82032	7
54	51354	85806	52844	84897	54317	83962	55775	83001	57215	82015	6
55	51379	85792	52869	84882	54342	83946	55799	82985	57238	81999	5
56	51404	85777	52893	84866	54366	83930	55823	82969	57262	81982	4
57	51429	85762	52918	84851	54391	83915	55847	82953	57286	81965	3
58	51454	85747	52943	84836	54415	83899	55871	82936	57310	81949	2
59	51479	85732	52967	84820	54440	83883	55895	82920	57334	81932	1
M	C.S.	S.	C.S.	S.	C.S.	S.	C.S.	S.	C.S.	S.	M
	59 Deg.		58 Deg.		57 Deg.		56 Deg.		55 Deg.		

M	85 Deg.		86 Deg.		87 Deg.		88 Deg.		89 Deg.		M
	S.	C.S.	S.	C.S.	S.	C.S.	S.	C.S.	S.	C.S.	
0	57358	81915	58779	80902	60182	79864	61566	78801	62932	77715	60
1	57381	81899	58802	80885	60205	79846	61589	78783	62955	77696	59
2	57405	81882	58826	80867	60228	79829	61612	78765	62977	77678	58
3	57429	81865	58849	80850	60251	79811	61635	78747	63000	77660	57
4	57453	81848	58873	80833	60274	79793	61658	78729	63022	77641	56
5	57477	81832	58896	80816	60298	79776	61681	78711	63045	77623	55
6	57501	81815	58920	80799	60321	79758	61704	78694	63068	77605	54
7	57524	81798	58943	80782	60344	79741	61726	78676	63090	77586	53
8	57548	81782	58967	80765	60367	79723	61749	78658	63113	77568	52
9	57572	81765	58990	80748	60390	79706	61772	78640	63135	77550	51
10	57596	81748	59014	80730	60414	79688	61795	78622	63158	77531	50
11	57619	81731	59037	80713	60437	79671	61818	78604	63180	77513	49
12	57643	81714	59061	80696	60460	79653	61841	78586	63203	77494	48
13	57667	81698	59084	80679	60483	79635	61864	78568	63225	77476	47
14	57691	81681	59108	8.552	60506	79618	61887	78550	63248	77458	46
15	57715	81664	59131	80644	60529	79600	61909	78532	63271	77439	45
16	57738	81647	59154	80627	60553	79583	61932	78514	63293	77421	44
17	57762	81631	59178	80610	60576	79565	61955	78496	63316	77402	43
18	57786	81614	59201	80593	60599	79547	61978	78478	63338	77384	42
19	57810	81597	59225	80576	60622	79530	62001	78460	63361	77366	41
20	57833	81580	59248	80558	60645	79512	62024	78442	63383	77347	40
21	57857	81563	59272	80541	60668	79494	62046	78424	63406	77329	39
22	57881	81546	59295	80524	60691	79477	62069	78405	63428	77310	38
23	57904	81530	59318	80507	60714	79459	62092	78387	63451	77292	37
24	57928	81513	59342	80489	60738	79441	62115	78369	63473	77273	36
25	57952	81496	59365	80472	60761	79424	62138	78351	63496	77255	35
26	57976	81479	59389	80455	60784	79406	62160	78333	63518	77236	34
27	57999	81462	59412	80438	60807	79388	62183	78315	63540	77218	33
28	58023	81445	59436	80420	60830	79371	62206	78297	63563	77199	32
29	58047	81428	59459	80403	60853	79353	62229	78279	63585	77181	31
30	58070	81412	59482	80386	60876	79335	62251	78261	63608	77162	30
31	58094	81395	59506	80368	60899	79318	62274	78243	63630	77144	29
32	58118	81378	59529	80351	60922	79300	62297	78225	63653	77125	28
33	58141	81361	59552	80334	60945	79282	62320	78206	63675	77107	27
34	58165	81344	59576	80316	60968	79264	62342	78188	63698	77088	26
35	58189	81327	59599	80299	60991	79247	62365	78170	63720	77070	25
36	58212	81310	59622	80282	61014	79229	62388	78152	63742	77051	24
37	58236	81293	59646	80264	61038	79211	62411	78134	63765	77033	23
38	58260	81276	59669	80247	61061	79193	62433	78116	63787	77014	22
39	58283	81259	59693	80230	61084	79176	62456	78098	63810	76996	21
40	58307	81242	59716	80212	61107	79158	62479	78079	63832	76977	20
41	58330	81225	59739	80195	61130	79140	62502	78061	63854	76959	19
42	58354	81208	59763	80178	61153	79122	62524	78043	63877	76940	18
43	58378	81191	59786	80160	61176	79105	62547	78025	63899	76921	17
44	58401	81174	59809	80143	61199	79087	62570	78007	63922	76903	16
45	58425	81157	59832	80125	61222	79069	62592	77988	63944	76884	15
46	58449	81140	59856	80108	61245	79051	62615	77970	63966	76866	14
47	58472	81123	59879	80091	61268	79033	62638	77952	63989	76847	13
48	58496	81106	59902	80073	61291	79015	62660	77934	64011	76828	12
49	58519	81089	59926	80056	61314	78998	62683	77916	64033	76810	11
50	58543	81072	59949	80038	61337	78980	62706	77897	64056	76791	10
51	58567	81055	59972	80021	61360	78962	62728	77879	64078	76772	9
52	58590	81038	59995	80003	61383	78944	62751	77861	64100	76754	8
53	58614	81021	60019	79986	61406	78926	62774	77843	64123	76735	7
54	58637	81004	60042	79968	61429	78908	62796	77824	64145	76717	6
55	58661	80987	60065	79951	61451	78891	62819	77806	64167	76698	5
56	58684	80970	60089	79934	61474	78873	62842	77788	64190	76679	4
57	58708	80953	60112	79916	61497	78855	62864	77769	64212	76661	3
58	58731	80936	60135	79899	61520	78837	62887	77751	64234	76642	2
59	58755	80919	60158	79881	61543	78819	62909	77733	64256	76623	1
M	C.S.	S.	C.S.	S.	C.S.	S.	C.S.	S.	C.S.	S.	M
	54 Deg.		58 Deg.		52 Deg.		51 Deg.		50 Deg.		

M	40 Deg.		41 Deg.		42 Deg.		43 Deg.		44 Deg.		M
	S.	C.S.	S.	C.S.	S.	C.S.	S.	C.S.	S.	C.S.	
0	64279	76604	65606	75471	66913	74314	68200	73135	69466	71934	60
1	64301	76586	65628	75452	66935	74295	68221	73116	69487	71914	59
2	64323	76567	65650	75433	66956	74276	68242	73096	69508	71894	58
3	64346	76548	65672	75414	66978	74256	68264	73076	69529	71873	57
4	64368	76530	65694	75395	66999	74237	68285	73056	69549	71853	56
5	64390	76511	65716	75375	67021	74217	68306	73036	69570	71833	55
6	64412	76492	65738	75356	67043	74198	68327	73016	69591	71813	54
7	64435	76473	65759	75337	67064	74178	68349	72996	69612	71792	53
8	64457	76455	65781	75318	67066	74159	68370	72976	69633	71772	52
9	64479	76436	65803	75299	67107	74139	68391	72957	69654	71752	51
10	64501	76417	65825	75280	67129	74120	68412	72937	69675	71732	50
11	64524	76398	65847	75261	67151	74100	68433	72917	69696	71711	49
12	64546	76380	65869	75241	67172	74080	68455	72897	69717	71691	48
13	64568	76361	65891	75222	67194	74061	68476	72877	69737	71671	47
14	64590	76342	65913	75203	67215	74041	68497	72857	69758	71650	46
15	64612	76323	65935	75184	67237	74022	68518	72837	69779	71630	45
16	64635	76304	65956	75165	67258	74002	68539	72817	69800	71610	44
17	64657	76286	65978	75146	67280	73983	68561	72797	69821	71590	43
18	64679	76267	66000	75126	67301	73963	68582	72777	69842	71569	42
19	64701	76248	66022	75107	67323	73944	68603	72757	69862	71549	41
20	64723	76229	66044	75088	67344	73924	68624	72737	69883	71529	40
21	64746	76210	66066	75069	67366	73904	68645	72717	69904	71508	39
22	64768	76192	66088	75050	67387	73885	68666	72697	69925	71488	38
23	64790	76173	66109	75030	67409	73865	68688	72677	69946	71468	37
24	64812	76154	66131	75011	67430	73846	68709	72657	69966	71447	36
25	64834	76135	66153	74992	67452	73826	68730	72637	69987	71427	35
26	64856	76116	66175	74973	67473	73806	68751	72617	70008	71407	34
27	64878	76097	66197	74953	67495	73787	68772	72597	70029	71386	33
28	64901	76078	66218	74934	67516	73767	68793	72577	70049	71366	32
29	64923	76059	66240	74915	67538	73747	68814	72557	70070	71345	31
30	64945	76041	66262	74896	67559	73728	68835	72537	70091	71325	30
31	64967	76022	66284	74876	67580	73708	68857	72517	70112	71305	29
32	64989	76003	66306	74857	67602	73688	68878	72497	70132	71284	28
33	65011	75984	66327	74838	67623	73669	68899	72477	70153	71264	27
34	65033	75965	66349	74818	67645	73649	68920	72457	70174	71243	26
35	65055	75946	66371	74799	67666	73629	68941	72437	70195	71223	25
36	65077	75927	66393	74780	67688	73610	68962	72417	70215	71203	24
37	65099	75908	66414	74760	67709	73590	68983	72397	70236	71182	23
38	65122	75889	66436	74741	67730	73570	69004	72377	70257	71162	22
39	65144	75870	66458	74722	67752	73551	69025	72357	70277	71141	21
40	65166	75851	66480	74703	67773	73531	69046	72337	70298	71121	20
41	65188	75832	66501	74683	67795	73511	69067	72317	70319	71100	19
42	65210	75813	66523	74664	67816	73491	69088	72297	70339	71080	18
43	65232	75794	66545	74644	67837	73472	69109	72277	70360	71059	17
44	65254	75775	66566	74625	67859	73452	69130	72257	70381	71039	16
45	65276	75756	66588	74606	67880	73432	69151	72236	70401	71019	15
46	65298	75738	66610	74586	67901	73412	69172	72216	70422	70998	14
47	65320	75719	66632	74567	67923	73393	69193	72196	70443	70978	13
48	65342	75699	66653	74548	67944	73373	69214	72176	70463	70957	12
49	65364	75680	66675	74528	67965	73353	69235	72156	70484	70937	11
50	65386	75661	66697	74509	57987	73333	69256	72136	70505	70916	10
51	65408	75642	66718	74489	68008	73314	69277	72116	70525	70896	9
52	65430	75623	66740	74470	68029	73294	69298	72095	70546	70875	8
53	65452	75604	66762	74451	68051	73274	69319	72075	70567	70855	7
54	65474	75585	66783	74431	68072	73254	69340	72055	70587	70834	6
55	65496	75566	66805	74412	68093	73234	69361	72035	70608	70813	5
56	65518	75547	66827	74392	58115	73215	69382	72015	70628	70793	4
57	65540	75528	66848	74373	68136	73195	69403	71995	70649	70772	3
58	65562	75509	66870	74353	68157	73175	69424	71974	70670	70752	2
59	65584	75490	66891	74334	68179	73155	69445	71954	70690	70731	1
60	65606	75471	66913	74314	68200	73135	69466	71934	70711	70711	0
M	C.S.	S.	C.S.	S.	C.S.	S.	C.S.	S.	C.S.	S.	M
	49 Deg.		48 Deg.		47 Deg.		46 Deg.		45 Deg.		

Text-Books in Astronomy

Bowen's Astronomy by Observation

By Eliza A. Bowen.

Boards, quarto, 94 pages.   Colored Maps and Illustrations $1.00

An elementary text-book for schools, and especially adapted for use as an atlas to accompany any other text-book in astronomy. Careful directions are given when, how and where to find the heavenly bodies, and the quarto pages admit star maps and views on a large scale.

Gillet and Rolfe's Astronomies

By J. A. Gillet and W. J. Rolfe.

First Book in Astronomy.   Short Course.   220 pages .   . $1.00
Astronomy.   415 pages 1.40

These books have been prepared by practical teachers and contain nothing beyond the comprehension of pupils in secondary schools.

Lockyer's Astronomies

By J. N. Lockyer, F.R.S.

Astronomy. (Science Primer Series.) 136 pages   .   35 cents
Elementary Lessons in Astronomy. 312 pages   .   . $1.22

The aim throughout these books is to give a connected view of the whole subject rather than to discuss any particular parts of it, and to supply facts and ideas founded thereon, to serve as a basis for subsequent study.

Ray's New Elements of Astronomy

By Selim H. Peabody, Ph.D., LL.D.

Cloth, 12mo, 352 pages $1.20

The elements of astronomy, with numerous engravings and star maps. In the revised edition, the scope and method of the original is retained, with the addition of all the results of established discovery. The book treats of the facts, principles, and processes of the science, presuming only that the pupil is acquainted with the simplest principles of mechanics and physics.

Steele's New Descriptive Astronomy

By J. Dorman Steele, Ph.D.   Cloth, 12mo, 338 pages . $1.00

This book is written in the same interesting and popular manner as other books of the Steele Series, and is intended for the inspiration of youth rather than for the information of scientific scholars. The book conforms to the latest discoveries and approved theories of the science. It supplies an adequate course in astronomy for all secondary schools and college preparatory classes.

Copies of any of the above books will be sent prepaid to any address, on receipt of the price, by the Publishers:

American Book Company

New York   •   Cincinnati   •   Chicago

Geology

Dana's Geological Story Briefly Told

By JAMES D. DANA. Cloth, 12mo, 302 pages . . . $1.15

A new edition of this popular work for beginners in the study and for the general reader. The book has been entirely rewritten, and improved by the addition of many new illustrations and interesting descriptions of the latest phases and discoveries of the science. In contents and dress it is an attractive volume either for the reader or student.

Dana's New Text-Book of Geology

By JAMES D. DANA. Cloth, 12mo, 422 pages . . . $2.00

A text-book for classes in secondary schools and colleges. This standard work has been thoroughly revised and considerably enlarged and freshly illustrated to represent the latest demands of the science.

Dana's Manual of Geology

By JAMES D. DANA.

Cloth, 8vo, 1087 pages. 1575 Illustrations $5.00

Fourth revised edition. This great work was thoroughly revised and entirely rewritten under the direct supervision of its author, just before his death. It is recognized as a standard authority in the science both in Europe and America, and is used as a manual of instruction in all the higher institutions of learning.

Le Conte's Compend of Geology

By JOSEPH LE CONTE, LL.D. Cloth, 12mo, 399 pages . $1.20

Designed for high schools, academies and all secondary schools.

Steele's Fourteen Weeks in Geology

By J. DORMAN STEELE, Ph.D. Cloth, 12mo, 280 pages . $1.00

A popular book for elementary classes and the general reader.

Andrews's Elementary Geology

By E. B. ANDREWS, LL.D. Cloth, 12mo, 283 pages . $1.00

Adapted for elementary classes. Contains a special treatment of the geology of the Mississippi Valley.

Nicholson's Text-Book of Geology

By H. A. NICHOLSON, M.D. Cloth, 12mo, 520 pages . $1.05

A brief course for higher classes and adapted for general reading.

Williams's Applied Geology

By S. G. WILLIAMS, Ph.D. Cloth, 12mo, 386 pages . $1.20

A treatise on the industrial relations of geological structure; and on the nature, occurrence, and uses of substances derived from geological sources.

Copies of any of the above books will be sent prepaid to any address, on receipt of the price, by the Publishers :

American Book Company

New York ♦ Cincinnati ♦ Chicago

(93)

Physical Geography

Appletons' Physical Geography

By JOHN D. QUACKENBOS, JOHN S. NEWBERRY, CHARLES H.
HITCHCOCK, W. LE CONTE STEVENS, WM. H. DALL, HENRY
GANNETT, C. HART MERRIAM, NATHANIEL L. BRITTON,
GEORGE F. KUNZ and Lieut. GEO. M. STONEY.

Cloth, quarto, 140 pages **$1.60**

Prepared on a new and original plan. Richly illustrated with engravings, diagrams and maps in color, and including a separate chapter on
the geological history and the physical features of the United States.
The aim has been to popularize the study of Physical Geography by
furnishing a complete, attractive, carefully condensed text-book.

Cornell's Physical Geography

Boards, quarto, 104 pages **$1.12**

Revised edition, with such alterations and additions as were found
necessary to bring the work in all respects up to date.

Hinman's Eclectic Physical Geography

Cloth, 12mo, 382 pages **$1.00**

By RUSSELL HINMAN. A model text-book of the subject in a new
and convenient form. It embodies a strictly scientific and accurate
treatment of Physiography and other branches of Physical Geography.
Adapted for classes in high schools, academies and colleges, and for
private students. The text is fully illustrated by numerous maps,
charts, cuts and diagrams.

Guyot's Physical Geography

Cloth, quarto, 124 pages **$1.60**

By ARNOLD GUYOT. Thoroughly revised and supplied with newly
engraved maps, illustrations, etc. A standard work by one of the ablest
of modern geographers. All parts of the subject are presented in their
true relations and in their proper subordination.

Monteith's New Physical Geography

Cloth, quarto, 144 pages **$1.00**

An elementary work adapted for use in common and grammar schools,
as well as in high schools.

---

*Copies of any of the above books will be sent prepaid to any address, on
receipt of the price, by the Publishers:*

American Book Company

New York   •   Cincinnati   •   Chicago

Physics

Appletons' School Physics

By JOHN D. QUACKENBOS, A.M., M.D., ALFRED M. MAYER, Ph.D., SILAS W. HOLMAN, S.B., FRANCIS E. NIPHER, A.M., and FRANCIS B. CROCKER, E.M.

Cloth, 12mo, 552 pages . . . $1.20

This book is a thoroughly modern text-book on Natural Philosophy, which reflects the most advanced pedagogical methods and the latest laboratory practice. It is adapted for use in the higher grades of grammar schools, and for high schools and academies.

Cooley's New Text-Book of Physics

By LE ROY C. COOLEY, Ph.D. Cloth, 12mo, 327 pages 90 cents

An elementary course in Natural Philosophy for high schools and academies. It is brief, modern, logical in arrangement, and thoroughly systematic.

Steele's Popular Physics

By J. DORMAN STEELE, Ph.D. Cloth, 12mo, 392 pages $1.00

This new work is a thorough revision of the popular text-book, "Fourteen Weeks in Physics," so long and favorably known. It presents a thoroughly scientific treatment of the principles of the science in such an attractive style and manner as to awaken and hold the interest of pupils from the first.

Stewart's Physics—SCIENCE PRIMER SERIES

By BALFOUR STEWART. Flexible cloth, 18mo, 168 pages 35 cents

This little book contains an exposition of the fundamental principles of Physics, suited to pupils in elementary grades or for the general reader.

Trowbridge's New Physics

By JOHN TROWBRIDGE, S.D. Cloth, 12mo, 387 pages . $1.20

A thoroughly modern work, intended as a class manual of Physics for colleges and advanced preparatory schools.

Hammel's Observation Blanks in Physics

By WILLIAM C. A. HAMMEL.

Flexible, quarto, 42 pages. Illustrated . . . 30 cents

A pupil's laboratory manual and note-book for the first term's work. Each pupil to make his own apparatus and then to perform the experiments as outlined. Blanks are left in which the pupil writes his observations and the principles illustrated. It is simple, practical, and inexpensive.

Copies of any of the above books will be sent prepaid to any address, on receipt of the price, by the Publishers :

American Book Company

New York • Cincinnati • Chicago

(90)

Zoölogy and Natural History

Burnet's School Zoölogy

By MARGARETTA BURNET.   Cloth, 12mo, 216 pages   .   75 cents

A new text-book for high schools and academies, by a practical teacher; sufficiently elementary for beginners and full enough for the usual course in Natural History.

Needham's Elementary Lessons in Zoölogy

By JAMES G. NEEDHAM, M.S.   Cloth, 12mo, 302 pages   .   90 cents

An elementary text-book for high schools, academies, normal schools and preparatory college classes.   Special attention is given to the study by scientific methods, laboratory practice, microscopic study and practical zoötomy.

Cooper's Animal Life

By SARAH COOPER.   Cloth, 12mo, 427 pages   .   .   . $1.25

An attractive book for young people.   Admirably adapted for supplementary readings in Natural History.

Holders' Elementary Zoölogy

By C. F. HOLDER, and J. B. HOLDER, M.D.
Cloth, 12mo, 401 pages $1.20

A text-book for high school classes and other schools of secondary grade.

Hooker's Natural History

By WORTHINGTON HOOKER, M.D.   Cloth, 12mo, 394 pages   90 cents

Designed either for the use of schools or for the general reader.

Morse's First Book in Zoölogy

By EDWARD S. MORSE, Ph.D.   Boards, 12mo, 204 pages   87 cents

For the first study of animal life.   The examples presented are such as are common and familiar.

Nicholson's Text-Book of Zoölogy

By H. A. NICHOLSON, M.D.   Cloth, 12mo, 421 pages   . $1.38

Revised edition.   Adapted for advanced grades of high schools or academies and for first work in college classes.

Steele's Popular Zoölogy

By J. DORMAN STEELE, Ph.D., and J. W. P. JENKS.
Cloth, 12mo, 369 pages $1.20

For academies, preparatory schools and general reading.   This popular work is marked by the same clearness of method and simplicity of statement that characterize all Prof. Steele's text-books in the Natural Sciences.

Tenneys' Natural History of Animals

By SANBORN TENNEY and ABBEY A. TENNEY.
Revised Edition.   Cloth, 12mo, 281 pages   .   .   . $1.20

This new edition has been entirely reset and thoroughly revised, the recent changes in classification introduced, and the book in all respects brought up to date.

Treat's Home Studies in Nature

By Mrs. MARY TREAT.   Cloth, 12mo, 244 pages   .   .   90 cents

An interesting and instructive addition to the works on Natural History.

Copies of any of the above books will be sent prepaid to any address, on receipt of the price, by the Publishers :

American Book Company

New York   •   Cincinnati   •   Chicago

(92)

Burnet's Zoölogy

FOR

HIGH SCHOOLS AND ACADEMIES

BY

MARGARETTA BURNET

Teacher of Zoölogy, Woodward High School, Cincinnati, O.

Cloth, 12mo, 216 pages.   Illustrated.   Price, 75 cents

This new text-book on Zoölogy is intended for classes in High Schools, Academies, and other Secondary Schools. While sufficiently elementary for beginners in the study it is full and comprehensive enough for students pursuing a regular course in the Natural Sciences.   It has been prepared by a practical teacher, and is the direct result of school-room experience, field observation and laboratory practice.

The design of the book is to give a good general knowledge of the subject of Zoölogy, to cultivate an interest in nature study, and to encourage the pupil to observe and to compare for himself and then to arrange and classify his knowledge.   Only typical or principal forms are described, and in their description only such technical terms are used as are necessary, and these are carefully defined.

Each subject is fully illustrated, the illustrations being selected and arranged to aid the pupil in understanding the structure of each form.

Copies of Burnet's School Zoölogy will be sent prepaid to any address, on receipt of the price, by the Publishers:

American Book Company

New York   •   Cincinnati   •   Chicago

Laboratory Physics

Hammel's Observation Blanks in Physics

By WILLIAM C. A. HAMMEL, Professor of Physics in
Maryland State School. Boards, Quarto, 42 pages.
Illustrated.   .   .   .   .   .   . **30 cents**

These Observation Blanks are designed for use as a
Pupil's Laboratory Manual and Note Book for the first
term's work in the study of Physics. They combine in
convenient form descriptions and illustrations of the apparatus required for making experiments in Physics, with
special reference to the elements of Air, Liquids, and Heat;
directions for making the required apparatus from simple
inexpensive materials, and for performing the experiments,
etc. The book is supplied with blanks for making drawings
of the apparatus and for the pupil to record what he has
observed and inferred concerning the experiment and the
principle illustrated.

The experiments are carefully selected in the light of
experience and arranged in logical order. The treatment
throughout is in accordance with the best laboratory practice
of the day.

Hon. W. T. Harris, U. S. Commissioner of Education,
says of these Blanks:

"I have seen several attempts to assist the work of
pupils engaged in the study of Physics, but I have never
seen anything which promises to be of such practical assist-
ance as Hammel's Observation Blanks."

*Specimen copies of the above book will be sent prepaid to any address,
on receipt of the price, by the Publishers:*

American Book Company

New York   •   Cincinnati   •   Chicago

Made in the USA
Lexington, KY
26 March 2012